MODERN ASPECTS OF ELECTROCHEMISTRY
No. 6

CONTRIBUTORS

A. J. ARVIA
Instituto Superior de Investigaciones
Facultad de Ciencias Exactas
Universidad Nacional de La Plata
La Plata, Argentina

J. O'M. BOCKRIS
Electrochemistry Laboratory
John Harrison Laboratory of Chemistry
University of Pennsylvania
Philadelphia, Pennsylvania

HAROLD L. FRIEDMAN
Department of Chemistry
State University of New York at Stony Brook
Stony Brook, New York

A. HICKLING
Department of Inorganic, Physical, and Industrial Chemistry
The Donnan Laboratory
The University of Liverpool
Liverpool, England

J. LLOPIS
Instituto de Química Física "Rocasolano"
C. S. I. C.
Madrid, Spain

S. L. MARCIANO
Instituto Superior de Investigaciones
Facultad de Ciencias Exactas
Universidad Nacional de La Plata
La Plata, Argentina

D. B. MATTHEWS
Union Carbide Australia, Ltd.
Chemicals Division
Rhodes, N. S. W., Australia

MODERN ASPECTS OF ELECTROCHEMISTRY

No. 6

Edited by

J. O'M. BOCKRIS

Electrochemistry Laboratory
John Harrison Laboratory of Chemistry
University of Pennsylvania, Philadelphia, Pennsylvania

and

B. E. CONWAY

Department of Chemistry
University of Ottawa
Ottawa, Ontario

℗ PLENUM PRESS • NEW YORK • 1971

Library of Congress Catalog Card Number 54-12732
ISBN-13: 978-1-4684-3002-8   e-ISBN-13: 978-1-4684-3000-4
DOI: 10.1007/978-1-4684-3000-4
© 1971 Plenum Press, New York

Softcover reprint of the hardcover 1st edition 1971

A Division of Plenum Publishing Corporation
227 West 17th Street, New York, N.Y. 10011

Preface

In the last decade, the evolution of electrochemistry away from concern with the physical chemistry of solutions to its more fruitful goal in the study of the widespread consequences of the transfer of electric charges across interphases has come to fruition. The turning of technology away from an onward rush, regardless, to progress which takes into account repercussions of techno-logical activity on the environment, and the consequent need for a reduction and then termination of the injection of CO_2 into the atmosphere (greenhouse effect), together with a reckoning with air and water pollution in general, ensures a long-term need for advances in a basic knowledge of electrochemical systems, an increased technological use of which seems to arise from the environmental necessities.

But a mighty change in attitude needs to spread among electro-chemists (indeed, among all surface chemists) concerning the terms and level in which their field is discussed. The treatment of charge transfer reactions has often been made too vaguely, in terms, it seemed, of *atom* transfer, with the electron-transfer step, the essence of electrochemistry, an implied accompaniment to the transfer of ions across electrical double layers. The treatment has been in terms of classical mechanics, only tenable while inadequate questions were asked concerning the behavior of the *electron* in the interfacial transfer. No process demands a more exclusively quantal discussion than does electron transfer. Objections may be made that electro-chemical systems are too complex for their quantum mechanical interpretation to be fruitful. But no other treatment is possible. And if discussions of biophysical processes can be made in quantum mechanical terms, there seems no argument against the application of quantum mechanical reasoning to the metal–solution interface. Quantal treatments already exist and need to be introduced immediately into *all* teaching in basic electrochemistry. But it is not enough to introduce a tunneling probability term into the derivation of the expression for the dependence of the rate upon the potential. The study of chemistry consists largely of attempts to understand bonding, and localized bonding forms a central part of the theory of electrode processes. This, too, must be treated quantum mechanically. And the applications of electrochemistry to the understanding of the properties of materials must be discussed in terms of solid-state physics, in the language of quantum chemistry. Such attitudes will increasingly be reflected in the future volumes of this series.

The quantum-mechanical treatments to which electrochemical processes have hitherto been subject are rudimentary. Moreover, they have been made by two somewhat different approaches which have not yet been related to each other. A general discussion of these two lines of quantum mechanical treatments given to charge-transfer reactions is presented in Chapter 4.

Well ahead of the theoretical treatment of charge transfer at interfaces is the statistical-mechanical treatment of ionic solutions given in Chapter 1, and here Dr. Friedman provides a discussion which shows that a break out from some of the restrictions of the Debye–Hückel model in this well-known area has been achieved.

Hydrodynamics is not a part of electrochemistry, but it is auxiliary to it. Electrochemical excursions here have been dominated by the Russian contributions, and these, and the more recent ones from other laboratories, are reviewed by Dr. Arvia in Chapter 2.

A number of modern discussions of the theory of the structure of the double layer are available. They are concerned largely with the statistical mechanics of the distribution of ions at the electrode interface. The experimental determination of the complex series of potentials which exist at interphases is little known, and is usefully discussed in Chapter 3 by Dr. J. Llopis.

Finally, in Chapter 5, by Dr. A. Hickling, a new type of electrolysis is discussed, in which the transfer of *energy*, in addition to electric charge, brings about chemical change.

J. O'M. Bockris
B. E. Conway

Breakers Club, January 1971

Contents

Chapter 4

THE MECHANISM OF CHARGE TRANSFER FROM METAL ELECTRODES TO IONS IN SOLUTION

Dennis B. Matthews and John O'M. Bockris

Chapter 5

ELECTROCHEMICAL PROCESSES IN GLOW DISCHARGE AT THE GAS–SOLUTION INTERFACE

A. Hickling

Computed Thermodynamic Properties and Distribution Functions for Simple Models of Ionic Solutions

Harold L. Friedman

Department of Chemistry
State University of New York at Stony Brook
Stony Brook, New York

I. INTRODUCTION

The aim of this review is to describe advances in the theoretical study of equilibrium properties of electrolyte solutions since the preceding review in this series by Falkenhagen and Kelbg.[1] It is intended for the reader with a special interest in electrolyte solutions, but not necessarily in the theoretical aspects. However, it is written with the philosophy that the role of theory in this area is to provide interpretation of experimental observations in terms of molecular interactions. On the other hand, providing equations for the representation and correlation of experimental data is regarded as a by-product of some value, but not an end in itself.

The molecular interactions of interest here may be called weak interactions, as contrasted with the strong interactions which result in chemical bonds within molecules. While our knowledge of the strong interactions has grown enormously in recent years, we know less about the weak interactions now than we did about the chemical bonds in the 19th century. Even the strongest of them, the hydrogen bond, is reasonably well characterized only in crystalline systems with regard to structural aspects,[2] while the energetic and kinetic aspects are largely undeveloped. In general, the weak interactions are distinguished from chemical bond interactions not so much because they are weaker but because they have short lifetimes in the situations in which they are encountered. One need only be reminded that the hydration energy of a sodium ion is about 100 kcal/mole but that the lifetime of a particular "complex" of a sodium ion with coordinated water molecules is of the order of only 10^{-11} sec in solution. It is the short lifetimes of the weak interactions which give studies of the molecular interpretation of solution properties and studies of the structure of the solutions their characteristic vagueness even when one employs those spectroscopic and radiation-scattering tools that have been so successful in elucidating chemical bonds. It results in the paradoxical situation that many of our most promising concepts of solution structure, at least for aqueous solutions, have been derived on the basis of thermodynamic studies[3] rather than spectroscopic or scattering studies.

This review deals almost entirely with the theory of the consequences of ion–ion interactions in the solutions, the theory of activity and osmotic coefficients, heats of dilution, and other thermodynamic excess functions. Although the advances in this particular area do lead back to problems in solvation (cases of purely ion–solvent interaction) and solvent structure, it seems that there is not enough advance in theory in these areas to merit a review, while the empirical aspects have been very recently reviewed.[4] The same may be said for the theory of the nonequilibrium properties, namely transport and relaxation coefficients, although in a more general sense, that is, without regard to molecular interpretation of electrolyte solution properties, there has been great progress in the theory of time-dependent phenomena in liquid systems.[5-7] There is also a rapidly growing body of experimental data concerning nuclear magnetic resonance relaxation, dielectric relaxation, and inelastic neutron scattering in ionic solutions which invite the application of these new theoretical methods.[8-11] Some of the connections between the material reviewed and that omitted are discussed in Section XI.

Finally, in the area remaining for review, emphasis is placed on a particular approach to the problem of the molecular interpretation of thermodynamic excess functions: the accurate computation of these coefficients for Hamiltonian models which characterize simplified hypothetical systems having some correspondence with the real systems which are of central interest. It is hoped to elucidate this statement in the following sections. It may be noted that it is only a somewhat less restricted point of view than that which was adopted by Falkenhagen and Kelbg,[1] p. 3.

II. THEORETICAL BACKGROUND

The molecular interpretation of the properties of ionic solutions is a problem which may be attacked by the methods of statistical mechanics. In this discipline, the general problem is, given the mechanical laws governing the microscopic particles which make up a system, to calculate those properties of the system which one may determine experimentally, especially thermodynamic properties, transport and relaxation coefficients, and other macroscopic properties.

In a sense, statistical mechanics lies between the other two principal theoretical disciplines, thermodynamics and quantum mechanics, which are most often needed for the elucidation of problems in chemistry. In thermodynamics, one is concerned with some relations among the macroscopic properties referred to above; namely those relations which are consequences of a few very general laws governing the behavior of physical systems. This description includes both equilibrium thermodynamics and the thermodynamics of steady-state processes. In quantum mechanics, one is concerned with calculating energy levels and transition probabilities for states of single molecules or assemblies of molecules, whether in isolation or interaction with various fields. In these calculations, one begins with a set of postulates governing the behavior of particles and their interaction with radiation.

These descriptions of three fields of theory should be regarded as giving only the focus of each field. In many problems, more than one field is basically involved and it does not seem very interesting to develop more precise definitions. For an example involving solutions, the study of liquid mixtures of ^3He and ^4He involves concepts associated with both statistical and quantum mechanics at a very deep level.[12] Also, the thermodynamic relations involving the phase transitions in this system may be employed to save having to derive the same relations from the particular mechanical laws for this system.[13]

For ordinary ionic solutions, which are the systems of central interest in this chapter, quantum mechanical effects are not dominant and we may take advantage of the great simplification possible in statistical mechanics when one may assume that the microscopic particles are governed by the laws of classical mechanics. However, we may expect that progress in understanding these systems will carry us to the point where this is no longer true. For example, we expect that the interaction of Na^+ with H_2O under conditions relevant to understanding solvation in aqueous NaCl solutions may be fairly well understood in terms of a classical picture[14a] in which one assumes a potential function with a repulsive core term together with terms describing the various interactions of fixed charges in H_2O with the charge in Na^+ as well as interactions of the charges with the polarizabilities of the two bodies, and in addition, a van der Waals interaction term. It is true that one must turn to quantum mechanics for more detailed understanding, and for the repulsive and van der Waals terms one even needs quantum mechanics to determine the correct function of distance, but this situation may be contrasted with the problem of understanding the forces *within* the water molecule. It is also true that a classical calculation of the $Na^+ H_2O$ interaction cannot be relied upon to elucidate the solvation of Na^+ in aqueous solutions before we have some quantum mechanical calculations for this and similar systems to "calibrate" our ideas about how to formulate the repulsive term and how to estimate the coefficient of the van der Waals term in the classical potential.

Turning now to the statistical-mechanical problem itself, the way one states the mechanical laws governing the particular microscopic particles which make up a system of interest is to specify the Hamiltonian function for the system,

$$H = \sum_i (p_i^2/2m_i) + U_N \qquad (1)$$

if the system is classical, or the corresponding Hamiltonian operator if the system is quantum mechanical. In equation (1), H is the sum of kinetic and potential energies for the system in a specified *phase*, i.e., at specified values of all of the location, orientation, and momentum variables required to give a microscopic description of the state of the system. The sums over all such variables, with p_i the momentum and m_i the mass associated with the ith variable. The potential energy U_N is determined by the location and orientation variables alone. In the absence of external fields, it comes just from the interactions among the molecules. For systems which are both classical and at equilibrium, and we shall consider only this restricted class of systems, the thermodynamic functions and distribution functions which characterize the interactions among the particles depend on U_N alone; therefore, for our purpose, specifying U_N is equivalent to specifying the Hamiltonian.

In quantum-mechanical problems, one generally knows the exact Hamiltonian operator for the system of interest, but may simplify the problem by treating a *model* system with a simpler Hamiltonian. In typical statistical-mechanical problems, the particles which make up the system are molecules rather than elementary particles, so one does not know exactly what their interactions are. In this case, we also treat a model system described by a Hamiltonian which is chosen to represent some attributes of the real system. It is chosen also to be simple enough so that by known methods one may calculate those properties of the model system which, for a real system, would be experimentally observable.

The word "model" is used in a number of ways in theoretical work nowadays. Sometimes, the term "Hamiltonian model" is used for a model of the kind described above. A different use of the term would be, for example, to speak of a model system for which the Poisson–Boltzmann equation of Debye–Hückel theory is an exact description, or a different model system to which the Debye-Hückel limiting law applies exactly over a finite range of concentrations. If one is primarily interested in the molecular interpretation of experimentally observable properties, then Hamiltonian models must be regarded in a different light than the others. Thus, in a Hamiltonian model, the molecular interactions are specified. If one calculates the observable properties of such a model and compares them with experimental properties of the corresponding real system, then those discrepancies that exceed the sum of the error in the calculation and the error in the experiment must result from differences in the molecular interactions, the Hamiltonian, of the model system and the real system. Often, from such a comparison, it is even possible to tell qualitatively what the difference in molecular interactions is. Then, by refining the model and trying again, we may hope to get model behavior which is closer to the behavior of the real system of interest and, presumably, more accurate knowledge of the molecular interactions. With non-Hamiltonian models, one can follow a procedure of refinement and comparison which superficially is similar to the one just described. However, then, the refinement is in an equation of state which becomes increasingly empirical, rather than in a representation of the molecular interactions.

The use of Hamiltonian models (henceforth in this chapter "model" means "Hamiltonian model") to refine our knowledge of molecular interactions by exercises of the form outlined above demands that we be able to calculate the model system's observable properties with controlled accuracy. Only those discrepancies in model behavior versus real-system behavior that exceed the sum of the calculational error for the model system and the experimental error for the real system can be interpreted in terms of changes needed in the model Hamiltonian to better represent the real system.

Another class of models which must be mentioned here may be called "association models" or "chemical models." In such a model, the chemical constituents of the system are specified together with the chemical reactions which may take place among them. The mixture of these constituents is assumed to be ideal. The observable properties of this model system are then calculated by applying the methods of chemical thermodynamics and chemical kinetics. Equilibrium and rate constants for the assumed processes may be fixed by comparing the computed observable properties with those of a real system that the model is supposed to represent.

Hamiltonian models and chemical models are complementary in an important way. The former provide detailed molecular interpretations of the properties of real systems, while the latter may be handled by more elementary methods and yet give interpretations which are satisfactory for many purposes, e.g., the prediction of macroscopic properties as yet unmeasured. On the other hand, much of the theoretical study of electrolyte solutions has been stimulated by the failure of chemical models for various ionic systems to yield thermodynamic and rate coefficients which are independent of composition, that is, the failure of the models to predict from data at one composition what will be observed at another. For example, it may be recalled that before the acceptance of the Debye–Hückel theory many investigators employed ion-association models to account for effects which later were attributed to the Debye-Hückel ion atmospheres. In other cases, chemical models have been very successfully used to treat systems which would seem to require extremely complicated Hamiltonian models. An example is Wyatt's treatment of concentrated sulfuric acid.[14b]

A feature of chemical models which is sometimes overlooked is that the thermodynamic association constants determined by comparing the model with observations are necessarily independent of the experimental method so long as, in reducing the data, the law of mass action is invoked. For example, if the same association constants are obtained from spectroscopic and colligative measurements, this is not necessarily verification that the model represents the real system of interest. This feature has been very clearly discussed by Orgel and Mulliken.[14c]

In some cases, for example the study of weak electrolytes, it may seem appropriate to investigate hybrid models in which some interactions among the component species are accounted for in the Hamiltonian while others are accounted for by assuming chemical reactions among the components. While such models are often very useful, they can involve troublesome inconsistencies, particularly if the effect of the Hamiltonian type of interactions is not quite small compared with the effect of the chemical reaction type of interactions.

There is an important question concerning the uniqueness of the molecular interpretation of the observable behavior of macroscopic systems. For clarity, we pose it for the special case of a classical system at equilibrium. Then, if we could determine all the thermodynamic properties of the system over all of the equilibrium states of the system, could we determine the U_N function $[U_N(r_1, r_2, \ldots, r_N)$, that is, U_N as a function of all of the coordinates of the particles of the system] from this information alone? The answer seems to be probably not, and certainly not if we allow any experimental error![15] Then, we may ask, if we determine the pair correlation functions (Section III) in the system of interest over all of the equilibrium states of the system, could we determine the U_N function from this information? Again the answer seems to be: not if there are experimental errors.[16]

To a chemist, this situation should not be dismaying since it is rather typical of the concepts with which we work. It can be described more sharply by considering models in which the direct potential U_N

is made up of pair contributions:

$$U_N = \tfrac{1}{2} \sum_{i=1}^{N} \sum_{j=1}^{N} u_{ij}(\mathbf{r}_i, \mathbf{r}_j) \qquad (2)$$

where $\mathbf{r}_i$ specifies the coordinates (location and orientation) of molecule i and $\mathbf{r}_j$ those of molecule j. Then, it is clear that the u_{ij} function may be investigated by comparison of model properties with experiment, as outlined above, but also, and more directly, by considering what we know about the interaction of particle i (now molecule i) with molecule j from more general considerations— molecular radii, van der Waals interactions, interactions of electric moments (charges, dipoles, etc.), and others. The most refined approach to this problem is the quantum mechanical calculation of the energy of the group of nuclei and electrons making up molecule i and the group of nuclei and electrons making up molecule j as a function of the separation and relative orientations of the two groups of nuclei. It seems that such calculations of nonbonded interactions are just becoming technically possible.

In summary, it seems that progress can be made in the molecular interpretation of the properties of ionic solutions by combining the two sources of information about molecular interactions:

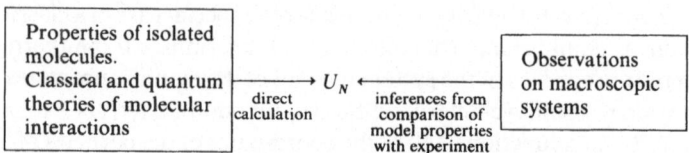

This chapter is mostly concerned with recent progress with the process on the right side of this diagram, using the methods of statistical mechanics. There has not been as much advance in the process on the left side in the case of ionic solutions. Incidentally, when one employs the McMillan–Mayer theory (Section V), the process on the left side also becomes one for which the methods of statistical mechanics are required, but there has been very little development of methods applicable to this sort of problem.

III. SOME RESULTS FROM STATISTICAL MECHANICS†

In order to provide a setting for the next part of the review, we summarize a few important results from statistical mechanics of classical systems. Let $A(N, V, T)$ be the Helmholtz free energy of a system of N molecules, possibly of several species, in a volume V at temperature T. Let $A^{id}(N, V, T)$ be the Helmholtz free energy of the corresponding ideal system, a hypothetical system with the same composition and the same internal molecular properties but with no interactions among the molecules. Then the interaction or "excess" part of the free energy of the real system

$$A^{ex} \equiv A - A^{id} \qquad (3)$$

is given by the equation‡

$$\exp(-A^{ex}/kT) = \int \exp[-U_N(\{N\})/kT]\, d\{N\} \Big/ \int d\{N\} \qquad (4)$$

where $\{N\}$ is an abbreviation for $\mathbf{r}_1, \mathbf{r}_2, \ldots, \mathbf{r}_N$, the coordinates that U_N depends upon. In each of the integrals, the coordinates take on all values consistent with the N molecules all being in the vessel of volume V. If the N molecules are all monatomic so that $\mathbf{r}_i$ is just the set of center-of-mass coordinates x_i, y_i, z_i of the ith molecule, then

$$\int d\{N\} = V^N \qquad \text{(monatomic molecules)}$$

but otherwise there is also a factor from the integration over orientational coordinates.

The other integral in equation (4) is called the configuration integral or, in the older literature, the phase integral. Clearly, if we are given a U_N function, then the configuration integral is determined, although evaluating it is not a trivial problem if N is of the order of

†Most of these results are derived by Hill,[17] Chapters 5 and 6, although, unfortunately, the notation is a little different. The equations here are written for the petit canonical ensemble but many of the references give only the grand ensemble theory because it is often more convenient to use even though the equations are more complicated. Hill discusses both. The references given for this section are selected for the reader who wishes to learn more about the results given; no effort is made to be complete or to indicate priority. However, the references given do contain bibliographical information.

‡Excess thermodynamic functions are sometimes called residual functions. See Rowlinson[18] for the thermodynamic relations. Also see Friedman,[19] where $-A^{ex}/VkT$ is called $\mathfrak{S}$.

10^{23}. However, if we can determine it, perhaps using an appropriate approximation method, then we can compute the excess free energy A^{ex} exhibited by a model for which U_N is the potential. We can obtain other interaction or excess thermodynamic functions either by computing A^{ex} for several states and applying the relevant thermodynamic equations, or by applying these equations to the evaluation of A^{ex} in terms of the configuration integral, thereby getting a somewhat different integral for each thermodynamic function. For example, for the excess energy, the energy of the interacting system minus that of the hypothetical system of the same composition but without interactions among the molecules, we have both [19,20]

$$E^{ex} = \left(\frac{\partial(A^{ex}/T)}{\partial(1/T)} \right)_{N,V} \tag{5}$$

and

$$E^{ex} = \frac{\int [\partial(U_N/T)/\partial(1/T)]\exp(-U_N/kT)\,d\{N\}}{\int \exp(-U_N/kT)\,d\{N\}} \tag{6}$$

If one begins with a model potential U_N and makes no approximations in the evaluation of the integrals in (4) and (6), then the two methods of calculating E^{ex} must agree exactly. In practice, such agreement is not so easy to obtain and the comparison of the two results for E^{ex} may be used as a *consistency test* which measures the quality of the approximation method actually employed.[20]

Of course, for large systems, A, A^{ex}, E^{ex}, etc. are proportional to the size of the system while the material properties, the quantities of intrinsic interest, are, for example, A^{ex}/N or A^{ex}/V, the corresponding quantities per molecule (or per mole) or per unit volume. Furthermore, such specific quantities are fixed by a choice of N/V and T and do not change if we vary N, V, T, while maintaining fixed values of N/V and T. In practical computations, we take advantage of this aspect of the behavior of the extensive thermodynamic variables for large systems, but it is simpler to present the fundamental relations in terms of the extensive variables themselves, just as in thermodynamics.

A second fundamental equation concerns the relation between the model potential U_N and the local concentrations in the system. Suppose we take a snapshot of the system of interest with a resolution in space and time such that we can see the individual molecules. Now, consider the chance that in a very small volume δv at a particular location in the system we find a molecule of species a. This chance, defined in terms of what we expect to see in a great many snapshots, is certainly $\rho_a \delta v$, where

$$\rho_a \equiv N_a/V \tag{7}$$

is the number of a molecules per unit volume of the system.

Now, suppose we reexamine the snapshots but, this time, in each one look for one molecule of species a in the very small volume δv, not in the same location in each snapshot, but at a chosen distance r from a molecule of species b. Now, we may express the chance of finding an a molecule in δv as $\rho_a g_{ab}(r) \delta v$, where the *correlation function* g_{ab} is a measure of the effect of the b molecule in repelling or attracting the a molecule to its neighborhood.† This factor did not appear when we chose δv without reference to the location of a b molecule because then, after many observations, the effect of the b molecules averages out.

By analogy with the barometric formula, we may write

$$g_{ab}(r) = \exp[-w_{ab}(r)/kT] \tag{8}$$

which *defines* the function w_{ab}, called the potential of average force. From the way we have defined g_{ab}, it is clear that when a is far from b. then $g_{ab} \to 1$. In this same limit, $w_{ab} = 0$. Then, again by analogy with the barometric formula, we see that $w_{ab}(r)$ is the reversible work required to bring a and b from infinite separation in the system to the distance r apart.

Unless molecules a and b are both monatomic, there will be a dependence of both g_{ab} and w_{ab} upon the orientation of one or both

†The notation of Falkenhagen and Kelbg[1] for this quantity is similar: they call it $g_{ab}(\mathbf{r}_a, \mathbf{r}_b)$, where $\mathbf{r}_a$ is a vector fixing the location of molecule a and $\mathbf{r}_b$ is a vector fixing the location of molecule b. In systems at equilibrium in the absence of external fields, g_{ab} depends on $|\mathbf{r}_a - \mathbf{r}_b|$ but not on the vectors $\mathbf{r}_a$ and $\mathbf{r}_b$ separately.

of the molecules a and b relative to the line between them. In this case, all of the above considerations are still valid except that in equation (8) one must regard r as denoting a list of numbers specifying the separation *and* all of the relevant angles. While it is easy to include such orientation-dependent effects in all of the equations in just this way, and while they are certainly important in many real systems of interest here, relatively little has been done in evaluating the observable properties of models with orientation-dependent forces. Therefore, it seems best to discuss explicitly only the simple cases in which there are no orientation-dependent forces, and, for the rest, not even indicate how these relations may be generalized.†

The fundamental equation relating g_{ab} to the model potential is

$$g_{ab}(r) = V^2 \frac{\int \exp(-U_{N+a+b}/kT)\,d\{N\}}{\int \exp(-U_{N+a+b}/kT)\,d\{N+a+b\}} \tag{9}$$

where, in the numerator, one integrates over all configurations of the N molecules in a vessel of volume V but keeps the specified a molecule and the specified b molecule at fixed locations in the vessel such that their separation is r. If we write $\exp(-w_{ab}/kT)$ in place of g_{ab} in this equation, the resulting expression is remarkably similar to equation (4), making it clear that w_{ab} is a local free energy just as $\rho_a g_{ab}$ is a local concentration.

All of these expressions are also valid if a and b are two different molecules of the *same* species, except that then a factor $(1 - 1/N_a)$ must be added on the right of (9). This is not important for considerations at the level of this chapter.

The integral in the denominator of (9) is the configuration integral which appears in equation (4) and the one in the numerator of (9) does not appear to be simpler, so it is remarkable that there seem to be more valuable approximation procedures which may be applied to equation (9) than to equation (4). In fact, the Debye–Hückel 1923 theory is an approximation procedure for evaluating g_{ab}. However, equation (9) is also important because the correlation

---

†For examples of the inclusion of orientation dependence, see Jepsen and Friedman.[21]

functions g_{ab} rather than the thermodynamic functions play a determining role in some experiments. In monatomic systems, $g_{aa}(r)$, the correlation function for two different molecules of the same species, may be determined by x-ray or neutron scattering experiments.[2] In systems of two atomic species a and b, the functions g_{aa}, g_{ab}, and g_{bb} may be determined by a combination of x-ray and neutron scattering, although not from scattering of just one kind of radiation.[23] However, in those polyatomic systems in which most of the scattering of a particular radiation is due to species a, one may determine g_{aa} after making more or less important corrections for the contributions of the other species to the scattering. These problems have been very clearly discussed by Pings and Waser.[23]

There are two independent ways to calculate the excess free energy functions from the correlation functions. One of these, the *compressibility equation*, has the form, for a one-component system,

$$(\partial P/\partial \rho)_{V,T} = kT/\{1 + \rho \int [g(r) - 1] d\mathbf{r}\} \qquad (10)$$

where

$$\int [g(r) - 1] d\mathbf{r} \equiv 4\pi \int_0^\infty [g(r) - 1] r^2 \, dr \qquad (11)$$

and where P is the pressure of the system. From $P(\rho)$, we may derive the excess free energy by the thermodynamic relation (Friedman,[19] p. 67).

$$A^{ex}(N, V, T) = N \int_0^{N/V} [P(\rho) - \rho kT] \rho^{-2} \, d\rho \qquad (12)$$

In order to use the compressibility equation to compute the pressure, one must first compute $g(r)$ at various densities, then use (10) to get the "compressibility" $\partial P/\partial \rho$ at various densities extending down to where $P \sim \rho kT$, and then integrate:

$$P(\rho) = \rho kT + \int_0^\rho [(\partial P/\partial \rho)_{at\rho'} - kT] \, d\rho' \qquad (13)$$

Then, in order to get the free energy, one must integrate again according to equation (12).

The other path to the free energy is through the *virial equation.* For a one-component system in which [see equation (2)]

$$U_N = \tfrac{1}{2} \sum u_{ij}(r) \tag{14}$$

this is

$$P(\rho) = \rho kT - \tfrac{1}{6}\rho^2 \int r(\partial u/\partial r)g(r)\,d\mathbf{r} \tag{15}$$

It can also be generalized to the case in which U_N is not merely a sum of pair contributions.[19] Both the compressibility and virial equations can be generalized to apply to systems of several species of molecules.[24]

If a computed $g(r)$ is exact for some model potential U_N, then the compressibility and virial equations must give the same A^{ex}, the same P, and the same $\partial P/\partial \rho$. In actual cases, comparison of the results of the two methods is a self-consistency test of the quality of the computed $g(r)$.[25] It is the same kind of test as was made in the classical Debye–Hückel theory by comparing the Debye charging process with the Guntelberg charging process. In fact, these classical charging processes provide paths for computing free-energy functions from correlation functions in addition to the compressibility and virial equations, although all four paths may not be independent. To use either charging process, one must calculate the correlation function for models in which the potential U_N' is related to the U_N in the model of interest by a scaling or charging parameter ξ which varies from 0 to 1. At $\xi = 1$, $U_N' = U_N$, while for smaller ξ, some or all of the terms in U_N' are ξ-fold smaller than the corresponding terms in U_N. These charging processes are powerful tools in solving problems in statistical mechanics by analytical mathematics,[17,26,27] but they have a serious drawback in numerical work: one must compute the correlation functions for model potentials U_N' having no intrinsic interest.

IV. APPROXIMATION TECHNIQUES

In this section, we review the methods which may be employed to estimate the excess Helmholtz free energy function A^{ex} which would be exhibited by model systems characterized by a potential function U_N if the systems could be studied experimentally. In some of the methods, one computes A^{ex} directly from U_N; in others, one obtains the correlation functions $g_{ab}(r)$ at an intermediate stage. In the interest of simplicity, the approximation techniques are discussed here only for systems subject to the following restrictions:

(a) U_N is a sum of pair potentials:

$$U_N = \tfrac{1}{2} \sum u_{ij} \qquad (16)$$

(b) The pair potential is just a function of separation,

$$u_{ij} = (|\mathbf{r}_i - \mathbf{r}_j|) \qquad (17)$$

(c) The N molecules are all of one species.

(d) If the system represented is a solution, the N molecules are just the solute molecules. The solvent molecules do not appear explicitly but enter into the u_{ij} function in a prescribed way.

(e) The pair potential $u_{ij}(r)$ decreases more rapidly than $1/r^3$ as r increases to infinity.

The greater part of the work on the approximation methods and their applications has been limited to models that are consistent with the above restrictions, although in many studies, including all of those summarized later in this chapter, one or more of these restrictions is lifted. It is possible but not easy to avoid restriction (a), but progress in this area has also been limited by lack of motivation: very little is known about what kind of nonpairwise model potential is physically reasonable.[19,28,29] We return to this problem in Section XI. It is easy to generalize both the theories and the computations to avoid restriction (c).[19] While it is easy to generalize the theories to avoid restriction (b), the computations become more difficult because of the increased number of variables.[21] Restriction (d) is imposed in order to make use of the McMillan–Mayer theory (Section V), which greatly simplifies the applications of all these methods to solutions. Restriction (e) was for a long time a stumbling block in applying systematic approximation procedures to ionic systems. While this problem was first overcome by Mayer in 1950,[30] some of the newer methods are indifferent to whether or not this condition is met.

Of course, a model to which all five of these restrictions applied would be a model that was, at best, suitable for studying a fluid consisting of a single species of rare-gas molecules, and not even quite adequate for this, since for He and Ne, quantum effects are important,[31] while for Ar and, presumably, Kr and Xe, there are observable effects due to the failure of the pairwise-additivity approximation, condition (a).[32]

1. Monte Carlo and Molecular Dynamic Methods

The most direct approximation technique is to begin with the integral in equation (4) for $\exp(-A^{ex}/kT)$ or the integral in equation (9) for $g_{ab}(r)$ and evaluate it numerically. A powerful procedure for doing integrals of this general type, having a great many variables, is the Monte Carlo scheme invented by Metropolis *et al.*[33]

In it, one divides the macroscopic model system of interest into cubic cells of volume V and containing N molecules each, so that N/V is the particle number density of the system of interest. One chooses an initial configuration $\{N\}$ of the particles in one cell and calculates the Boltzmann factor $B_N \equiv \exp[-U_N(\{N\})/kT]$ for this configuration. Then, a new configuration is selected by the following process: One of the N molecules is selected at random and moved an amount $\Delta r = \Delta x, \Delta y, \Delta z$, where each of the components is selected at random, subject to the condition that it is in the range between $-\Delta_{max}$ and $+\Delta_{max}$. After the particle is moved, we have a new configuration and the Boltzmann factor B_N is calculated for this configuration. If the new B_N is in a certain relation to the old one, the new configuration is recorded and the process of moving one molecule at random is repeated. If not, the old configuration is recorded a second time and the same process is repeated. This is continued until a list of perhaps 100,000 configurations of the N molecules has been constructed. The list is called a Markov chain of configurations because of the random way in which it is generated.

If the Markov chain is long enough, then the statistical-mechanical integral of interest may be evaluated by merely averaging the integrand over the Markov chain. The art comes in getting answers with relatively short chains. That is, if the cell edge is length L, and if the x, y, and z coordinates of each molecule can be located at any one of 10^8 values along this length, then there are about 10^{24N} possible configurations; a rather large number if $N \simeq 100$. The problem is to make the 100,000 or so configurations that one can afford to list the representative ones so that the chain will indeed be "long enough." This is done by choice of the initial configuration, by choice of the parameter Δ_{max}, as well as choice of the smallest amount a particle may be moved, and, especially, by choice of the selection procedure applied to the successive Boltzmann factors. Commonly, if the new B_N is larger than the preceding one, the configuration is accepted; if it is smaller, then the new configuration is rejected or accepted according to some probability rule so that it is easy for the developing chain to stay in regions of high B_N and hard for it to go into regions of low B_N.

The Monte Carlo procedure is described in more detail and very clearly by Brush et al.[34] in their study of a model representing point positive charges in a uniform negative background. They also discuss very carefully what steps may be taken to include the effects of interactions between cells, and give some very detailed information about the progress of the computations, as shown in Fig. 1, for

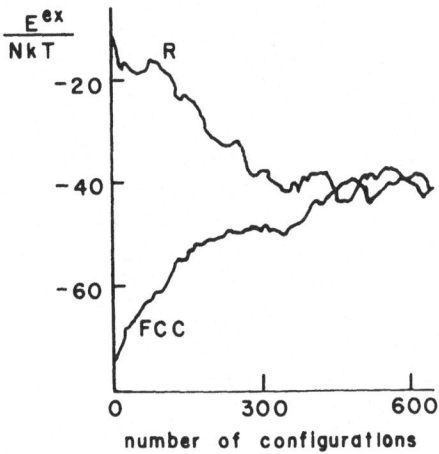

Figure 1. Change in the excess energy of the configuration as one follows a Markov chain in the Monte Carlo method.[34] In this model (point charges of one sign), $N = 108$ and $\Gamma \equiv (e^2/kT) \times (4\pi\rho/3)^{1/3} = 0.1$, corresponding to an aqueous 1–1 electrolyte at 10^{-3} M. The R chain started with the ions in a random configuration. The FCC chain started with the ions on a face-centered cubic lattice.

example. However, models for point charges in a uniform background, while a favorite for developing and intercomparing approximation methods, may be very misleading if used to represent systems of more or less equivalent positive and negative species and their results do not seem very helpful for the purposes of this chapter.

Closely related to the Monte Carlo (MC) method is the molecular dynamic (MD) method, in which one starts with an initial configuration $\{N\}$ and an initial set of momenta $\mathbf{p}_1, \ldots, \mathbf{p}_N$ and then calculates the new configuration which develops in a short interval Δt of time later as a consequence of the laws of classical mechanics together with the assumed masses and model potential U_N.[35] The MD method is even more demanding of computer power than the MC method, and so is restricted to even smaller values of N for a given computing facility. However, since it follows the dynamics of the model system, the MD method provides data from which the transport and relaxation coefficients of the model system may also be calculated.

The MC and MD results are of limited accuracy for several reasons. First, there is the limitation to cells with N up to 100, or, in a few cases, 1000 molecules. Real systems of only 1000 molecules have large fluctuations which affect the sharpness of the thermodynamic properties and also these properties have large contributions from surface effects.[36] In the standard way of doing MC and MD calculations, one approximately corrects for these effects by imposing periodic boundary conditions on the cells and by estimating the interactions with molecules in other cells; this reduces and intermingles the surface and fluctuation effects so they are difficult to estimate. However, they may be studied empirically, as in Fig. 2, which suggests that the effects are small for $N > 30$ (even with a long-range potential!), but not negligible compared to experimental errors in determining the properties of the corresponding physical systems.

It also seems reasonable that the errors due to remaining fluctuation effects would be smaller at larger densities because the intermolecular forces reduce the number of configurations with reasonably large Boltzmann factors that are accessible to the system and this effect is relatively more important at large densities (molecular concentrations). This expectation seems to be consistent with many of the reported results.[34,35]

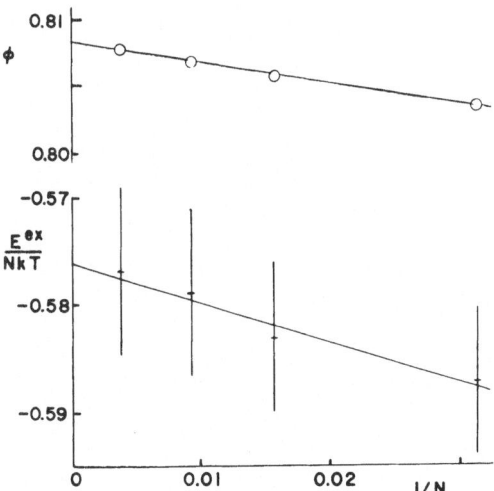

Figure 2. Osmotic coefficient and excess energy given by
the Monte Carlo method as a function of the reciprocal
of the number of ions in the cell.[34] The error bars shown
for E^{ex} are $\pm$ the standard deviation calculated on the
basis that pieces of the Markov chain 5000 steps long are
random samples. The total chain generated was over
50,000 steps long for each N. In this model, $\Gamma = 1.0$,
corresponding to an aqueous 1–1 electrolyte at 1 M.

A further source of error is the limited length of the Markov
chain actually generated, since it is only infinitely long chains which
exactly represent the system in thermodynamic equilibrium.
Another is substituting a discrete set of configurations for the
continuum appropriate to a physical system. Finally, there are,
of course, errors due to rounding-off at every stage of the computations.

As a result of careful analysis of all of these errors, together with
a study of the dependence of the results upon N, it is now generally
agreed that MC or MD results for a model system, when carefully
obtained, have the same reliability as experimental results for a real
system. These results have been obtained for models in which U_N
is a sum of pair contributions [equation (2)]. An example in which
u_{ij} is appropriate to a hard-sphere model for a one-component fluid,

$$u_{ij} = \infty \qquad \text{if} \quad r_{ij} < a$$
$$u_{ij} = 0 \qquad \text{if} \quad a < r_{ij} \tag{18}$$

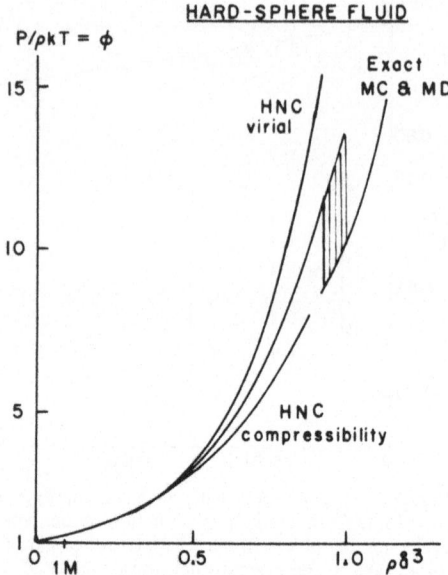

Figure 3. The equation of state of a system of hard spheres obtained by several approximation methods.

is shown in Fig. 3.[37] For comparison with the corresponding figures for electrolyte solutions, it is helpful to note that pV/NkT for the fluid corresponds to the osmotic coefficient ϕ for an electrolyte solution, while the reduced concentration ρa^3 is 0.08 for 1 M solution of a 1–1 electrolyte whose ions are charged hard spheres of diameter 4 Å. This correspondence will be made clear later in the discussion of the McMillan–Mayer theory.

Some MC results are also available for models for ionic solutions. They are presented below in comparison with other results.

The MC and MD results are so demanding of computational facilities that it seems wise to regard them as methods for obtaining accurate results for the experimentally observable properties of key model systems rather than for the exhaustive study of model systems that seems to be needed in the interpretation of solution properties in terms of intermolecular forces. Then, we may use these results for key systems to test other approximation methods which are much easier to use. This is already the practice in the study of simple fluids (e.g., Fig. 3) and it seems likely to become accepted in the statistical-mechanical study of more complicated systems involving ions, dipolar molecules, etc. This way of testing approximation methods is enormously superior to comparing the results of approximate calculations from a model with experimental observations on a real system: then one cannot know whether discrepancies are due to the limitations of the approximation method or the difference between U_N for the model and U_N for the real system. Unless the discrepancy is certainly due to differences in U_N, one cannot learn anything about the real U_N by the comparison.

2. Other Approximation Methods

Now, in turning to the various other approximation methods, it must be remarked that most of them have been derived from exact equations of statistical mechanics by more than one technique. In the following discussion, the aim is to give the simplest elementary presentation of each. Extensive references to the literature are given in a review by Brush[38a] and an article by Stell.[38b]

Another general remark is that no effort is made in the following summary of approximation methods to give the historical context of each, not even to the point of indicating how each was in fact first derived. It seems more helpful, at the level of this chapter, to merely indicate *one* of the methods by which the approximation may be derived. While this affords the reader one view of the relations among the approximation methods (they do not form a simple hierarchy), it is no more than one view of a complex and fascinating subject.

(i) Cluster-Expansion Methods

We may think of A^{ex} as a function of $\rho \equiv N/V$, the concentration. If this functional relation corresponds to a simple curve which may be graphed as in Fig. 4, then $A^{ex}(\rho)$ has a Taylor series expansion. That is, one can express $A^{ex}(\rho_2)$ in terms of $A^{ex}(\rho_1)$ and all the derivatives of A^{ex} with respect to ρ at ρ_1. It is particularly useful to take $\rho_1 = 0$ because we know $A^{ex}(0) = 0$. This Taylor series expansion is often written in the form

$$A^{ex}(\rho) = -VkT(B_2\rho^2 + B_3\rho^3 + B_4\rho^4 + \cdots) \qquad (19)$$

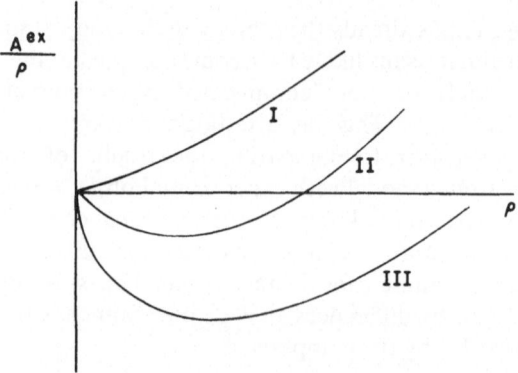

Figure 4. Schematic representation of A^{ex}/ρ for systems I and II having a cluster expansion in powers of ρ and a system III not having one. System I might be a nonionic gas above its Boyle temperature, system II the same below its Boyle temperature, and System III a plasma. For solutions, A^{ex} might be the McMillan–Mayer excess free energy of (I) a nonionic solution showing positive deviation from ideality, (II) a nonionic solution showing negative deviation from ideality at low concentration only, and (III) an ionic solution. In III, the slope of the curve is $-\infty$ at $\rho = 0$.

The coefficients B_n, called virial coefficients, may be calculated from Eq. (4) and are functions of temperature, but not of ρ. Therefore, it is quite easy to calculate the corresponding equation for the pressure in view of the thermodynamic relation

$$P = -(\partial A^{ex}/\partial V)_{N,T} + \rho kT \qquad (20)$$

or, what is equivalent but clearer if we think of A^{ex}/N as the dependent variable and ρ as the independent variable as in Fig. 4,

$$P = \rho^2[\partial(A^{ex}/N)/\partial\rho]_T + \rho kT \qquad (21)$$

This result is the well-known virial equation of state,[†]

$$P = \rho kT - kT[\rho^2 B_2 + 2\rho^3 B_3 + 3\rho^3 B_4 + \cdots] \qquad (22)$$

hence the name for the B_n coefficients, but equation (22) is the more useful form, particularly if one is interested in solutions or mixtures.

†According to strict definition, $-kTB_2$, $-2kTB_3$, $-3kTB_4$, etc. are the virial coefficients. It seems advantageous to use this nomenclature loosely as we do here and to indicate in other ways which virial coefficient is meant.

The virial equation does show that $B_1 = 0$, otherwise the infinitely dilute fluid would not obey the ideal-gas law.

It is quite a difficult mathematical problem to use equation (4) to find the expression for the B_n in terms of the model potential U_N.[17,19] The results for B_2 and B_3 follow. We define first the cluster function (or Mayer f-function)

$$f(r) = e^{-u(r)/kT} - 1 \tag{23}$$

Then we have

$$B_2 = \tfrac{1}{2} \int f(r)\, d\mathbf{r} = 2\pi \int_0^\infty f(r) r^2\, d\mathbf{r} \tag{24}$$

$$B_3 = \tfrac{1}{6} \int f(r_{ij}) f(r_{jk}) f(r_{ki})\, d\mathbf{r}_i\, d\mathbf{r}_j \tag{25}$$

where r_{ij} is the distance between point i, at $\mathbf{r}_i$, and point j, at $\mathbf{r}_j$. Expressions for the higher B_n become increasingly more complicated. Thus, there are 3 terms in B_4 and 10 terms in B_5.[40] However, as in the examples given, the virial coefficients of every order are explicitly determined by the pair potential. With very few exceptions, if one begins with a physically interesting model potential $u(r)$, it is found that the integrals B_n, even for $n = 2$, cannot be performed by analytical mathematics and must be evaluated numerically.† For large n, where other numerical methods fail, one may use the Monte Carlo method,[39,41] but in this connection, even large n means n in the range 5–10, so the integrals are much less formidable than the entire configuration integral!

For the hard-sphere fluid, which has been most extensively investigated, keeping the first n virial coefficients and dropping the others gives a good fit (i.e., within 1% in $P/\rho kT$) up to a particular reduced concentration as follows ($\mathring{a}$ is the sphere diameter):

n	2	3	4	5	6
$\rho \mathring{a}^3$	0.05	0.16	0.29	0.41	0.53

Recalling that, for a 1 M solution of a 1–1 electrolyte whose ions are 4 Å in diameter, we have $\rho \mathring{a}^3 = 0.08$, it is seen that there is some

†See Ree et al.[39] and Ree and Hoover.[41] The behavior of the second virial coefficient for nonionic systems is often of value in interpreting the properties of ionic systems. Extensive graphs and tables showing the behavior of B_2 for various models for nonionic systems are given by Hirschfelder et al.[51]

reason to hope that a cluster expansion in which one keeps only virial coefficients B_n for n less than 4 or 5 will give an accurate value for A^{ex} for such a solution.

The cluster-expansion method is not limited to expansions for A^{ex}. It is also particularly useful when applied to the correlation function $g(r)$. One finds that at $\rho = 0$ equation (9) reduces to

$$g(r) = e^{-u(r)/kT}, \qquad \rho = 0 \tag{26}$$

and that a Taylor series expansion for $g(r)$ as a function of the concentration ρ has the form

$$g(r) = e^{-u(r)/kT}(1 + \rho g_1 + \rho^2 g_2 + \cdots) \tag{27}$$

Each coefficient g_n is related to the model potential in a way similar to B_{n+2}. The first one is

$$g_1(r) = \int f(r_{ij})f(r_{jk})\,d\mathbf{r}_j, \qquad |\mathbf{r}_i - \mathbf{r}_k| \equiv r_{ij}$$

Even for this simplest g_a one cannot usually evaluate the integral except by numerical methods.

There is an important difference between the cluster expansion for A^{ex}, equation (19), and that for $g(r)$: In the first case, for the limit $\rho = 0$, $A^{ex} = 0$ has no information in it about the model. The effect of the model potential first comes in the B_2 term. Thus, the first n terms of the g expansion might be expected to lead to a more accurate A^{ex} than the first n terms of the expansion for A^{ex}. Of course, to make use of the g expansion, it is also necessary to calculate A^{ex} from the approximate $g(r)$ via the virial equation (14) or the compressibility equation (10), but these numerical computations are generally quite easy compared with the evaluation of the cluster integrals; i.e., the expressions for the B_n and the g_n in terms of the model potential.

This difference in accuracy of the A^{ex} and $g(r)$ expansions, truncated at a given ρ^n, is found in actual calculations. It is emphasized here in order to lay the groundwork for the discussion of a more powerful expansion technique to be described shortly.

First, we consider the application of the above cluster-expansion methods to ionic solutions. To do this in a realistic way requires going over to a more elaborate notation, suitable to systems of many components. This would obscure many really important aspects which are not that complicated. Therefore, we use an artifice widely employed in statistical-mechanical studies of this problem: We consider a one-component system in which the pair potential is

$$u(r) = e^2/\varepsilon r \qquad (28)$$

(model for fluid of point ions of a single charge). In the resulting A^{ex} there appears a "self-energy" term due to the fact that the model system is not electrically neutral; this is readily identified and omitted. It may be surprising to students of the Debye–Hückel theory that this model exhibits the ion-atmosphere effects and Debye-shielding $e^{-\kappa r}$ characteristic of neutral electrolyte solutions, but this is so. It is the model treated by Brush et al.[34] among others.

In employing this model, one finds that the cluster integrals all become infinite when they are evaluated in the limit of infinite volume, the limit one needs to study to obtain the thermodynamic properties such as A^{ex}. This may be easily seen in one case: Consider the last integral in equation (24) as the sum of two terms, one in which the range of integration is $0 < r < L$ and the other in which the range is $L < r < \infty$. If L is chosen big enough, and we may choose whatever we please for it, then, in the second term of the integral, the exponential function may be represented accurately by the first two terms of its series expansion, and we have

$$\int_L^\infty f(r)r^2 \, dr \underset{e^2/\varepsilon kTL \ll 1}{=} \int_L^\infty (-e^2/\varepsilon kTr)r^2 \, dr \qquad (29)$$

which apparently diverges. This happens for neutral solutions as well and was known to Debye and Hückel, who invented another approach to the problem. The direct solution to the problem was found by Mayer[30] in 1950, who showed how to combine the infinite parts of all of the cluster integrals B_n, for $2 \leq n \leq \infty$, so that they mutually cancel, as one expects they must if the statistical-mechanical theory is correct.

The final result of Mayer's intricate mathematical analysis may be represented as follows.[19,42] The expressions for B_2, B_3, and the other B_n, as well as g_1 and the other g_n, have the same form as for the system without ionic interactions except that we replace

$$u(r) = e^2/\varepsilon r \qquad \text{by} \qquad e^2 e^{-\kappa r}/\varepsilon r \qquad (30)$$

in the definition of the f-functions, equation (23). The screening factor κ which results from this analysis is defined by the equation

$$\kappa^2 = 4\pi e^2 \rho/\varepsilon kT \tag{31}$$

It is the one-component form of the screening factor in the Debye–Hückel theory. The screening factor also appears in exponential $u(r)$ in equation (27), i.e., one must replace the given $u(r)$ as indicated in equation (30). The resulting expression for the excess free energy is

$$A^{\mathrm{ex}} = -VkT[(\kappa^3/12\pi) + B_2(\kappa)\rho^2 + B_3(\kappa)\rho^3 + \cdots] \tag{32}$$

where now the B_n are functions of concentration because the screening factor appears in the f-function. (There are some other differences between the B_n and g_n for ionic and nonionic systems besides the one specified here. They are a little complicated to describe and it seems that omitting them does not at all mislead the reader unless he tries to do further work with the equations given here. Reference should be made to more complete presentations [19,42–45] for these details.)

Mayer's analysis works equally well for models in which the potential is[19,42–45]

$$u(r) = u^*(r) + e^2/\varepsilon r \tag{33}$$

where u^* represents repulsive and other non-long-range terms in the potential of the interaction of two ions. The process of removing the infinite parts of the cluster integrals replaces $u(r)$ by $u^*(r) + e^2 e^{-\kappa r}/\varepsilon r$ in all of the f-functions and in the exponential factor in equation (27). The limiting-law term in A^{ex} is not affected. The u^* potential enters into determining A^{ex} and $g(r)$ in the same way as the total pair potential in nonionic systems. Thus, if one begins with the ionic equations and lets the charges go to zero, the nonionic equations are recovered exactly.

The cluster expansion is closely related to the other approximation methods to be described and has in fact been derived by Kelbg[43] by the method of collective coordinates and by Ebeling et al.[44] from the BGY hierarchy. The cluster-expansion method can be used even for quantum mechanical ionic systems.[45]

The κ^3 term in equation (32) is the Debye–Hückel limiting law for A^{ex}. One gets the familiar limiting-law expression for $\ln \gamma_{\pm}$ from this just by applying the appropriate thermodynamic transformation.

This way of deriving the Debye–Hückel theory leaves it imbedded in an exact statistical mechanical expression. It is not easy to see how one can extract the familiar extended forms of the Debye–Hückel theory from this exact expression. It is not even clear how one gets the extended form

$$g(r) = \exp[-e^2(\exp -\kappa r)/\varepsilon r(1 + \kappa \acute{a})kT] \qquad (34)$$

where $\acute{a}$ is the diameter of an ion, now taken as a charged hard sphere.

(ii) Functional Expansions and Integral Equations†

Rather than thinking of A^{ex} as a function of the concentration, as in the cluster-expansion methods, perhaps it is more natural to think of its dependence on the pair potential function $u(r)$. When $u(r) = 0$ for every r, then $A^{ex} = 0$, so we may hope to find an expansion of the Taylor series form

$$A^{ex}(u) = 0 + u(\partial A^{ex}/\partial u)_{u=0} + \tfrac{1}{2}u^2(\partial^2 A^{ex}/\partial u^2)_{u=0} + \cdots$$

Of course, this is not so easy, since u, unlike ρ, is not simply a number which takes on increasing values as we go from the $A^{ex} = 0$ reference state to the A^{ex} of the state of interest. That is, $A^{ex}(u)$ cannot be represented by a graph similar to that in Fig. 4. However, we may get a Taylor series by the following procedure.

First, we think of r, the distance between two particles, as having only discrete values $r_1, r_2, \ldots, r_i, \ldots$, and we suppose that A^{ex} is determined just by $u(r_1), u(r_2), \ldots, u(r_i), \ldots$ rather than by $u(r)$ for all r from 0 to ∞. In this case, we would have a Taylor series expansion of the form

$$A^{ex}(\mathbf{u}) = 0 + \sum_i \left[\frac{\partial A^{ex}}{\partial u(r_i)}\right]_{\mathbf{u}=0} u(r_i)$$

$$+ \frac{1}{2}\sum_i \sum_j \left[\frac{\partial^2 A^{ex}}{\partial u(r_i)\,\partial u(r_j)}\right]_{\mathbf{u}=0} u(r_i)u(r_j) + \cdots \qquad (35)$$

where the independent variables are $u(r_1)$, $u(r_2)$, etc., and where $\mathbf{u}$ represents the *set* of all of these variables and $\mathbf{u} = 0$ means that all of them are zero. One could evaluate the derivatives appearing here in terms of the model potential by using the relation between A^{ex} and U_N given in equation (4).

† This section is derived from Percus.[46]

In the case of interest, in which the spacing of $r_1, r_2, \ldots$ is infinitesimal, we may replace the sums in equation (35) by integrals. Then we have

$$A^{ex}(\mathbf{u}) = 0 + \int \left[\frac{\delta A^{ex}}{\delta u(r)} \right]_{\mathbf{u}=0} u(r) \, d\mathbf{r}$$

$$+ \frac{1}{2} \int \left[\frac{\delta^2 A^{ex}}{\delta u(r) \, \delta u(r')} \right]_{\mathbf{u}=0} u(r) u(r') \, d\mathbf{r} \, d\mathbf{r}' + \cdots \quad (36)$$

This is called a *functional expansion*. The coefficients are now "functional derivatives," with the operator δ, rather than partial derivatives, with the operator ∂. The δ operation may be worked out by examining the very limiting process we have just described; it is rather similar to the ∂ operation.†

Like the cluster-expansion technique, the functional expansion may be applied to $g(r)$ as well as to A^{ex}. Then, we obtain

$$g(r, \mathbf{u}) = g(r, 0) + \int [\delta g(r, u)/\delta u(r')]_{\mathbf{u}=0} u(r') \, d\mathbf{r}' + \cdots \quad (37)$$

We have $g(r, 0) = 1$; the correlation function is unity for a system of noninteracting particles. This may be verified by equation (9). Using this same equation to evaluate the functional derivative in (37), and then rearranging, we obtain

$$g(r) - 1 = [-u(r)/kT] - (\rho/kT) \int [g(r') - 1] \, d\mathbf{r}' \, u(|\mathbf{r} - \mathbf{r}'|) + \cdots$$

$$\quad (38)$$

where we have written $g(r)$ for $g(r, \mathbf{u})$ since both g functions in *this* equation pertain to the model $\mathbf{u}$, not to $\mathbf{u} = 0$. Equation (38) results from investigating how $g(r)$ changes from its ideal-system value, $g(r) = 1$, as the interaction potential $u(r)$ is "turned on," just as one may consider that equation (27) exhibits how $g(r)$ changes from its infinite-dilution value as the *effect* of the interaction potential is "turned on" by increasing the concentration.

The omitted terms in equation (38) represent second-order and higher-order effects of turning on the interaction potential. If they

†See Percus.[46] The introductory part of the account of this method by its inventor is very elementary; see Volterra.[47]

are omitted, the remaining equation is an example of an *integral equation*. It is a relation between the known function $u(r)$ and the function it determines, $g(r)$. In this example, it is possible by further mathematical manipulation to rearrange the equation so that $g(r)$ is on the left and a complicated series of operations involving only $u(r)$ is on the right, thus "solving" the equation. In most cases, one can only solve integral equations by an iterative method: e.g., by choosing a trial form for the unknown function, here $g(r)$, evaluating the integral with this and the known function, here $u(r)$, and thus obtaining a refined value approximation for the unknown function. This is inserted in the integral and the process is repeated until the results of the process converge and the $g(r)$ that comes out is the same as the $g(r)$ that goes in. The computational procedures needed to evaluate the integrals appearing here and in the integral equations which follow are the same as those required to evaluate B_3 in equation (19) or g_1 in equation (27). Furthermore, the iteration procedure is computationally straightforward. Therefore, it is not much harder to obtain numerical solutions to one of these integral equations than to evaluate A^{ex} directly through the third virial coefficient or $g(r)$ directly through the g_1 coefficient.

Equation (38) is called the linearized Debye–Hückel integral equation. In the special case that $u(r)$ is the Coulomb potential [equation (28)], one can show that equation (38) is mathematically equivalent to the linearized Debye–Hückel equation, i.e., equation (34) without the $\kappa \mathring{a}$ term.

An unsatisfactory feature of equation (38) is that at low concentration ($\rho = 0$) it gives $g(r) = 1 - [u(r)/kT]$. Comparing with equation (27), we see that this is only a good approximation if $u(r)/kT \ll 1$. This is not true at all r for any model of a physical system, and the difficulties are avoided in the linearized Debye–Hückel case only by restricting the application to models at concentrations ρ so small that configurations in which the particles come close to each other do not make an appreciable contribution to the observable properties.

A much better result may be obtained by starting in the same way as in equation (37) but by making a functional expansion of $\ln g$ rather than of g. The result is

$$-kT \ln g(r) = w(r) = u(r) + \rho \int \lfloor g(r') - 1 \rfloor \, d\mathbf{r}' \, u(|r - r'|) + \cdots \quad (39)$$

The equation as written, with higher-order terms omitted, is the Debye–Hückel integral equation (not linearized!) If one takes $u(r)$ to be the Coulomb potential [equation (28)], then it may be converted mathematically to the Poisson–Boltzmann equation. Kaneko[49] derived the integral equation from the Poisson–Boltzmann equation.† It has also been obtained by simplifying the Falkenhagen–Kelbg equation[50] (discussed below) apparently without recognition of its relation to the Poisson–Boltzmann equation.

Guggenheim[51] has obtained numerical solutions of the Poisson–Boltzmann equation, equivalent of course to numerical solutions of the Debye–Hückel integral equation.

It is obvious how to proceed to get the "next term" in the Debye–Hückel integral equation; one must continue the functional expansion. It is less obvious how to do the equivalent thing with the Poisson–Boltzmann form, but some progress in this direction has been made by Outhwaite.[52]

The functional expansions for $g(r)$ and $w(r)$ which we have just considered would necessarily give equivalent results if extended to second, third, and higher orders, but if both are truncated after the first-order term, the $w(r)$ expansion is expected to be a good approximation to much higher concentrations than the $g(r)$ expansion, for the reason given. This observation may suggest that perhaps there is some other function whose expansion in $u(r)$, truncated after the first-order term, will work still better; perhaps it will give accurate results at *any* concentration. Another thought is to choose a different expansion function in place of $u(r)$. Thus, one might argue that $\exp[-u(r)/kT]$, which can have a range only between 0 and 1, ought to be a better expansion function than $u(r)$ which, in systems of physical interest, varies from very large positive values to very large negative values. Just as a truncated Taylor series expansion is more satisfactory for a small change of the independent variable, a truncated functional expansion is expected to be more accurate if the expansion function does not vary too much from its reference form.

This line of reasoning (with an additional important refinement‡), when applied to the expansion of the function $g(r)e^{u(r)/kT}$,

†His integral equation involved the Debye–Hückel approximation for w_{ab} in terms of the electrical potential.[48]
‡This refers to the definition of the direct correlation function as a functional derivative.[46]

leads to the Percus–Yevick integral equation:

$$g(r)\exp[u(r)/kT] = 1 - \rho \int g(r')\{\exp[u(r')/kT - 1]\}\,d\mathbf{r}'$$

$$\times [g(|\mathbf{r} - \mathbf{r}'|) - 1] \tag{40}$$

For deep reasons which are apparent in some of the details omitted here, the Percus–Yevick (PY) equation is particularly accurate for systems of hard-sphere particles. It is remarkable that the PY equation has been solved *analytically* for such systems,[53,54] even for the many-component case.[55] The resulting algebraic expressions for the thermodynamic functions of such systems are very close to those obtained earlier by the "scaled particle" method of Reiss, Frisch, and Lebowitz (see Reiss[56]). These results are of great importance for the study of systems with more physical potentials because, quite generally, the repulsive core interactions play a dominant role in the properties of fluids and solutions at high concentrations. The hard-sphere fluid gives a first-order picture of these effects which is expected to be of great use in treating more realistic models.[57]

The same expansion procedure which leads to the Percus–Yevick integral equation may be applied to $w(r)$, in which case one obtains the hypernetted chain (HNC) integral equation

$$w(r) = u(r) - \rho kT \int C(r')\,d\mathbf{r}'\,[g(|\mathbf{r} - \mathbf{r}'|) - 1]$$

$$C(r) \equiv g(r) - 1 - [u(r) - w(r)]/kT \tag{41}$$

The name describes the way it was first derived.† In this case, we also give the many-component form, since many results from the HNC equation will be presented. For a many-component system with species $a, b, \ldots, s, \ldots$, the HNC equation is

$$w_{ab}(r) = u_{ab}(r) - kT \sum_s \int C_{as}(r')\rho_s\,d\mathbf{r}'\,[g_{sb}(|\mathbf{r} - \mathbf{r}'|) - 1] \tag{42}$$

with C_{as} given by the expression for C in (40) with the "as" species subscript on each function.

†Sometimes, this is called the convolution hypernetted chain equation (CHNC).

The physical significance of these equations seems only to be clear in the framework of the functional expansion theory : The HNC equation expresses the potential of average force in terms of the model potential $u(r)$ together with the first-order effect of "turning on" the pair correlations due to $u(r)$. However, another view which may be helpful can be obtained in terms of cluster expansions in the following way.

As pointed out above, the coefficients in the cluster expansion of the excess free energy

$$A^{ex}(\rho) = -VkT(B_2\rho^2 + B_3\rho^3 + B_4\rho^4 + \cdots)$$

may be expressed in terms of the model potential, e.g., equations (24) and (25). For the present purpose, it is instructive to consider also the corresponding expression for the fourth virial coefficient :

$$B_4 = (1/24) \int d\mathbf{r}_i \, d\mathbf{r}_j \, d\mathbf{r}_k \, [3f_{ij}f_{jk}f_{kl}f_{li} + 6f_{ij}f_{jk}f_{kl}f_{li}f_{ik}$$

$$+ f_{ij}f_{jk}f_{kl}f_{li}f_{ik}f_{jl}]$$

where we have abbreviated $f(r_{ij}) = f_{ij}$. Like B_5 and all the higher virial coefficients, this one consists of several terms with different numbers of f bonds. Since $f(r_{ij})$ tends rapidly to zero as molecules i and j move farther apart, the integrand with six f-bonds tends to be zero unless each of the four molecules i, j, k, and l is close to *all* of the others, as in a close tetrahedral configuration. On the other hand, the integrand with just four f-bonds is nonzero over a much wider range of configurations. So it is expected that the term with four f-bonds dominates B_4, the term with five f-bonds makes a smaller contribution, and the term with six f-bonds contributes relatively little. A similar discussion may be made of the higher B_n but the proportion of terms which have nearly the minimum number of f-bonds quickly gets small as n gets large.

Now, from the approximate equation (41) together with the virial equation (15), one can derive an approximate virial expansion

$$A^{ex}(\rho) = -VkT(B_2'\rho^2 + B_3'\rho^3 + B_4'\rho^3 + \cdots)$$

and then compare the approximate B_n' with the exact B_n. One finds

$B'_2 = B_2$ and $B'_3 = B_3$ but B'_4 is equal to B_4 less the integral on the term with six f-bonds. Also, B'_5 has the terms of B_5 with fewer f-bonds, but not the others. In view of the discussion in the preceding paragraph, one expects that B'_4 will be a good approximation to B_4, B'_5 will be a less good approximation to B_5, etc.

In this way, one finds that the hypernetted chain equation "generates" the correct second and third virial coefficients, an approximate fourth virial coefficient, and poorer values of the higher coefficients. Thus, it ought to be a useful approximation tool in a concentration range above that in which the second virial coefficient alone is satisfactory. Similar statements are valid for the PY and some of the other integral equations, but the approximation obtained for a given B_n beyond B_3 depends on the integral equation and whether one uses the virial or compressibility equation in deducing the free energy from the correlation function.

Some results for the HNC equation applied to a model for a system of hard-sphere particles are shown in Fig. 3. In a similar comparison, the PY equation gives accurate results to higher concentration than the HNC equation, but both are quite satisfactory at concentrations corresponding to the solute concentration in aqueous electrolyte solutions up to 1 M.

If one expands the functions in the HNC equation in powers of the density and keeps only the lowest-order terms, the linearized Debye–Hückel equation is recovered. Therefore, the HNC equation applied to ionic systems is consistent with the Debye–Hückel limiting law. The same is true of the PY equation.

Allnatt[59] has shown how modification of the cluster-theory derivation of the PY equation, following Mayer's analysis described above in connection with equation (29), leads to a slightly modified equation which is advantageous for numerical computations. He has made the analogous modification of the HNC equation which results in a form that is mathematically identical to the original HNC equation, but which again is more convenient for numerical work. This is the HNC form used in the work described below, except as noted.

If in the HNC equation, equation (41), one makes the approximation, appropriate at low concentration,

$$C(r) \simeq e^{-u(r)/kT} - 1 \qquad (43)$$

then equation (42) becomes the integral equation investigated by Falkenhagen and Kelbg[50] and Kelbg.[60] They obtained it by combining a surprising and unjustified version of Kirkwood's superposition approximation with an integral equation due to Kirkwood and Salsburg.[61]

The functional expansion method had its genesis in a pioneering contribution by Bogoliubov[62] which displayed the essential advantages of the method for equilibrium and nonequilibrium theory. Among other things, Bogoliubov obtained the Debye–Hückel limiting result for the potential of average force. His method was applied by Glauberman and Juchnowski[63] (see also Falkenhagen and Kelbg[1]) to get a result useful at higher concentrations by including a representation of the u^* part of the pair potential; this work is referred to again in Section IX.

The approximation method used by Bogoliubov, Glauberman, and Juchnowsky is the same as the BGY (Born, Green, Yvon) integral equation. This and a somewhat similar integral equation due to Kirkwood[64,65] are discussed in some detail, together with methods of solution, by Hill.[17] Both the BGY and Kirkwood integral equations are readily derived by the functional expansion method.[46]

Other first-order integral equations may be advantageous.[66] There has recently been some study of integral equations of higher order, called PY2, HNC2, and BGY2, in which one keeps one more term in the functional expansion[46,67,68] or does something different but roughly equivalent to this in order to get a more accurate treatment of a model required to be applicable to concentrations higher than $\rho \mathring{a}^3 \sim 0.3$ (see Fig. 3). Such more elaborate integral equations have not been applied to ionic solution problems but may be useful at the point where models are introduced in which the solvent molecules are explicitly represented rather than in the way employed in the McMillan–Mayer treatment.

(iii) Other Methods

Here, brief mention is made of several theoretical developments which do not fit easily into any of the major groups considered above.

Edwards[69] has devised an approximation method for ionic systems which makes special use of the properties of Gaussian integrals, similar to earlier contributions by Kramers[70] and Berlin and Montroll.[71] Edwards carries the method considerably further and is able to treat systems with realistic short-range potentials.

A new development by Stell and Lebowitz[72] relates the properties of an ionic system to those of the corresponding uncharged system having all the same intermolecular potentials except for the Coulomb term. If e is the ionic charge and $\mathring{a}$ the ionic diameter, which one can define even for ions which are not charged hard spheres, then the Bjerrum ratio

$$q = e^2/\varepsilon k T \mathring{a} \qquad (44)$$

plays a central role in this theory. The free energy and the potential of average force in the ionic system are expressed as series in q where the coefficients depend only on properties of the uncharged system. If the uncharged system is an assembly of hard spheres, then its known equation of state and correlation functions can be used with this theory to get the properties of the ionic system at any concentration. This development may be contrasted with a cluster expansion, which can be thought of as a power series in $\rho \mathring{a}^3$.

When one applies the mathematical operation of Fourier transformation to the configuration integral, the unusual difficulties associated with the long range of the Coulomb potential become less important. This is the basis of the method of collective coordinates, which was first devised for the study of the electron gas but which has been applied by Kelbg[43,73] and then by Eisenthal and McMillan[74] to models for ionic solutions. When one tries to go beyond the limiting law and get an accurate account of the effects of the core and other contributions to the pair potential, the theory tends to look the same whether Fourier transforms were used at the outset or only at a late stage[30]; this is particularly clear from the work of Kelbg.

Stillinger and Lovett[75-78] have considered the consequences of *defining* the "particles" in a solution of an n–n electrolyte to be each a pair of ions of opposite signs. The precise definition of a pair corresponds approximately to that in Fuoss's refinement of Bjerrum's theory[79,80]; roughly speaking, each ion is paired to the nearest ion of opposite charge which is not still nearer to another ion of opposite charge to it. The configuration integral is written in terms of the pairs, each of which has a center-of-mass location, an orientation, and a separation variable. Pair–pair correlation functions g_{ab} are also defined and may be studied by standard techniques, for example cluster expansions. Particularly interesting results are obtained by considering the dielectric behavior of the system, e.g., its response to an applied electric field. The response to a field which is constant in time but varies periodically in space, the "wavelength-dependent dielectric constant," is used to derive a new integral relation for the ordinary ion–ion pair correlation function (Section VI), as well as an expression for the free energy as a function of concentration, which can be approximately evaluated by analytical methods.

Polyelectrolytes. While polyelectrolytes are outside the scope of this review, it seems of interest to point out that some of the newer approximation methods are finding application in this field, as evidenced by the work of Manning on cluster expansions.[82] Perhaps this should be expected; Kirkwood and Mazur numerically solved the BGY integral equation to elucidate a polyelectrolyte problem ten years ago.[82]

3. Summary

In a model for a one-component system of interest, one must specify $u(r)$, the potential for the interaction of a pair of molecules. Then, the approximation methods reviewed here enable the correlation function $g(r)$ for a pair of molecules in the system to be estimated at the concentration of interest. This correlation function may be compared with experiment, in some cases, or, more often, can be used to calculate thermodynamic functions. An example is shown schematically in Fig. 5.

The methods given, which are described for one-component systems, may be generalized to apply to many components. Except for aspects of the McMillan–Mayer theory (Section V), these additional steps are regarded as details outside the scope of this review.

V. McMILLAN–MAYER THEORY[83]

This is a theory† having a generality and exactness equivalent to those of thermodynamics and which enables us to obtain

†The complete derivation is also given by Friedman.[19] A simple derivation is given by Hill.[17,18]

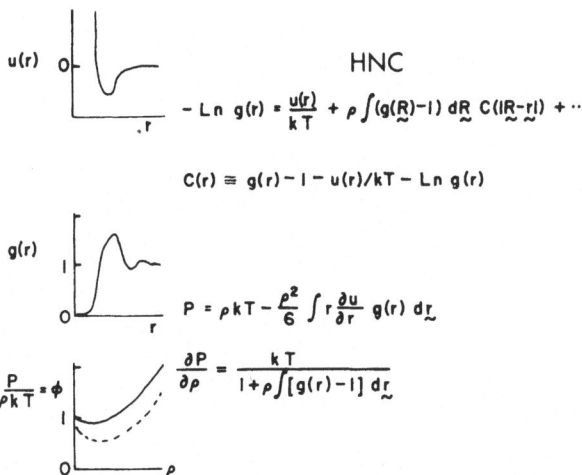

Figure 5. Schematic representation of the application of a statistical mechanical approximation method to find the observable properties of a model system. In this example, one begins with the pair potential, uses the HNC equation to find the correlation function which would result at a given ρ and T, and then uses the virial or the compressibility equation to find the pressure or osmotic coefficient. The solid curve in the last figure represents a nonionic system, the dashed curve an ionic system.

statistical mechanical theories for solutions in a particularly convenient way. For example, all of the approximation methods of the preceding section, which were formulated there for one-component systems, may be applied to solutions of one solute species by reinterpreting each of the functions in a particular way.

In Section IV, in the interests of simplicity, the equations were mostly written for systems of only one component, but of course the simplest electrolyte solution has two solute components, and for application to such cases, it is necessary to write the approximation equations in a more general form. An example of this which was given [equations (41) and (42)] is quite representative of how the equations are changed for the multicomponent case. In discussing the McMillan–Mayer theory, we shall explicitly consider a system of ns components labeled† $a, b, \ldots, s, \ldots, ns$. It seems helpful to

†The components here are the various species of particles or molecules which make up the system. Thus, in NaCl(aq), one has $ns = 3$: H_2O, Na^+, and Cl^-, if we neglect H^+ and OH^-.

Table 1
Definitions for McMillan–Mayer Theory

Function	Simple fluid	Solution
A	Helmholtz free energy of system	Helmholtz free energy of system
A^{id}	A for hypothetical system of same composition but with no interactions between the component particles: thus, for an ideal gas	A for hypothetical system of same composition but with no interactions of the solute components with each other; solute–solvent interactions same as in real system at infinite dilution of the solute: thus, for an ideal solution[a]
A^{ex}	$A - A^{id}$; excess compared to ideal gas	$A - A^{id}$; excess compared to ideal solution[a]
E, E^{id}, E^{ex}	Energies, as for A; hence, $$E^{ex} = [\partial(A^{ex}/T)/\partial(1/T)]_{N,V}$$ is $-\Delta E$ for the process in which the simple fluid is allowed to expand isothermally into a vacuum	Energies, as for A; hence, $$E^{ex} = [\partial(A^{ex}/T)/\partial(1/T)]_{N,V}$$ is $-\Delta E$ for the process in which the solution is diluted isothermally with an infinite amount of solvent
N_s	Number of particles of species s	Number of solute particles of species s
ρ_s	N_s/V; concentration of species s	N_s/V; concentration of solute species s
P	Pressure of system on a piston	Pressure of system on a piston that is permeable to the solvent particles, but not to the solute: the osmotic pressure[b]
P^{id}	For the noninteracting system $$P^{id} = kT \sum \rho_s$$ $$P^{ex} = P - P^{id} = -(\partial A^{ex}/\partial V)_{N,T}$$ is the contribution to the pressure from the interactions among the particles	For the system with no solute–solute interactions $$P^{id} = kT \sum \rho_s$$ $$P^{ex} = P - P^{id} = -(\partial A^{ex}/\partial V)_{N,T}$$ is the contribution to the osmotic pressure from the interactions of the solute particles with each other
$\rho_a g_{ab}(r)$	The local concentration of an a particle at a distance r from a b particle in the simple fluid	The local concentration of a solute particle of species a at a distance r from a solute particle of species b in the solution
$U_N(\{N\})$	The potential of interaction of N molecules at the locations $\{N\}$; $U_N = 0$ when the molecules are all far from each other	The potential of interaction of N solute molecules at configuration $\{N\}$; $U_N = 0$ when all the solute molecules are far from each other in a body of the pure solvent

Table 1—(*continued*)

$u_{ab}(r)$	The potential of the interaction of a molecule of species a with one of species b when they are at a separation r in the vacuum; $u_{ab}(\infty) = 0$	The potential of the interaction of a solute molecule of species a with a solute molecules of species b when they are at a separation r in an infinite body of pure solvent; $u_{ab}(\infty) = 0$

[a]One can define an ideal solution in many ways. The appropriate definition here is one that is ideal on a molarity scale, not a mole fraction scale. See Friedman,[19] Section 16 or Friedman.[81a]

[b]For solutions, one needs also another pressure variable P_0 which is the pressure on the pure solvent in osmotic equilibrium with the solution.

coin a name, a "simple fluid," for a system comprising just these ns components. This is to be distinguished from a "solution," in which there are ns *solute* components, with the labels given, and in addition one *solvent* component.†

Now, before stating the McMillan–Mayer theory, we require certain definitions, which we make in parallel for the simple fluid and for the solution in Table 1. This list is not exhaustive but it seems long enough so that the reader can extend it as he finds necessary.

It will be noted that the van't Hoff equation for the osmotic pressure of an ideal solution has just the same form as the equation of state of an ideal simple fluid, the ideal gas. The McMillan–Mayer theory is just a generalization of this. It may be stated as follows:

Any exact relation among the functions for a simple fluid is equally valid in terms of the functions for a solution when these are defined as in Table 1.

To derive this result, one treats the solution itself as a simple fluid, i.e., by including the solvent in the list of species. Then, it is found that in the most general statistical mechanical relations among the observable properties, the correlation functions, and the model potentials for such a system, one can group the terms in such a way that the above result is produced, quite independently of the nature of the model potential for the entire system.

†The theory is not limited to a one-component solvent, but this restriction is made here for simplicity.

The force of this result may be appreciated by considering an observation made by McMillan and Mayer.[83] In the nineteenth century, it was widely assumed that the space between material bodies was occupied by a fluid called the ether which could penetrate any body. If this were the case, any pressure we measure would be an osmotic pressure, with the ether as the solvent because it would permeate the piston or membrane of the measuring instrument. Any experiments we did to determine the potentials between particles would be determinations of the kind relevant to U_N and u_{ab} in the solution column of Table 1, again with the ether as the solvent. Then, it follows from the McMillan–Mayer theory that there is no possibility of investigating the existence of the ether by use of equilibrium statistical mechanics, since the equations have the same form whether the ether is there or not!

On a more mundane level, the use of the McMillan–Mayer theory does require attention to certain matters in addition to those which are relevant to the theory of simple fluids. An important requirement is that, in specifying the state of the solution, one specifies the chemical potential of the solvent in addition to the variables $\rho_a, \rho_b, \ldots, \rho_s, \ldots, T$, and V needed for the simple fluid. It is most often convenient to specify the chemical potential of the solvent by specifying P_0, the pressure of the solvent in osmotic equilibrium with the solution of interest. Then, in treating changes of state corresponding to changes of concentration or of temperature of the solution, the simplest results are obtained if one keeps P_0 fixed. Now, if one picks $P_0 = 1$ atm, corresponding to the usual laboratory situation, then the pressure on the solution of interest is not 1 atm, but is larger by an amount equal to the osmotic pressure, about 50 atm for a 1 M solution of a 1–1 electrolyte. In comparison with experiment one must of course bring the experimental and calculated quantities to apply to the same states by correcting one or the other.

As a simple example of the way the solvent enters into the model potential, consider the pair potential for an ionic system, which we may write in the form

$$u_{ab}(r) = u_{ab}^*(r) + (e_a e_b / \varepsilon r) \tag{45}$$

where the part of the potential with the $1/r$ dependence is written out while the u_{ab}^* term contains all other contributions, including the possible effect of the field-strength dependence of the dielectric constant as well as effects due to the mutual repulsion of the ion cores, effects of the molecular structure of the solvent, and effects of van der Waals or chemical interaction of the ions with each other. Now we consider the question : What is ε in this equation? It turns out that if we use the McMillan–Mayer theory as stated above, then ε *must* be the macroscopic dielectric constant of the pure solvent at P_0 and at the same temperature as the solution.†

To understand this, we recall that u_{ab} for a pair of ions in the solution is the potential of their interaction at separation r in the pure solvent in osmotic equilibrium with the solution. Hence, T is the temperature of the solution and P_0 must be kept fixed, independent of the concentration of the solution, as described above. Furthermore, it is essential that the $e_a e_b / \varepsilon r$ term in equation (45) be the exact asymptotic form of u_{ab} at large r. In other words, $u_{ab} - (e_a e_b / r \varepsilon)$ must tend to zero faster than $1/r$ as r approaches infinity. This mathematical condition results from the fact that the $1/r$ term tends to generate infinities in all of the integrals [of equation (29)]. This behavior, said to result from the long range of u_{ab} for an ionic system, is significant even when we contrive to make the infinities mutually cancel. In fact, it gives rise to the Debye–Hückel limiting law, the $e^{-\kappa r}$ shielding, and other consequences of the ion atmosphere. Therefore, we must treat it accurately.

It follows that ε in equation (45) must accurately describe the shielding effect of the solvent as $r \to \infty$. In this limit, it is only the solvent far from both ions, where the electric field due to the ions is very weak, which enters. Hence, ε is the ordinary dielectric constant of the solvent. In fact, if we denote the entire pair potential for ions a and b in the gas phase by u_{ab}° and in a solvent phase by u'_{ab}, then

$$\varepsilon \equiv \lim_{r \to \infty} [u_{ab}^\circ(r)/u'_{ab}(r)] \tag{46}$$

is as good an operational definition of the dielectric constant of the solvent as any other.[86]

Since the dielectric constant of a real ionic solution is a function of the concentration, it is important to ask : How is this variation

---

†Again in the interests of simplicity this description does not do full justice to the McMillan–Mayer theory, which may also be used in a more general way, for example, in the treatment of the Donnan equilibrium. See, e.g., Hill.[85]

consistent with the McMillan–Mayer theory? In order to be as clear as possible and, especially, to avoid any detailed consideration of the very difficult molecular theory of dielectric phenomena, this question will be discussed here only in terms of a very simple model. Fortunately, this is enough to show just how the dielectric behavior enters into the McMillan–Mayer type of theory.

The model we consider (Fig. 6) is that used by Hasted et al.[87] to interpret their measurements of the dielectric response of ionic solutions in water. Each ion and the surrounding annular region of solvent comprise a complex in which the dielectric constant is ε_c, while outside this complex it is ε, the dielectric constant of the pure solvent. They concluded that if the annular regions are about one water molecule thick, then $\varepsilon_c \sim 5$ for aqueous electrolyte solutions. Further details of their conclusions need not be of concern here.

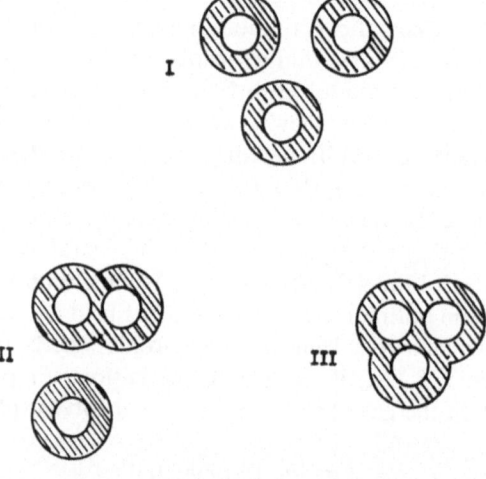

Fig. 6. Ionic cospheres (shaded). Configuration I contributes to those properties of the dilute solution to which the ions make additive contributions, e.g., limiting partial molar volume, linear change of dielectric constant with concentration, solvation energies, etc. In configuration II, the cospheres contribute to the excess functions in a way which may be taken into account by the methods described in this review. In configuration III, there is likely to be a contribution to the failure of condition (a) in Section IV.

To further define the model, assume that the charge of each ion is fixed in its center and that the ion cores, the inner spheres in the figure, are hard spheres. For this model, one can write

$$u_{ab} = u_{ab}^{hc} + u_{ab}^{el}$$

where u^{hc} is the hard-sphere potential [equation (18)] and u^{el} is the electric contribution which may, in principle, be calculated from classical electrostatics for this model. Levine[88–90] has done this for nonoverlapping complexes† and one can see from his results how u^{el} is explicitly a function of ε_c. In fact, he obtains†

$$u_{ab}^{el} = \frac{e_a e_b}{\varepsilon r} + \frac{1}{2\varepsilon r^4} \frac{\varepsilon - \varepsilon_c}{2\varepsilon + \varepsilon_c}(e_a^2 R_b^3 + e_b^2 R_a^3) + \cdots \qquad (47)$$

where the omitted terms have larger negative powers of r and where R_a and R_b are the radii of the complexes. This important problem has also been studied by Bellemans and Stecki[91–93] and Marcus.[94,95]

For this model, then, the dielectric response of the solution is reflected in the pair potential. It will also enter in another way. Consider now an assembly of three ions and their potential of mutual interaction in the solvent $U_3(\mathbf{r}_a, \mathbf{r}_b, \mathbf{r}_c)$. We may write

$$U_3(\mathbf{r}_a, \mathbf{r}_b, \mathbf{r}_c) = u_{ab}(|\mathbf{r}_a - \mathbf{r}_b|) + u_{bc}(|\mathbf{r}_b - \mathbf{r}_c|) + u_{ca}(|\mathbf{r}_c - \mathbf{r}_a|)$$

$$+ u_{abc}(\mathbf{r}_a, \mathbf{r}_b, \mathbf{r}_c) \qquad (48)$$

If, as is usually assumed, U_N is a sum of pair contributions [see equations (2) and (16)], then the term u_{abc} must vanish. This remainder term, which may be defined by equation (48), since each of the other terms is independently defined, is an example of a function, called a higher component potential, which appears if one tries to improve on the approximation that U_N comprises only pairwise components. Now, for the model considered here, it may be shown that u_{abc} does not vanish; neither will the still higher component potentials. We shall not elaborate on these statements except to point out that these many-body interactions have the same origin as in a simpler problem which has been carefully studied: The potential of interaction U_N of a gas of polarizable spheres in an electric field is not made up of just pair contributions.[96]

†See also Friedman,[19] Section 11 for further discussion. The $1/r^4$ term has also been obtained by Bellemans and Stecki[91] and Jepson and Friedman.[86]

We may conclude that, in general, local variations in the dielectric behavior of an ionic solution will be reflected both by the pair potential u_{ab} and by nonvanishing higher component potentials. The latter effect, if important, must greatly limit what can be learned from models in which one assumes that U_N comprises only pair components. While it is conceivable that one may mimic the effects of the higher component potentials in a model in which they are absent but in which, for example, one assumes a concentration dependence for ε in equation (45), a theoretical framework to show how this sort of procedure is embedded in exact theory is not at hand.

A word of caution is necessary here. Since the objective of the studies emphasized in this review is to get a molecular interpretation of thermodynamic and other equilibrium data, the use of the McMillan–Mayer theory is only appropriate to the extent that we can use it with the data to learn about U_N, i.e., the process in the right side of the diagram at the end of Section II. The objective cannot be reached until we also have some understanding of the process on the left side of the diagram. This is much more difficult for a McMillan–Mayer U_N than for a simple fluid U_N, since in the former case the calculation involves averages over the coordinates of the solvent molecules and in all cases of interest these are present at high concentration, so that few of the approximation methods summarized in Section IV can be expected to work. Now, if we make U_N still more complicated by making it a sum of concentration-dependent pair potential terms with the use of some theory that has yet to be devised, we may say that in terms of the diagram at the end of Section II, we move U_N closer to the data, in terms of accessibility, but farther from the elementary molecular interactions.

VI. THERMODYNAMICS

All of the approximation methods described in Section IV are formulated for systems with the independent variables V, T, $N_1, N_2, \ldots, N_{ns}$ and the thermodynamic potentials directly obtained are those appropriate in a thermodynamic sense to these variables, namely the Helmholtz free energy A, the thermodynamic energy E, etc. Moreover, in calculations using the McMillan–Mayer theory, one obtains these thermodynamic potentials for the solution

at a pressure $P_0 + P$, where P is the osmotic pressure and P_0 is the pressure on the pure solvent in osmotic equilibrium with the solution. We call these "thermodynamic functions in the McMillan–Mayer system."

Thermodynamic data for ionic solutions are generally gathered and formulated with a different choice of independent variables, namely T, P', and the solute molalities $m_1, m_2, \ldots, m_s, \ldots, m_{ns}$, where P' is the external pressure, typically not more than 1 atm. The appropriate thermodynamic potentials are normally used: the Gibbs free energy G, the enthalpy H, etc.

The conversion from one system to another has been investigated† and it is found that for a given sample of solution we have

$$G^{ex}/(RT \sum m_s) = (A^{ex}/VkT \sum \rho_s) - R_1(m_1, \ldots, m_{ns}) \qquad (49)$$

where G^{ex} is the molal-scale excess free energy per kilogram of solvent for the solution at pressure P_0 (say, 1 atm), A^{ex} is the excess Helmholtz free energy of the solution at pressure $P_0 + P$ (see Table 1), and R_1 is a rather complicated thermodynamic function with the main part

$$R_1 = \ln[V(m_1, \ldots, m_{ns})/V(0, \ldots, 0) + \cdots] \qquad (50)$$

just a correction for the difference in concentration scales, m_s compared to ρ_s. Here, $V(m_1, \ldots, m_{ns})$ is the volume of a solution of the indicated molal composition and containing 1 kg of solvent.

In this review, comparisons of model calculations are presented in the McMillan–Mayer system except as otherwise noted, with the concentration scales usually given in molarity or molar ionic strength rather than particles per unit volume. A typical comparison of G^{ex} and A^{ex} is shown in Fig. 7.

Some additional problems are encountered when one studies models for mixed electrolyte solutions. This area is discussed here only in the experimental system ($P', T, m_1, \ldots, m_{ns}$ as independent variables). There is a completely analogous formulation in the McMillan–Mayer system and one requires both to compare model calculations with experimental data.[19]

In a mixture, there are many thermodynamic coefficients competing for our attention. It is proposed that the principal

†Friedman[97] and Poirier[98]; also see Friedman,[19] Section 16.

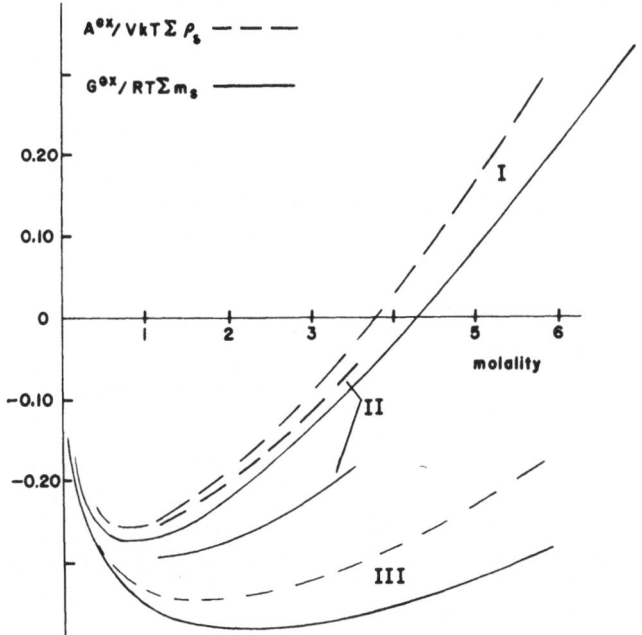

Figure 7. Comparison of McMillan–Mayer and practical excess
free energies in typical cases.[114] Aqueous solutions (25°C) of (I)
LiCl; (II) NaI; and (III) NaCl.

coefficients are those which characterize the process of forming
the mixture from solutions of the single electrolytes at the same
ionic strength. There is a systematic way to do this for any mixture,
as pointed out by Reilly and Wood,[99–101] but for this discussion,
we limit ourselves to the case (Fig. 8) where the mixture is formed
from y parts of electrolyte solution A (ion species 1 and 3) and
$1 - y$ parts of electrolyte solution B (ion species 2 and 3). These
"parts" are defined so that y is the fraction of the ionic strength of
the mixture due to electrolyte solution A. The enthalpy change
$\Delta_m H(y, I)$ in this process may be measured directly in a calorimeter,
while the volume change $\Delta_m V(y, I)$ in this process may be measured
directly in a dilatometer. The change in excess Gibbs free energy
$\Delta_m G^{ex}$ is more difficult to determine experimentally; it plays a
central role because it (or $\Delta_m A^{ex}$) comes most simply from model

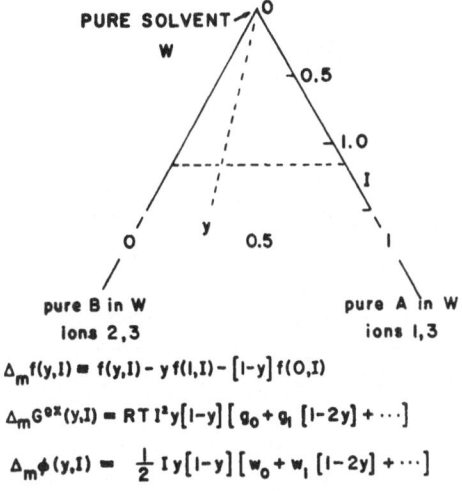

Figure 8. Coordinates and coefficients for mixtures of two electrolytes with a common ion. I is the molal ionic strength and y is the fraction of the ionic strength due to electrolyte A.

calculations and because its thermodynamic derivatives yield $\Delta_m H$ and $\Delta_m V$.

It is convenient to express $\Delta_m G^{ex}$ in terms of expansion coefficients $g_0, g_1, \ldots$ which depend upon I but not upon y. These relations are shown in Fig. 8 with the corresponding relation for the expansion coefficients of ϕ. Just as we have

$$\partial(G/T)/\partial(1/T) = H \qquad (51)$$

so we have

$$\partial(g_n/T)/\partial(1/T) = h_n, \qquad n = 0, 1, 2, \ldots \qquad (52)$$

where the h_n are the expansion coefficients of $\Delta_m H$. The theory of the expansion coefficients for the partial molar quantities is somewhat more complicated, the expansion for $\Delta_m \phi$ in Fig. 8 being correct only for mixtures of 1–1 electrolytes, but in general we have

$$w_n = \partial(Ig_n)/\partial I \qquad (53)$$

if the w_n are correctly defined.† The Harned coefficients[19,102] are the expansion coefficients for the mean ionic activity coefficient of each of the electrolytes; for example,

$$\log_{10} \gamma_A(y, I) = \log_{10} \gamma_A(1, I) - \alpha_A[1 - y] - \beta_A[1 - y]^2 - \cdots \quad (54)$$

For a mixture of 1–1 electrolytes, we have[19]

$$g_0 = -2.303\{\alpha_A + \alpha_B + I[\beta_A + \beta_B] + \cdots\} \quad (55)$$

$$g_0 = -(2.303/3)I[\beta_A - \beta_B] + \cdots \quad (56)$$

Although the Harned coefficients are important in the historical development of the theory of mixed electrolyte solutions and are needed in certain applications, it is rather tricky to use them because they are interrelated with other data in complicated ways. For example, $\alpha_A - \alpha_B$ is completely determined by the properties of the pure A and B solutions.[102]

In comparing model calculations with the results of experiments, there are difficulties because one does not usually have all of the data to perform the required Gibbs–Duhem integrations‡ needed to get g_0 from the measured osmotic or activity coefficients. It would be best to compute w_0 to compare with the experimental data, but this has not yet been done for the primitive model. Guggenheim[103] and Scatchard[104] have proposed special assumptions about the I dependence of the mixing coefficients which do enable the Gibbs–Duhem integrations to be carried out. These were employed by Harned and Robinson.[105] However, the above assumptions are not consistent with the known limiting law for mixtures (Section VII) and it is not easy to estimate the resulting error. In any case, if one has osmotic coefficient data, the approximation $g_0 \sim w_0$ is apparently correct within a factor of two and this is enough to learn something by comparing the mixture data with the primitive model results.

VII. RESULTS WHICH ARE INDEPENDENT OF THE MODEL

Here, we summarize some results for ionic systems which are independent of all aspects of the model except that the pair potential

†The definition is given by Friedman,[19] equation (18.46), but this should be modified by inserting a minus sign after the equal sign; then, (18.47) and (18.48) should be similarly modified.
‡These are described in Friedman,[19] Section 16.

has a Coulomb term and a core repulsion term, for example,

$$u_{ab}(r) = u_{ab}^*(r) + e_a e_b/\varepsilon r \tag{57}$$

with u_{ab}^* large and positive at small r so u_{ab} does not go to $-\infty$ at small r even if a and b have charges of opposite sign. No other restriction is placed on the model except that u_{ab}^* is assumed to decrease faster than $1/r^3$ at large r. In particular, there may be nonpairwise terms in U_N.

We shall call results which are independent of the model, except as specified, "general results." They are especially valuable because they can be applied with confidence to real systems, the U_N for which are generally unknown except for a few general characteristics like those specified above. There is another class of results called in statistical mechanics "exact results." Exact results may be obtained for any sort of model; they are characterized by complete mathematical rigor. This tends to be somewhat a matter of taste, but the results of the approximation methods in Section IV are generally not accepted as exact because of unresolved mathematical questions concerning the convergence of series or integrals. However, some exact results obtained by the cluster-expansion method are described by Groeneveld[106] and Lebowitz.[107] One exact result, due to Onsager, is relevant to ionic solutions. It is a lower bound on the energy of a system of charged hard spheres.

An example of an exact and general result for ionic systems concerns the mathematical problems arising from the long range of the Coulomb potential, which one usually resolves with the aid of the assumption that ionic systems do have thermodynamic properties in the usual sense. From the point of view of statistical mechanics, this is an extraneous assumption! Recently, Lieb and Lebowitz[108] have proved that ionic systems do indeed have thermodynamic properties.

That is the newest general result. The oldest is the familiar Debye–Hückel limiting law, which may be expressed as follows:

$$\lim_{\kappa \to 0} \sum A^{ex}/\kappa^3 = 1/12\pi \tag{58}$$

where

$$\kappa^2 \equiv (4\pi/\varepsilon kT) \sum \rho_s e_s^2 \tag{59}$$

This is a general result for the asymptotic form of A^{ex} at low concentration: its functional dependence on concentration as the concentration goes to zero. The more familiar corollaries in terms of $\ln \gamma_\pm$, the osmotic coefficient, the heat of dilution, the partial molar volume, etc., are all readily obtained from this simplest form by thermodynamic operations.

Another general result of the same kind is

$$\lim_{\rho_s = 0} \frac{A^{ex} - (\kappa^3/12\pi)}{\kappa^4 \ln(\kappa L)} = \sum \rho_s e_s^3 \tag{60}$$

a higher-order limiting law first found by Gronwall.[109,110] In this equation, L is *any* constant with the units of length. The reason for this is as follows. We have, for a given electrolyte solution or other ionic system,

$$A^{ex} = \alpha\kappa^3 + \beta\kappa^4 \ln(\kappa L) + \gamma\kappa^4 + \cdots \tag{61}$$

where the coefficients α and β are given by the preceding limiting laws. Now, if L' is another length, then we also have

$$A^{ex} = \alpha\kappa^3 + \beta\kappa^4 \ln(\kappa L') + [\gamma + \ln(L/L')]\kappa^4 + \cdots \tag{62}$$

with the same α and β coefficients. There is no result for γ which is general in the same sense; it depends on the functional form of the pair potentials [see equation (33) and its discussion]. Therefore, there does not seem to be any natural way to choose the length L, although a convenient choice is the Bjerrum-type length† :

$$b = e^2/\varepsilon kT \tag{63}$$

where e is the electronic charge.[113,114]

It is possible to overcome this difficulty in comparing the higher-order limiting law with experimental data.[19] However, the effect is not easy to find in the experimental range and it is often neglected. It does vanish for single, symmetrical electrolytes: 1–1, 2–2, etc. It makes a *large* contribution to the deviations from the ionic strength principle (i.e., to the Harned coefficients) in dilute mixtures of electrolytes of charge type such that $\sum \rho_s e_s^3$ changes with composition at fixed ionic strength.

Another general result pertains to mixtures of two electrolytes with a common ion (Harned's rule mixtures) in which $\sum m_s e_s^3$ does not change when the mixing fraction y is changed at constant molal ionic strength (see Section VI). The simplest coefficient

---

†The Bjerrum length is often taken as one-half this. See, for example, Table 7.1 (Appendix) of Robinson and Stokes.[113]

characterizing the change in excess free energy on mixing at constant I, namely g_0 in Fig. 8, is then governed by the limiting law[19,114]

$$\lim_{\kappa \to 0} (1/g_0) \, dg_0/d\kappa = e_1^2/\varepsilon kT \qquad (64)$$

There are at present no data at sufficiently low concentrations that this limiting-law effect is dominant, but the result should be used in the extrapolations needed when one makes Gibbs–Duhem integrations of data obtained at higher concentrations.

In mixtures for which $\Sigma \, m_s e_s^3$ changes on mixing at constant I, the limit in equation (64) does not exist; there is a logarithmic infinity in g_0 as $I \to 0$ instead.[19] Of course, like the Debye–Hückel limiting law, the Gronwall limiting law and the mixture limiting law may be converted by the thermodynamic operations from laws for the total excess free energy to laws for the activity and osmotic coefficients, heats of dilution, partial molar volumes, etc.

These results for the thermodynamic properties of ionic systems all arise from the long range of the Coulomb potential. This also leads to some general results for the ion–ion correlation functions $g_{ab}(r)$ in such systems.

One is the local electroneutrality condition, better called the zeroth-moment condition†:

$$e_a + \sum \rho_s e_s \int [g_{as}(r) - 1] \, d\mathbf{r} = 0 \qquad (65)$$

Another such exact result is the second-moment condition due to Stillinger and Lovett[75,77]

$$6 + (4\pi/\varepsilon kT) \sum_s \sum_s \rho_s e_s \rho_{s'} e_{s'} \int [g_{s,s'}(r) - 1] r^2 \, dr = 0 \qquad (66)$$

These equations are very useful for checking the accuracy of computed correlation functions, especially since they are very sensitive to errors in the correlation functions at large r, where the computations are least reliable.[20]

Possible additional applications of the second-moment condition have been pointed out by Stillinger and Lovett.[77]

†Stillinger and Lovett.[75] See, e.g., Hill,[117] equation (53b).

The zeroth-moment condition requires that the local charge concentration of the atmosphere of ion a,

$$CH_a(r) \equiv \sum_b e_b \rho_b \, g_{ab}(r) \tag{67}$$

when integrated over all space, is equal to minus e_a. It is of interest to inquire about the local charge concentration itself and, in particular, whether it invariably is opposite in sign to e_a. In the Debye–Hückel limiting-law region, where one has

$$g_{ab} = \exp[-e_a e_b(\exp -\kappa r)/\varepsilon kTr] \tag{68}$$

it is well known that $CH_a(r)$ is opposite in sign to e_a at all r, while in a crystalline salt, $CH_a(r)$ oscillates in sign : it becomes alternately positive and negative as r increases. Therefore, one may ask whether, as an electrolyte solution is concentrated, one comes to a critical concentration above which $CH_a(r)$ oscillates in sign. We shall call κ^* the value of the Debye κ at the critical concentration.

This problem has so far only been studied in the context of a charged hard-sphere model with two ionic species of the same size and opposite charge. It is discussed here with the general results because it seems that κ^* ought to be roughly independent of the details of the model.

Kirkwood pointed out on the basis of his integral equation theory[116] that a critical concentration exists, and Kirkwood and Poirier,[65] using the same method, found $\kappa^* \mathring{a} \sim 1.03$. Outhwaite[52] has extended this method by refined estimates of the "fluctuation terms" which appear when one imbeds the Poisson–Boltzmann equation in exact theory and finds $\kappa^* \mathring{a} \sim 1.495$. Stillinger and Lovett[77] find from the second-moment condition that $\kappa^* \mathring{a} \leq \sqrt{6}$. It may be remarked that for an aqueous 1–1 electrolyte with hard-sphere ions 4 Å in diameter, one has $\kappa^* \mathring{a} = 1.495$ at 0.53 M. It is not clear whether the critical concentration discussed here has any thermodynamic consequences since neither the experimental systems nor accurately studied model systems give any evidence of transitions in thermodynamic properties in this range.†

---

†This is a controversial point. See, e.g., Vaslow.[116a]

VIII. PROPERTIES OF THE PRIMITIVE MODEL

By far the greatest number of studies have been made on this model for ionic systems. It corresponds to a system of charged hard spheres (the ions) in a structureless dielectric medium (the solvent). The ions are each made of a material with the same dielectric constant as the solvent and each has its charge at its center.

More precisely, the primitive model is defined by the equations

$$U_N(\{N\}) = \sum_{\text{pairs}} u_{ij}(|r_i - r_j|) \tag{69}$$

$$u_{ab}(r) = \text{COR}_{ab}(r) + e_a e_b/r \tag{70}$$

$$\text{COR}_{ab}(r) = \infty \quad \text{if} \quad r < r_a^* + r_b^*$$

$$\text{COR}_{ab}(r) = 0 \quad \text{if} \quad r_a^* + r_b^* < r \tag{71}$$

In equation (69), i and j are the labels of two of the N molecules. There are various pair-potential functions corresponding to the species a of molecule i and the species b of molecule j, as exhibited in equation (70). In this model, the pair potential has a repulsive core term COR and a Coulomb term which is written out. The COR term is defined in equation (71) (Fig. 9). Here, r_a^* is the radius of an

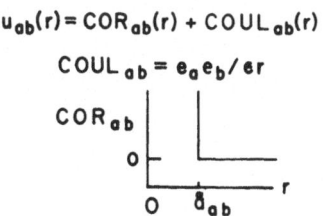

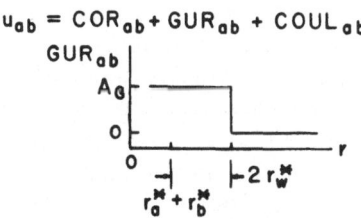

Figure. 9. Potential functions for some discontinuous models.

ion of species a, and r_b^* that of an ion of species b. These radii are the only parameters appearing in this model and, in most uses of the model, are regarded as adjustable parameters. The dielectric constant ε must, according to the McMillan–Mayer theory, be taken as that of the pure solvent in the model.

The simplest system of this sort one has but one species a of ion present at concentration ρ_a. Of course, the system is then not electrically neutral, but it is not hard to recognize and omit the self-energy terms arising from the net charge. Computations for this model, with the additional restriction $r_a^* = 0$, have been made by Carley[127] using the HNC equation and by Brush et al.[34] by the Monte Carlo method with reasonable agreement in the range of Bjerrum length $e_a^2/\varepsilon k T$ and concentration ρ_a that is relevant to the properties of aqueous electrolyte solutions up to a few molar in concentration. This comparison tends to support the hypothesis that the HNC equation may yield reasonably accurate properties of models for electrolyte solutions of practical interest.

We now turn to the primitive model with two ionic components; $a = +$ and $b = -$, corresponding to a solution of a single electrolyte. In this case, we define the parameter

$$\mathring{a} = r_+^* + r_-^* \tag{72}$$

which is the collision diameter for a $(+ -)$ pair (Fig. 9). It is convenient to discuss the calculations which have been done in terms of the dimensionless quantities $q \equiv |e_+ e_-|/\varepsilon k T \mathring{a}$, $c \equiv (\rho_+ + \rho_-)a_0^3$, r_-^*/r_+^*, and $|e_+/e_-|$. Most calculations have been made with the last two quantities set at unity: $r_+^* = r_-^*$ and $e_+ + e_- = 0$. Therefore, the first two, the Bjerrum length divided by the distance of closest approach, and the reduced concentration, are used to organize our discussion. In Fig. 10, are indicated several domains of q, c values that are of special interest. The dotted lines indicate where one finds models representative of systems of practical interest, namely quite far from both axes on the scale represented here.

The region of low c includes, extremely near the $c = 0$ axis, the region of validity of the Debye–Hückel limiting law, the Gronwall limiting law, and the limiting law for mixtures.

At somewhat higher concentrations, it is expected that the cluster expansion, equation (32), truncated after the ρ^2 term, will give accurate results. Some computations for this approximation,

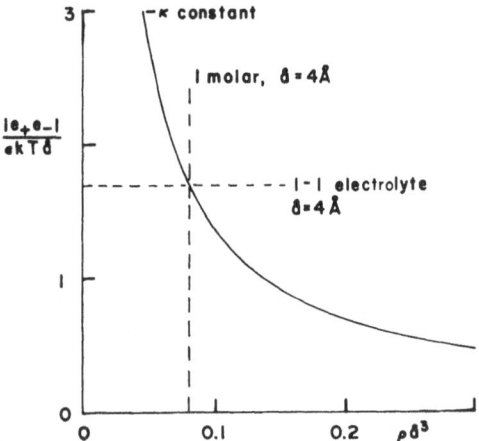

Figure 10. Reduced variables for discussing ionic solution theory. The 1-1 electrolyte line is for aqueous
solutions at 25°C and the κ constant curve is for variation
of $\mathring{a}$ at fixed εT.

which is the collision diameter for a $(+ -)$ pair (Fig. 9). It is con-
electrolytes, with $r_+^* = r_-^*$, have been made by Poirier[98,117] with
the results presented in Fig. 11. The abscissa scale has been chosen to
show that the computed values are at least as nearly linear in the *cube*
root of the concentration as the experimental data. This observation
tends to demolish the familiar argument that such cube-root
dependence shows there is some sort of lattice formed by the ions in
the solution, since there is nothing else about the model or these
computations which suggests that such a lattice is formed. The
computed values are apparently in quite good agreement with
experiment. However, the significance of this is not so clear, since
we have no independent evidence that the cluster expansion, truncated in this way, gives an accurate treatment of the model over this
range of b and c.

   In work of this kind, one adjusts $\mathring{a}$ to get the best agreement of
computation and experiment. The bars in Fig. 11 represent the points
at which the computed $\ln \gamma_\pm$ was fitted to the experimental result by
adjusting $\mathring{a}$. To prevent confusion, a name is needed for ion size
parameters deduced by comparing computations with experiment in
this way ; we shall call them Debye diameters or Debye radii, as may

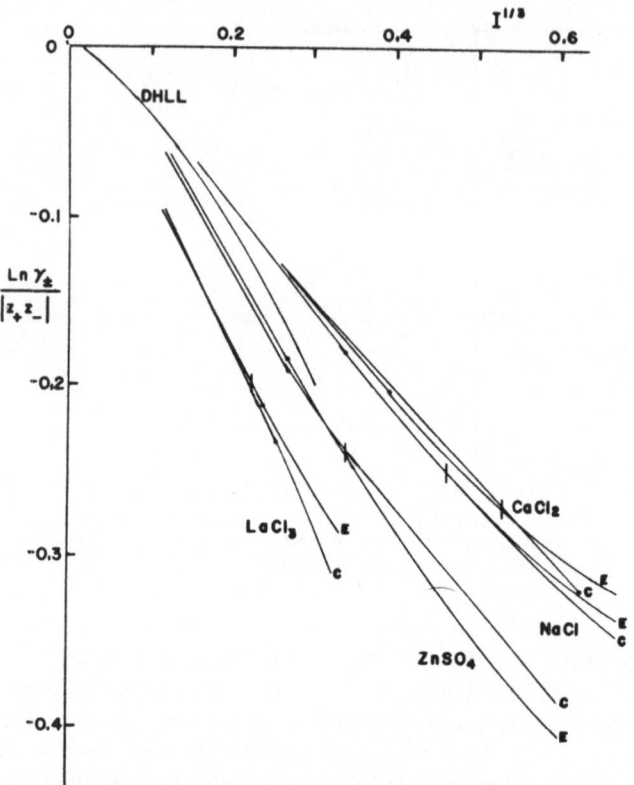

Figure 11. Primitive-model results for aqueous solutions at 25°C obtained by Poirier[117] plotted against the cube root of the molar ionic strength. DHLL is the Debye–Hückel limiting law on this scale. For each of the salts, the experimental data quoted by Poirier lie on curve E, while his computed results lie on curve C. The deviations of the data or computed results from the respective curves are not visible on the scale shown here. The bars show the points at which the model was fitted to the experimental systems by adjusting $\mathring{a}$. The heavy dot on each curve marks the high-I end of a straight-line segment.

be appropriate. They may be compared with ion size parameters taken from other sources, for example, radii taken from Pauling's compilation based on the crystallography of the alkali halides.[118] We shall call these Pauling radii. There are strong arguments for the superiority of other sets of crystal radii, but, as we shall see later, these distinctions probably have no significance for the molecular interpretation of properties of electrolyte solutions because they are not relevant for more realistic core potential functions!

It is of interest that the $\mathring{a}$ values deduced by Poirer for 1–1 and 1–2 electrolytes are very close to those obtained in a similar way but using extended forms of the Debye–Hückel equation.[102] Thus, in this range, several different approximation methods give nearly the same results for the primitive model, which is some evidence that they are all accurate within, say, 0.001 in $\ln \gamma_+$ in this range. In this region, the Stillinger–Lovett theory, which is based on an ion-pair formulation of the particle components of the ionic system, is found to give[76]

$$\ln \gamma_\pm = -A\rho^{1/2} + B\rho + C\rho^{4/3} + \cdots$$

for the primitive model with both ionic species having the same sphere diameter and the same magnitude of charge. The coefficient A is the usual Debye–Hückel limiting-law coefficient, the coefficient B depends on the ion size parameter as usual, but the nonzero coefficient C seems to be a new feature not found in other theories. According to the cluster-expansion method,[19] the next term after $B\rho$ in this case is expected to be a term proportional to ρ^2. It will be interesting to see how this difference is resolved.

In the region of low c but at higher q, that is, for 2–2 electrolytes, Poirier's computations exhibit a feature of the experimental data, namely approach to the limiting-law line from the lower (attractive) side which cannot be obtained from various extended forms of the Debye–Hückel theory without assuming a $(+-)$-complex with an association constant K_a which is treated as an adjustable parameter. This suggests that in the region of low c the cluster expansion truncated after the ρ^2 term is accurate to higher values of the Bjerrum parameter than various extended forms of the Debye–Hückel limiting law. Again, there is another approximation method which gives similar results; Guggenheim's numerical solution of the full Poisson–Boltzmann equation.[51] Several interesting features of this comparison are exhibited in Fig. 12.

Here, the function plotted is chosen to emphasize the deviations from the limiting law. It also has an advantage over functions of $\ln \gamma_\pm$ in comparison with experiment: in this case, the experimental data are osmotic coefficients and additional errors are introduced in the Gibbs–Duhem integration needed to get activity coefficients. In this plot, the Debye–Hückel limiting law is a straight line DHLL.

For $\mathring{a} = 3.47$ Å, Guggenheim's computation is the curve G, the extended limiting-law expression

$$\ln \gamma_\pm = -Az^2 I^{1/2}/(1 + B\mathring{a}I^{1/2}) \tag{73}$$

gives curve F, and the cluster expansion truncated after the ρ^2 term gives curve P.[119] We remark that in this plot introduction of more attraction into the model raises a point, while introduction of more repulsion lowers it. Thus, at low concentration, the data[120] tend to exhibit more attraction than the limiting law (tendency to association), while at high concentration, they exhibit more repulsion. The P and G curves seem to show that this may be accounted for by more accurate treatment of the model. Moreover, their mutual consistency indicates that both are reasonably accurate up to $I = 0.04\ M$. Then, clearly, curve F is the result of an inaccurate treatment of the model and it does not seem to offer any basis for the molecular interpretation of these data.

There is a considerable body of work (e.g., Davies[121]) in which association constants for electrolytes are determined by a process which is equivalent to taking curve F in this figure as the behavior of a model for a strong electrolyte and attributing the difference between this and experimental data to ion association. There is a basis for such procedures in the Bjerrum–Fuoss theory[79] of combined ion-atmosphere and association effects which is doubtless accurate when the latter are strong and the former weak, corresponding roughly to the data points being all far above the DHLL line in Fig. 12, but there seems to be no reason to assume that the procedure is also valid in the converse case, represented by the data in this figure. It must be emphasized that such procedures cannot be justified by their empirical success in comparison with experimental data, since the U_N for the real systems is different from that for the primitive model. However, the empirical success of these procedures does invite a search for a theoretical basis.†

The use of the cluster expansion in equation (22) truncated after the second term requires a method to estimate the cluster integrals $B_2(\kappa)$.

Tables given by Poirier enable one to estimate either $B_2(\kappa)$ or slightly modified integrals appropriate for calculating $\ln \gamma_\pm$ or ϕ.

†For a review of qualitative aspects, see Swarc.[122]

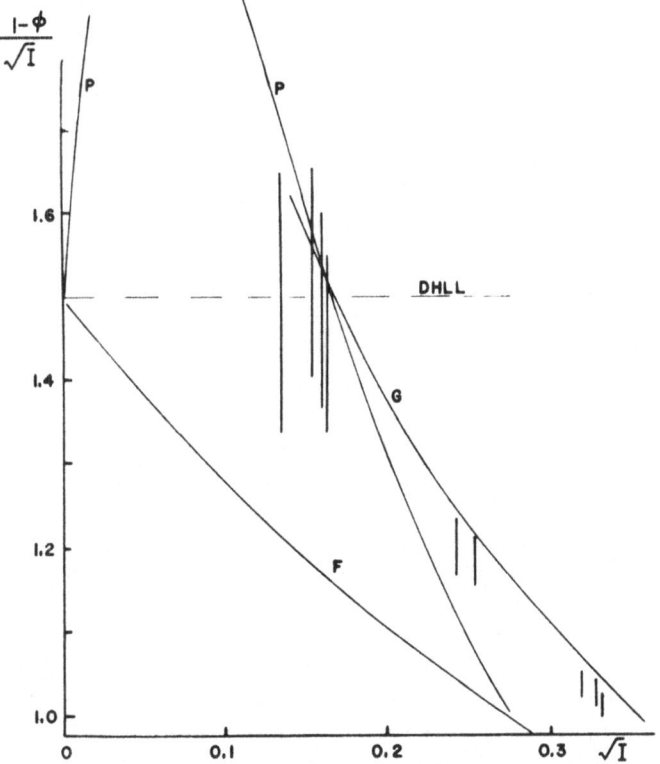

Figure 12. Calculations and experiment for aqueous 2–2 electrolytes at 0°C.
The vertical lines represent experimental data points taken with the authors'
uncertainty of 0.002°C in the freezing-point determination.[120] The curve P has
a maximum off the scale of this figure.

Other estimation procedures are described by Friedman,[19] who also
gives tables of a dimensionless form of $B_2(\kappa)$. However, experience
has shown that such tables are often not as convenient as a computer
program which numerically evaluates the integrals for any given
parameters.[123]

On the other hand, Scatchard[124] has shown how certain
tabulated solutions of the Poisson–Boltzmann equation[125] may be
used to calculate activity coefficients in ionic solutions of simple
electrolytes. His results are in good agreement with those of Guggenheim.[51]

Ebeling, Kelbg, and Kreinke[126a-c] have investigated the behavior of $B_2(\kappa)$ at low κ and large Bjerrum ratio q. If q is large enough and κ small enough, the cluster term reduces to Bjerrum's expression for the ion–ion association constant provided an assumption is made concerning the thermodynamic definition of the latter. The authors discuss the various assumptions which may be made concerning the association constant and show how Bjerrum's theory goes into the cluster expression as one "turns down" q.

The region of low b in Fig. 10 is natural for the application of the methods of Stell and Lebowitz,[72] but this has yet to be done. However, Carley[127] has computed osmotic coefficients for some primitive model systems in this region by means of the PY and HNC equations without the Allnatt modifications.

We turn now to computed results for aqueous 1–1 electrolytes up to the molar concentration range. Monte Carlo computations for such systems have been reported by Vorontsov-Veliaminov, and coworkers,[128,129a,129b] and some of their results are shown in Fig. 13. They studied Markov chains about 50,000 steps long and their results seem accurate within a few percent at the higher concentrations. They also have results for some other models; reference will be made to these later. Concordant Monte Carlo results have recently been obtained by Card and Valleau.[130] Monte Carlo results for the primitive model had already been reported by Poirier,[131] but only for one system at about 0.01 M.

Extensive computations for primitive models for 1–1 electrolytes up to 1 M have been made by the HNC method by Rasaiah and Friedman.[20,24] The results obtained by this method agree quite well with the Monte Carlo results for the same model, as shown in Fig. 13. Good agreement was also obtained for a model in which the ions sizes were taken unequal. The accuracy of the HNC method for treating models for aqueous 1–1 electrolyte solutions has also been examined by a variety of self-consistency tests and other quality tests with very encouraging results.[20] The results of primitive-model computations by this method which follow have all been subjected to various self-consistency tests.[20]

First, we note that the HNC results for the osmotic coefficients for the primitive model that are to be described are concordant, in the small range of parameters in which a comparison is possible, with results obtained by Carley[24,127] using the HNC and PY equations

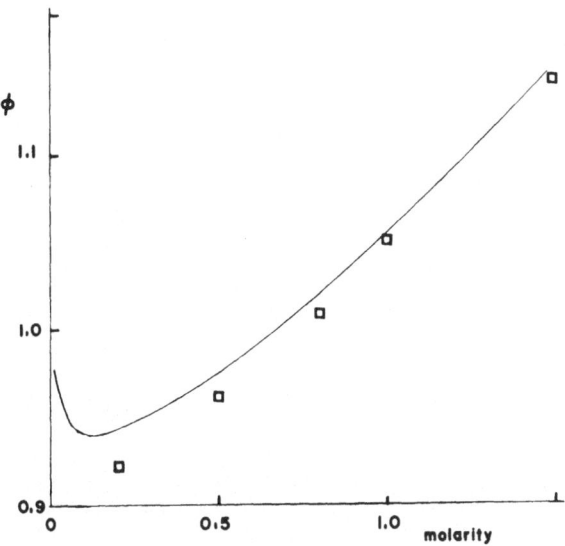

Figure 13. Comparison of Monte Carlo and HNC approximations for the primitive model of an aqueous 1–1 electrolyte solution at 25°C, with both ion diameters equal to 4 Å. Solid curves: HNC results. Squares: Monte Carlo results.

without the Allnatt[59] modifications and by Möller using the BGY equation. The latter results are published only in an article by Falkenhagen and Kelbg.[50] At 0.4 M the curve of ϕ as a function of $å$ for aqueous 1–1 electrolytes with $r_+^* = r_-^*$ which may be constructed from Möller's results[50] agrees with that from the HNC equation[20,26] within 0.02 for 2.5 Å $\leq å \leq$ 4.7 Å.

In Fig. 14, we have the function $A^{ex}/kT(\rho_+ + \rho_-)$, which is qualitatively similar to $\ln \gamma_\pm$, as a function of ionic strength. The experimental curve is for aqueous LiBr, and we shall discuss primitive-model computations in which the parameters are chosen for this system and, in particular, $å$ is the Debye diameter chosen to give a good fit to experiment at low ionic strength.

In this figure, the curve labeled DHLL + B_2 is obtained by the same approximation as the results in Fig. 11, namely, equation (32) truncated after the ρ^2 term. This also corresponds rather closely to what is obtained from extended forms of the Debye–Hückel limiting law, such as equation (73). Proceeding on the assumption that

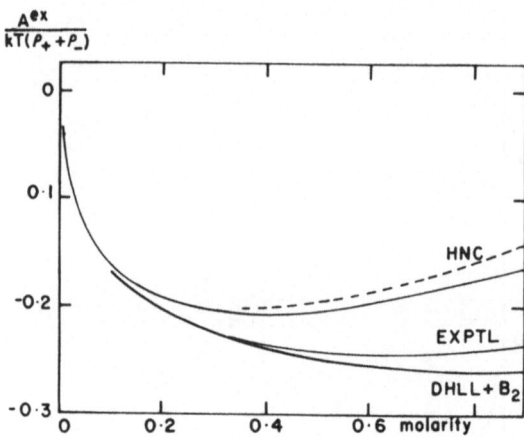

Figure 14. Result of HNC calculation for the primitive model for aqueous LiBr at 25°C compared with experiment and with another approximation. Ion size parameters: solid curves, $r_+^* = r_-^* = 2.3\,Å$; dashed curve, $r_+^* = 1.8\,Å$, $r_-^* = 2.8\,Å$.

DHLL $+$ B_2 was an accurate representation of the behavior of the primitive model, a number of authors have undertaken to interpret the difference between this curve and the experimental one in terms of effects omitted from the model. Although the comparison in this figure is only for LiBr, the experimental curve also lies above the DHLL $+$ B_2 curve for most other alkali halides, so the effect omitted by the primitive model has been taken to be something which is effectively a repulsion, i.e., an effect to raise A^{ex}, ϕ, or $\ln \gamma_\pm$. While these contributions to our interpretation of the properties of ionic solutions have not been shown to be imbedded in exact theory, in the sense of the theories discussed in Section IV, they have had remarkable success in producing equations which, with one or two adjustable parameters, fit the free-energy data for aqueous solutions of single electrolytes.[113,132]

The DHLL $+$ B_2 curve in Fig. 14 is computed for a model in which $\mathring{a} = 4.6\,Å$, is chosen to give a good fit to experiment at low concentrations; it is a Debye diameter. The HNC computation for the same model yields the curves labeled HNC in the figure. The solid HNC curve is for $r_-^*/r_+^* = 1$, while for the other HNC curve, it is 3.6 (or 1/3.6; this model does not distinguish). The HNC

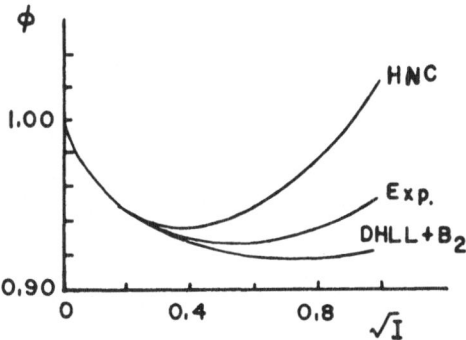

Figure 15. Calculations for the primitive model for aqueous NaCl at 25°C compared with experiment. Ion size parameters: $r_+^* = r_-^* = 1.95 \text{ Å}$.

computations are believed to be accurate within one or two widths of the lines in the figure. Apparently the primitive model with the Debye $\mathring{a}$ exhibits too much repulsion compared with the experimental system and a better model needs an additional effect which reduces this repulsion, contrary to the conclusion reached earlier on the basis of the DHLL + B_2 approximation or its equivalents. The figure also shows that this conclusion is not changed if we consider models with radius ratios different from unity.

It is of great interest to make a comparison like that in Fig. 14 (for LiBr) for the rest of the alkali halides. One other example, for NaCl, is given in detail in Fig. 15.[24,139] For the rest, it is sufficient just to compare the free energies at 1 M, since the curves have all about the same shape. Of the free-energy quantities that are relevant, A^{ex}, ϕ, and $\ln \gamma_\pm$, it is ϕ that is the easiest to obtain from the computations, and this is the function for which the comparison is made in Table 2.[139] In all these computations, the radius ratio is unity and $\mathring{a}$ is the Debye diameter; i.e., it is chosen to give a good fit at low concentration. We must conclude that the primitive model, with the Debye diameter, uniformly exhibits more repulsion than that corresponding to the behavior of the aqueous alkali halides at higher concentrations, so that the situation depicted in detail for LiBr (Fig. 14) is relevant to all of these systems.

The primitive model may also be employed with distance parameters taken from other data. Some HNC investigations have

Table 2
Comparison of Osmotic Coefficients of 1 M Aqueous Alkali Halides at 25°C with HNC Calculations for the Primitive Model with Debye Diameters[a]

		Li$^+$	Na$^+$	K$^+$	Rb$^+$	Cs$^+$
Cl$^-$	$\dot{a}$	4.25	4.0	3.8	3.6	3.0
	HNC Δ	91	47	15	-13	-84
	Expt. Δ	37	-47	-77	-85	-108
Br$^-$	$\dot{a}$	4.3	4.1	3.84	3.55	2.93
	HNC Δ	102	64	21	-20	-92
	Expt. Δ	61	-18	-61	-83	-109
I$^-$	$\dot{a}$	5.1	4.2	3.94	3.5	2.87
	HNC Δ	289	82	37	-27	-97
	Expt. Δ	119	27	-31	-76	-104

[a] In each case, $r_+^* = r_-^*$. Here, $\Delta = 1000(\phi - 1)$. For the same model except with no charges, $\Delta = 288$ for $\dot{a} = 4.6$ Å.

been made with Pauling radii [defined above, following equation (72)]. An example is given in Section IX.

We turn now to the study of the primitive model for electrolyte mixtures. First, we note a result which was obtained by the cluster-expansion method[19] and which does not depend on the model; namely, that very near zero ionic concentration, the mixing coefficient g_0 (cf. Fig. 8) is given by

$$g_0 = - [1.2046 \times 10^{-3}/z_1^2(z_1 - z_3)^2 V_w]$$

$$\times \int [2\psi_{12} - \psi_{11} - \psi_{22}] \exp[-e_1^2(\exp - \kappa r)/\varepsilon k Tr] \, d\mathbf{r}$$

(74)

where e_a is the charge on an ion of species a, z_a is the same in electronic units, V_w is the volume in liters of a kilogram of the solvent, r is in angstrom units, and

$$\psi_{ab} \equiv \exp\{ -(1/kT)[u_{ab}(r) - (e_a e_b/\varepsilon r)] \}$$

(75)

is the Boltzmann factor for the non-Coulomb part of the potential between a particle of species a and one of species b. This equation is relevant to a mixture of electrolyte A (ions of species 1 and 3) with electrolyte B (ions of species 2 and 3), so it is clear that g_0 near $I = 0$

depends on differences among the 1–1, 2–2, and 1–2 interactions, i.e., among the interactions of ions of *like* charge in the mixture. Furthermore, for r larger than, say 5 Å, the bracketed function in the integral in equation (74) must vanish so the integral is not likely to exceed 500 Å^3, which gives a bound $|g_0| < 1.5$ for mixtures of 1–1 electrolytes. The only primitive-model computations made for electrolyte mixtures employ the cluster expansion, equation (32), truncated after B_2. Comparisons of g_0 obtained from models in this way, with the model ionic radii r_a^* taken as various factors times the Pauling radii for particular ions, are compared with experimental data in Fig. 16. The results clearly indicate that the primitive model with Pauling radii gives effects much smaller than those in the experimental systems, while double the Pauling radii fits much better.

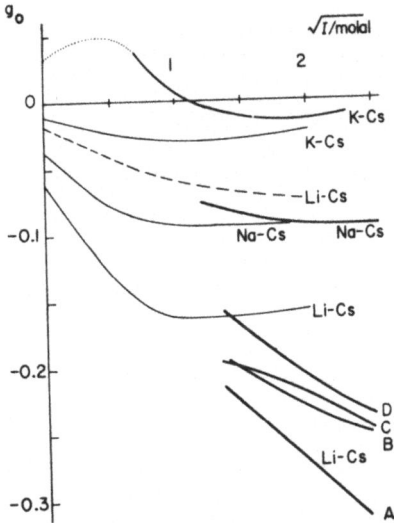

Figure 16. Calculations for the primitive model for mixed aqueous alkali chloride solutions at 25°C compared with experiment. Thin solid curves: computed from the primitive model, each $r_a^* = 2 \times$ Pauling radius. Dashed curve: computed from primitive model, each $r_a^* = 1.5 \times$ Pauling radius. Heavy solid curves: g_0 estimated from the experimental data for w_0. The results of four methods of estimation are shown for Li–Cs. Method A, assuming $g_0 = w_0$, was used for the other systems.

This is taken to be a measure of the failure of the primitive model, since the ions cannot really be this large; this point of view will be developed in later sections. It is noted, however, that these computations reproduce the qualitative behavior of the g_0 function of real systems. This agreement is improved if one uses the cluster expansion truncated after the ρ^3 term, although only if ions with double the Pauling radii are used.[19]

A computation with a model corresponding to a non-symmetrical mixture (ions 1 and 2 having different charges) has been made.[19] It exhibits a divergence of g_0 as $I \to 0$, which is a reflection of the same term in A^{ex} that gives rise to Gronwall's limiting law. This divergence of g_0 is in agreement with experiment.[19]

In summary, the primitive-model calculations for mixtures which are described above, all at a lower level of approximation than the HNC equation, offer interpretation of some of the features of the g_0 data, but for the rest, they mainly show that the primitive model cannot quantitatively account for the data with model parameters having molecular significance.

IX. SOME OTHER MODELS WITH DISCONTINUOUS POTENTIALS

A simple example of a nonprimitive model with a discontinuous potential is the one treated by Eisenthal and McMillan.[74] It has the pair potential

$$u_{ab}(r) = 0 \qquad \text{if} \quad r < \mathring{a}$$
$$u_{ab}(r) = e_a e_b / \varepsilon r \qquad \text{if} \quad \mathring{a} < r \tag{76}$$

We note that this corresponds to equation (70) with

$$COR_{ab} = -e_a e_b / \varepsilon r \tag{77}$$

which is repulsive for a $(+ -)$-pair but attractive for a pair of like-charged ions. Therefore, it can only give physically interesting results at the lowest concentrations where the short-range interactions of $(+ +)$- or $(- -)$-pairs have only negligible effects on the solution properties. A fundamental difficulty with models which, like this one, lack a core repulsion in the pair potential has been pointed out by Ruelle and Fisher, as reported by Eisenthal and McMillan.[134] The model determined by equation (77) was introduced in order to carry the collective-coordinate theory to the point of yielding numerical results for an ionic solution model while using only analytical mathematics except in the evaluation of the final integral. This is achieved with the help of approximations whose effects are hard to estimate. They obtain results for $\ln \gamma_{\pm}$ as a function of concentration for various $\dot a$ which are very close to those obtained by Poirier[120] with a different model and a different approximation procedure (Section VII). It seems particularly clear in the present case, because of the nonphysical nature of the model, that the agreement of these results with experiment cannot be used as a step toward our understanding of the molecular interactions in electrolyte solutions. It may also be remarked that the collective-coordinate method was independently studied by Kelbg[73] but with a better physical model, to be described below.

One may make the pair potential in equation (70) more flexible by adding a term to represent a possible non-Coulomb interaction of the two particles at separations somewhat larger than the sum of of their radii. Gurney[135] pointed out that when two ions are close enough together so that their cospheres overlap, there may be expected to be an additional force between them because of this. The cosphere, in this usage, is the region about an ion in which the solvent properties are altered in a nonlinear way by the presence of the ion.† Figure 6 may be taken to represent, in a very crude way, the mechanism by which cosphere overlap contributes to the pair potential: When two ions approach close enough, the sum of their cosphere volumes is reduced by overlap, assuming the cospheres are fixed as shown in Fig. 6. The volume difference represents solvent which must return to its normal state, and the free-energy change for this process is a contribution to the pair potential. We call such a contribution a Gurney (GUR) potential. It is expected to be larger in many cases than other effects, such as van der Waals interaction, charge-polarizability interaction, covalent interaction, etc. which we have also neglected. Frank has pointed out that the same process may make an important contribution to the entropy of dilution as well.[136] For further discussion of cosphere overlap, reference may be made to an earlier review by Desnoyers and Jolicoeur in this series.[4]

---

†Linear effects could be accounted for in other ways. For example, the linear part of the dielectric response of the solvent to the electric fields of the ions is accounted for by the dielectric constant factor ε in the denominator of the Coulomb term.

Now we consider the potential

$$u_{ab}(r) = COR_{ab}(r) + GUR_{ab}(r) + (e_a e_b / \varepsilon r) \qquad (78)$$

$$GUR_{ab}(r) = (A_g)_{ab} \qquad \text{if} \quad r < r_a^* + r_b^* + 2r_w^*$$
$$GUR_{ab}(r) = 0 \qquad \text{if} \quad r_a^* + r_b^* + 2r_w^* < r \qquad (79)$$

where r_w is the radius of a solvent molecule and the other symbols have the same meaning as in equation (71). Thus, in this model, the non-Coulomb part of the potential has the form shown in Fig. 9. We note that this model offers three parameters: core diameter, cosphere thickness, and A_g for each species pair ab. Thus, there are nine parameters for a solution of a single electrolyte!

Computations with this potential were first made by Kelbg[60] for comparison with the properties of dilute solutions of tetraalkyl-ammonium halides. The approximation method he used is based on

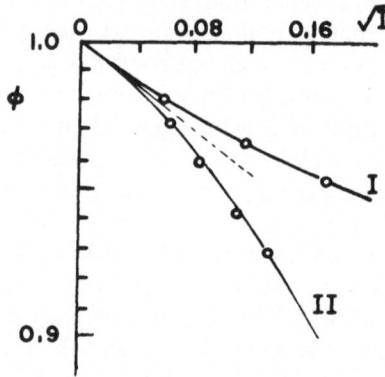

Figure 17. Calculations for the model comprising charged hard sphere plus square well, compared with experiment for aqueous tetralkylammonium halides at 0°C.[60] (I) $(C_4H_9)_4NCl$. (II) $(C_4H_9)_4NI$. Circles: experiment. Solid curves: computed. Dashed line: DHLL. For both salts, the model has the cospheres touching at $r = 14$ Å and the cores touching when $r = 7$ Å, where r is the center-to-center separation for a $(++)$-, $(--)$-, or $(+-)$-pair of ions. For the chloride, $(A_g)_{ab} = -0.16kT$, while for the iodide, $(A_g)_{ab} = -0.37kT$.

the Falkenhagen–Kelbg integral equation but, at the level of approximation employed, rather looks like a cluster expansion truncated after the ρ^2 term, and indeed gives practically the same computed values from the model.[137] The results of two of Kelbg's computations are compared with experimental data in Fig. 17. In these computations, Kelbg used three adjustable parameters, the core diameter, the cosphere thickness, and the mound height (or minus the well depth) A_g, which he assumed were the same for $(+\,+)$ and $(-\,-)$ as for $(+\,-)$ interactions. This does not seem particularly reasonable from the point of view of molecular interpretation of the data[136] and can only be defended if it does not make any difference what one chooses for the $(+\,+)$ and $(-\,-)$ potentials. At sufficiently low concentrations, it is always true that A^{ex}, ϕ, and $\ln \gamma_{\pm}$ are determined mainly by the $(+\,-)$ interaction,[42] but the concentration range in this comparison is not low enough. It is found[137] by direct calculation that $\phi-\phi_{DHLL}$, the simplest quantity with information about the intermolecular forces, is very sensitive to whether one takes $(A_g)_{++} = (A_g)_{--} = 0$ or $(A_g)_{++} = (A_g)_{--} = (A_g)_{+-}$. This surprising result is presumably not due to failure of the approximation method employed, which ought to be satisfactory at these low concentrations, but rather to the large values taken for the distance parameters, e.g., for $(n\text{-}C_3H_7)_4NCl$, $r_+^* = r_-^* = r_w^* = 2.89$ Å. It may be concluded that the use of this model to obtain a molecular interpretation of the excess thermodynamic properties of the tetraalkylammonium halides is promising but that further work is required. In particular, one needs to use data for mixed electrolyte solutions to get independent evidence about the $(+\,+)$ and $(-\,-)$ interactions.

More recently, the same model has been used with the HNC approximation to find the behavior of the model at concentrations up to 1 M.[138,139] In this work, all of the distance parameters were calculated from Pauling radii and the parameters in the calculations were just the $(A_g)_{ab}$ values, or mound heights. There is an argument for this procedure, namely that a great deal is known about distance parameters from crystallography and other structural investigations while what is really needed is a comparison of model behavior with thermodynamic data to elucidate the energy parameters in the molecular pair-potential functions.

Some computations of this kind are shown in Fig. 18, where the distance parameters are all chosen to correspond to an aqueous

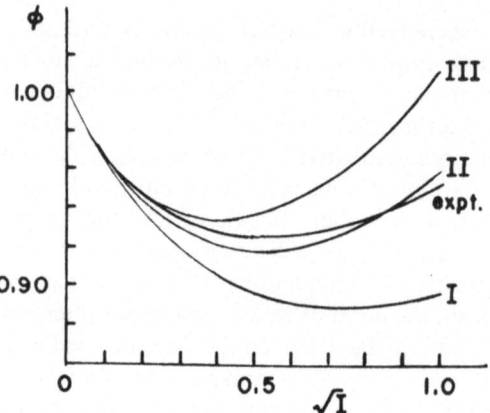

Figure 18. HNC calculations for the model comprising charged hard sphere plus square mound, compared with experiment for aqueous NaCl at 25°C. Parameters in computation: $r^*_+ = 0.95$ Å, $r^*_- = 1.81$ Å; $(A_g)_{++} = (A_g)_{--} = 0$; and $(A_g)_{+-} = 0$ for (I), $= \frac{1}{4}kT$ for (II), $= \frac{1}{2}kT$ for (III).

solution of NaCl. In these computations, A_g for the $(++)$ and $(--)$ interactions was taken as zero. With $(A_g)_{+-} = 0$, we have just the primitive model and, as noted above, we find that the model system with Pauling radii exhibits much less repulsion than does the experimental system. When we include a $(+-)$ Gurney potential of only $\frac{1}{2}kT = 300$ cal/mole in the model, we find that it exhibits much more repulsion than the experimental system, while $(A_g)_{+-} = \frac{1}{4}kT = 150$ cal/mole is about right.

Now let us compare this result, in which we find a model whose observable behavior is in accord with that of an experimental system, with the result of a more *ad hoc* sort of theory, in which the assumptions concern distributions or even thermodynamic relations in the actual solution rather than the intermolecular forces in the model. To be explicit, we consider a Robinson–Stokes type of theory[113] in which the parameters are $\mathring{a}$ and h, a hydration number parameter. Consider the implications for two different kinds of experiment. In interpreting rates of chemical reaction between two species, or effects of one species in inducing the NMR relaxation of another,† it is found that the correlation function $g_{ab}(r)$ of the two species at

†See Hertz.[8] Some experiments relevant to models for aqueous 1–1 electrolytes are given by Eisenstadt and Friedman[140a] and by Woessner *et al.*[140b]

contact is especially important. The reason is that the forces responsible for chemical reaction or for NMR relaxation are rather short-range and it is mainly the concentration of pairs in contact $\rho_a \rho_b g_{ab}(r_a^* + r_b^*)$ which controls the rate of the process of interest. In the model in Fig. 18, the separation at contact is $0.95 + 1.81 = 2.76$ Å, while if one uses the Debye diameter, as in the Robinson–Stokes formula, it is 3.9 Å. A short-range interaction, like those mentioned, falls off enormously in passing from 2.8 to 3.9 Å. Furthermore, the correlation function at contact is not greatly affected by the Gurney potential required to bring the osmotic coefficient of the model in agreement with that of the experimental system (Fig. 19). Therefore, for these important applications of what we can learn about the molecular interactions in ionic systems, it makes all the difference whether it is believed that it is the model discussed here or the Robinson–Stokes theory that is representing what really goes on in the system.

The H^{ex} functions, heats of dilution as a function of concentration, of electrolyte solutions are even more varied and specifically characteristic of the ionic species than the free-energy functions. In the model represented by equation (78), the E^{ex} ($\sim H^{ex}$) comes from the temperature dependence of u_{ab} as well as from u_{ab} itself; an exact equation corresponding to equation (15) is

$$E^{ex} = \tfrac{1}{2}\rho^2 \int \{\partial[u(r)/T]/\partial(1/T)\}g(r)\,d\mathbf{r} \tag{80}$$

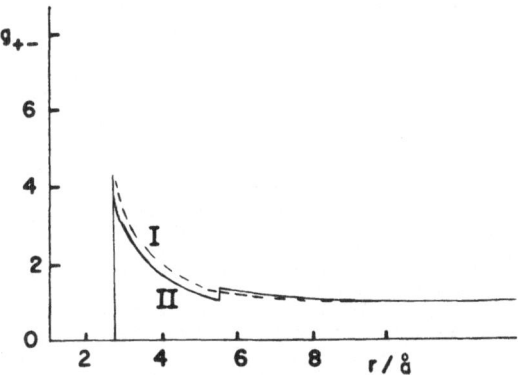

Figure 19. Computed pair correlation functions at 1 M for two of the models in Figure 18.

and it is readily generalized to apply to many-component solutions. If we assume that the only temperature dependence in the model potential enters through the dielectric constant, then it is expected that the E^{ex} curves for various models will tend to form a family which is ordered in a way corresponding to the order of the excess free-energy curves for the same models, contrary to what is observed in the experimental systems. If, in addition, we allow the temperature dependence of A_g to be a parameter, corresponding to an energy of the cosphere overlap different from its free energy, i.e., a nonvanishing entropy effect, then the differences in correlation function near contact from one system to another will be strongly reflected in the E^{ex} functions.

A comparison of this sort for several models for aqueous NaCl is shown in Fig. 20.[139] It is clear that a rather small $\partial A_g/\partial T$ is adequate to account for E^{ex} of NaCl(aq).

The results in Figs. 18 and 20 are typical of the success of the same model in representing the behavior of all of the alkali halides

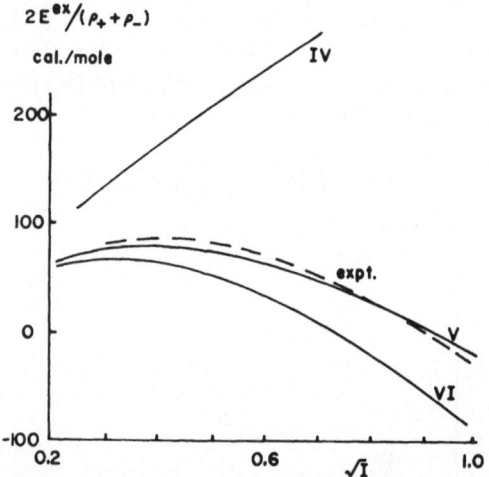

Figure 20. Excess energy of aqueous NaCl at 25°C. Comparison of computation for model comprising charged hard sphere plus square mound and experiment. Model parameters same as in model II in Figure 18, and in addition, $\partial(A_g)_{++}/\partial T = \partial(A_g)_{--}/\partial T = 0$; and $\partial(A_g)_{+-}/\partial T = 0$ for curve IV, $= -1.1k$ for curve V, and $= -1.3k$ for curve VI.

(fluorides omitted) according to a comprehensive investigation by Rasaiah[139] of the application of this model to 1–1 aqueous electrolytes. It should be emphasized that in all of these studies the HNC approximation has been used together with numerous self-consistency tests which tend to verify its accuracy in these applications.[132]

The success of the simple model represented by equation (78) in comparison with experimental systems is reason enough to be interested in it, but not reason enough to accept it as a very good representation of the molecular interactions in the real systems. There are several reasons for this. The discontinuous potentials are not physical: real potential functions do not have corners and kinks. One might think that smooth potentials with approximately the same behavior will give the same observable properties, but some recent results cited in Section X indicate this to be only very roughly true. The choice of $(A_g)_{++} = (A_g)_{--} = 0$ is not natural; comparison with data for mixed electrolyte solutions is needed to elucidate the effect of these parameters. Finally, we have the usual problem that in the model U_N is made up solely of pair contributions, while this is not true in the real systems with which the model results are compared. While it may conceivably turn out that none of these objections is significant, they all require further study.

X. MODELS WITH CONTINUOUS POTENTIAL FUNCTIONS

One motivation for investigating continuous potentials is that they are more realistic. Another is that the computational method employed for solving the integral equations[24] is simplified when the potentials are continuous rather than discontinuous.

Glauberman[63a] and Juchnowski[63b] investigated the model[70] in which the pair potential is

$$u_{ab}(r) = (1 - e^{-\alpha r})e_a e_b / \varepsilon r \qquad (81)$$

It is instructive to note that this corresponds to equation (70) with

$$COR_{ab} = -e_a e_b e^{-\alpha r} / \varepsilon r \qquad (82)$$

which is repulsive for a $(+ -)$-pair but attractive for a pair of like-charged ions. Therefore, it has the difficulties associated with the model in equation (76). This model is peculiarly advantageous for mathematical analysis because the potential has a simple Fourier transform. However, this does not seem reason enough to explore it as a basis for the molecular interpretation of solution properties, although it might yet be useful for testing approximation techniques. The results obtained with this model were reviewed by Falkenhagen and Kelbg.[1]

At the opposite extreme in terms of physical content are models in which $u_{ab}(r)$ is derived by finding the *pseudopotential*, the quantum mechanical equivalent of $u(r)$. If a and b are fundamental charged particles, their kinetic energies are large at small separation, which produces a core repulsion in the pseudopotential, while at large separation, the dominant term is the Coulomb interaction. These theories, due to Morita,[141] Kelbg,[73,142] and Ebeling,[143] are of course of more interest for the treatment of gaseous ionic media, namely plasmas, than electrolyte solutions.

Ramanathan and Friedman[144] have made extensive calculations with the HNC approximation using the following pair-potential function:

$$u_{ab} = COR_{ab} + GUR_{ab} + CAV_{ab} + e_a e_b / \varepsilon r \qquad (83)$$

Here, the core potential is an inverse power function in r:

$$COR_{ab} = B_{ab}[(r_a^* + r_b^*)/r]^n \qquad (84)$$

with the coefficient B_{ab} as well as the distance parameters determined by crystal data using Born's approximation for the lattice energy of an ionic crystal as a sum of repulsive and Coulombic terms.[118] Then, we have

$$B_{ab} = Fe^2/n[r_a^* + r_b^*] \qquad (85)$$

where F is the ratio of Madelung's constant to coordination number for the crystal and e is the electronic charge. Various values of n have been used; for the results described here, we use Pauling's value, $n = 9$.[118] This elementary theory really only yields B_{+-} but they use the same formula for B_{++} and B_{--}.

The Gurney potential in equation (83) is taken to be [cf. equation (79)].

$$GUR_{ab} = (A_g)_{ab} V_w^{-1} V_{mu}(r_a^* + R_a^*, r_b^* + R_b^*, r) \qquad (86)$$

where V_{mu} is the mutual volume of the cospheres of ions a and b at a separation r. The thicknesses of the cospheres are R_a^* and R_b^*. Here, V_w is the molar volume of the pure solvent in the same volume units as V_{mu}. Then, A_g is the free-energy change per mole of the cosphere solvent as it reverts to the normal solvent state. This is simply a more detailed representation of the Gurney potential described in Section IX.

The cavity potential (CAV) arises as a consequence of the fact that each ion is in a region of low dielectric constant compared with that of the bulk solvent. If nothing else, the ion itself is a region where $\varepsilon \sim 2$. While CAV may also be a contribution from the cosphere, it is considered to be a part of the Gurney potential term. Expressions for the cavity potential have been derived by a number of authors, especially Levine and his co-workers.[89,90] The longest-range term is

$$\text{CAV}_{ab} = \frac{1}{2\varepsilon r^4} \frac{\varepsilon - \varepsilon_c}{2\varepsilon + \varepsilon_c} (e_a^2 r_b^{*3} + e_b^2 r_a^{*3}) \tag{87}$$

where ε_c is the dielectric constant of the cavity with the ion in it and ε is the dielectric constant of the pure solvent. This term is always repulsive, as are the higher-order terms as long as r is larger than the sum of the ion radii. The general shape of the various contributions to u_{+-} which arise in this model are represented in Fig. 21.

Some computed results for this model, with parameters chosen to correspond to aqueous NaCl, are shown in Fig. 22. All of these computations have been tested by several self-consistency criteria and there is no reason to doubt that the results represent the behavior of the model as accurately as the HNC approximation applied to the discontinuous models. In all cases, $(A_g)_{++} = (A_g)_{--} = 0$.

Comparing curves IV and VII in Fig. 22, we see the enormous effect of the soft core upon the computed osmotic coefficient. Although this has not been anticipated, it is understandable in view of functions shown in Figs. 19 and 21. The soft core differs from the hard one especially in the range of r where $g_{+-}(r)$ reaches large positive values. The population of pairs $\rho_+ \rho_- g_{+-}$ near contact, each in a potential well due to the other, makes a strong attractive contribution to the free energy which tends to be reduced when we go from the hard- to the soft-core potential. This conclusion is contrary to what might be concluded from the common statement that ions are the same size in solution as in the crystal,[145] but it seems inescapable.

One curve in Fig. 22 is given for $n = 15$ to show that, even with such a large n, the difference from a hard-core model is very large.

Adding the CAV term, which is at least an approximation to an effect in the real systems, makes the difference from the hard-core model greater.

The computed osmotic coefficients compare very poorly with the experimental ones for the real system if $(A_g)_{+-}$ is taken as zero, but also if it is adjusted to make the computed and experimental curves intersect at some point, as may be seen from Fig. 22.

It can be concluded from this study that the model, at least with the Gurney terms for $(++)$ and $(--)$ terms set equal to zero, has properties quite different from those of the experimental system with which it is compared. What is needed to improve the agreement with experiment is a change in the model which produces a markedly larger attractive effect (or smaller repulsive effect) at high concentrations compared to low. Negative Gurney terms for the like-charged interactions might do this since the $(++)$ and $(--)$ interactions

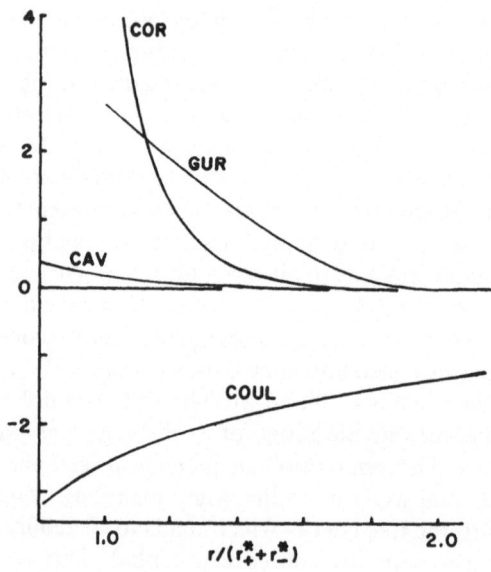

Figure 21. Terms of the model potential for a $(+-)$-pair in units of kT for an aqueous 1–1 electrolyte with $r_+^* + r_-^* = 2.76$ Å, $\varepsilon_{CAV} = 2$, and $(A_g)_{+-} = 100$ cal/mole.

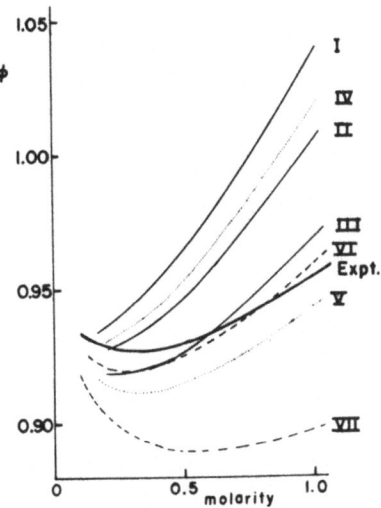

Figure 22. Computed results for models for NaCl(aq) compared with experiment. $r_+^* = 0.95$ Å, $r_-^* = 1.81$ Å. Solid curves: model of equation (83) with $n = 9$, $(A_g)_{++} = (A_g)_{--} = 0$; and $(A_g)_{+-} = 0$ for I, $= -25$ cal/mole for II, and $= -50$ cal/mole for III. Dotted curves: model of equation (83) without CAV term, with all $(A_g)_{ab} = 0$, and with $n = 9$ for IV, $n = 15$ for V. Dashed curve: model of equation (78) with $(A_g)_{++} = (A_g)_{--} = 0$; and $(A_g)_{+-} = 150$ cal/mole for VI, $= 0$ for VII.

become markedly more important as the concentration increases, but it does not seem wise to adjust these except by comparison with data for electrolyte mixtures. The needed change in the model might also be attained by introducing an attractive term in the pair potential u_{+-} at small r, since the population of $(+ -)$-pairs at small r also increases markedly with concentration. This hypothesis is especially attractive since the effect has, in essence, been predicted by Levine and his co-workers in their studies of what we call the cavity potential.[90]

Essentially, what happens is that, when two oppositely charged ions are very close together, the lines of force from one to the other travel much more through the ionic material than through the dielectric medium, giving rise to a large attractive interaction at very small r compared to anything represented by the terms of equation (83). In order to represent this effect, an effective dielectric constant may be *defined* by the equation

$$u_{ab} = e_a e_b / r \varepsilon_{ab}(r) \tag{88}$$

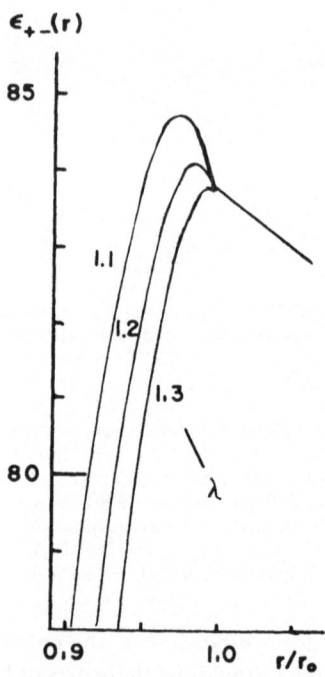

Figure 23. The local dielectric constant $\varepsilon_{+-}(r)$ computed for a model of interpenetrating hydration spheres by Levine and Rozenthal.[90] λ is a measure of the deformation of the spheres as they come together. The right branch where the three curves lie together is due to the cavity potential for nonoverlapping hydration spheres, essentially the same effect which gives rise to the CAV curve in Figure 21. In this model, $\varepsilon = 80$ and $\varepsilon_{CAV} = 5$.

With this definition, $\varepsilon_{ab}(r)$ at $r = \infty$ is exactly the dielectric constant
for the solvent and this may be used as the basis of the statistical-
mechanical theory of the dielectric constant.[86] The behavior of $\varepsilon_{ab}(r)$
at smaller r is not always facile to understand. For example,
$\varepsilon_{+-}(r)$ becomes larger if there is a repulsive effect in u_{+-}, as may
arise, e.g, from the COR or CAV terms, but $\varepsilon_{++}(r)$ becomes smaller
if there is a repulsive effect in u_{++}. In Fig. 23, one of Levine and
Rozenthal's[93] calculated curves for $\varepsilon_{+-}(r)$ is shown. It is obtained
from a model in which ions are regarded as being located in cavities
in the dielectric medium. The region where $\varepsilon_{+-}(r)$ is larger than
$\varepsilon = 80$ corresponds to nonoverlapping cavities; this is the effect
accounted for by the CAV term in equation (83). The region in
which $\varepsilon_{+-}(r)$ becomes very small corresponds to overlapping
cavities. In the Levine–Rozenthal model, these are interpenetrating
cospheres. The effect would equally well arise with ions with soft
cores, like those considered in the model of equation (83), when they
are close enough to deform each other. It can easily be seen from Fig.
21 that doubling the Coulomb potential in the region of separation r
corresponding to moderate ion deformation would tend to cancel
the effect of the soft-core potential.

While this effect seems to be identified, both empirically through
Fig. 22 and through the model calculations, it is risky to do more with
it without the benefit of a statistical-mechanical theory at the
molecular level of the interaction of two ions in a solvent. The dielectric models are neither very easy to study nor very trustworthy,
except in a qualitative sense, when applied to phenomena on a
molecular scale. The comparison with experiment in Fig. 22 is likewise not definitive since it is possible that the deviation of U_N from
pairwise additivity in the real system is significant even below 1 M.

Computations have also been made for mixed 1-1 electrolyte
solutions with a common ion using the model of equation (83). Some
results are presented (Fig. 24) for a model with parameters appro-
piiate to an aqcuous solution of mixed 1-1 electrolytes at 25°C with
$r_1^* = 0.60$ Å, $r_2^* = 1.69$ Å, $r_3^* = 1.81$ Å; thus, electrolyte A (ions of
species 1 and 3) corresponds to LiCl, electrolyte B to CsCl. Also,
these computations are for a model in which $n = 9$ and $(A_g)_{ab} = 0$ for
all pairs.

Computations of the properties of this model were made using
the HNC approximation. The self-consistency tests and other

criteria for the accuracy with which the computed properties represent the model are as good as for computations for a single electrolyte. The computed quantity of greatest interest is shown in Fig. 24. This is obtained from a suitable combination of the compressibility integrals [equation (10)]:

$$\int [g_{ab}(r) - 1] \, d\mathbf{r}; \qquad a = 1, 2, 3; \quad b = 1, 2, 3$$

at each composition y, I for which the model properties are computed. If Harned's rule is obeyed, i.e., the β and higher coefficients in equation (57) are negligible, then these lines are horizontal. Straight lines in Fig. 24 correspond to nonzero Harned α and β coefficients. The computed results in this figure correspond to about the same fraction g_1/g_0 that is found in many real systems. The small curva-

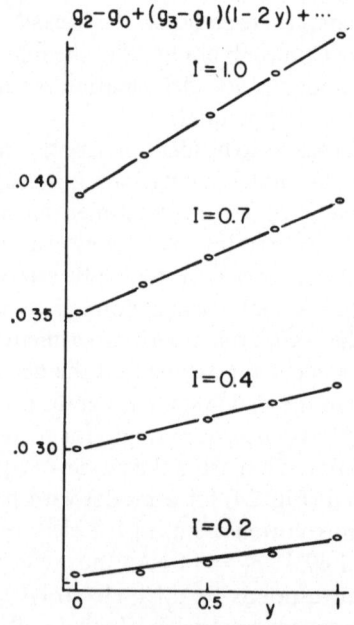

Figure 24. Computed free energy of mixing coefficients for a model of equation (83) with all $(A_g)_{ab} = 0$, $r_1^* = 0.60$ Å, $r_2^* = 1.69$ Å, and $r_3^* = 1.81$ Å.

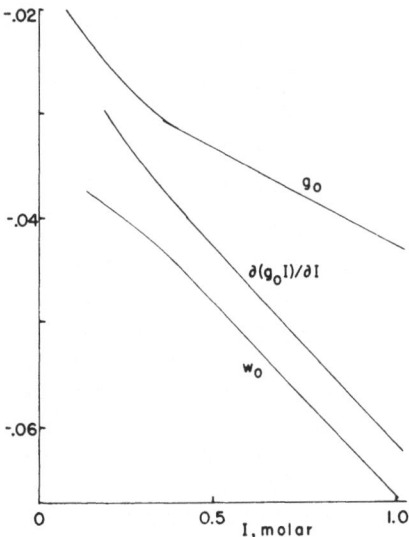

Figure 25. Additional results for model
described in Figure 24.

ture in the computed results at lower I corresponds to nonvanishing
Harned γ coefficients.

The model g_0 results are shown as a function of ionic strength in
Fig. 25. The curve $\partial(g_0 I)/\partial I$ is calculated from the g_0 curve: it is a
thermodynamic expression for w_0 in terms of the g_0 function. The w_0
curve is obtained from the same computed correlation functions, but
by calculating the osmotic coefficients by the virial equation
[equation (15)] and taking differences among $\phi(y, I)$ values at the
same I to get w_0 according to Fig. 8. Thus, the degree to which the w_0
and $\partial(g_0 I)/\partial I$ curves agree is a self-consistency test in which one looks
at variations along an $I = $ const line in Fig. 8, whereas the self-
consistency tests studied earlier come from variations along a y
$= $ const line in Fig. 8.

It is noteworthy that the mixture computations satisfy this new
self-consistency test very well : The observed discrepancy of 0.005 in
w_0 corresponds to a discrepancy of less than 0.00063 in ϕ.

The long linear regions in the curves in Fig. 25 are remarkable.
They also correspond to similar linearity found in experimental

systems, but here it is especially clear that they result from some numerical coincidence because the model functions, when plotted this way, are sharply curved at lower concentrations, in accord with the mixture limiting law (Section VI).

XI. MODEL CALCULATIONS IN SOME RELATED AREAS

One may study the effects of core repulsive potentials and of cosphere overlap in solutions without the complicating effects of the Coulomb potential by studying solutions of nonelectrolytes. The data for aqueous solutions of many nonelectrolytes have been treated by Kozak et al.[146] to extract the McMillan–Mayer B_2 and B_3 [equation (19)], which were then studied further to learn something about the model potential U_N for the system.

Closely related are solutions containing two solutes, one an electrolyte and the other nonionic, in which case the salting-out coefficient (Setschenow coefficient) is proportional to the McMillan–Mayer second virial coefficient[147] for the interaction of the two solutes. In this case, some aspects of model calculations have been worked out,[147,148] but there has been little progress toward a molecular interpretation except in a negative sense: The failure of the Debye–McAuley theory to account qualitatively for the observations in general shows that the dominant effect is not the particular dielectric phenomenon they considered. Stern's recent measurements of the heat effect associated with the Setschenow coefficients[149] tend to support the hypothesis that cosphere-overlap effects are dominant here. Another interesting aspect is that the experimental Setschenow coefficients are often independent of solute concentration up to concentrations over $1\,M$. It is of great interest to learn whether this linear behavior is real, i.e., due to the B_3 and higher terms in the cluster expansions being negligible, or whether it is merely apparent: If the experiments could be carried out down to lower concentrations, would a change to a different constant value of the Setschenow coefficient be found? Then the linearity in the experimental range presently accessible would be one of the accidents to which workers in this field have become accustomed (e.g., Masson's law).

It is of enormous importance to investigate the nonequilibrium properties of the various models which may be used in the study of ionic solutions. Then, comparison with experimental data for transport and relaxation coefficients as well as thermodynamic coefficients would allow a more rigorous test of the success of the model in representing the real system of interest. Unfortunately, there seems to be a great difficulty in this program which has not been generally recognized.

The ionic solution models which have been studied treat the solvent as a medium which contributes to the solute–solute interactions rather than as an assembly of molecules. For equilibrium calculations, one can use McMillan–Mayer theory to interpret this treatment more fully and to make it rigorous, as discussed in Section V. One can generalize the McMillan–Mayer derivation to apply to nonequilibrium processes in the solutions,[150] at least in the framework of a Kubo theory[5] for the transport and relaxation coefficients, but then U_N for the McMillan–Mayer system (Table 1) no longer provides enough information to specify the model. In fact, to calculate the nonequilibrium properties, one also needs a function for each solute species characterizing the fluctuating motion of a solute particle in the solvent (roughly speaking, its Brownian motion) and another function for each pair of solute species characterizing momentum transfer from one solute particle to another through the solvent (roughly speaking, a hydrodynamic force).[150,152]

In the classical theories of nonequilibrium properties of ionic solutions, attention is focused on the balance of forces and flows in a stationary state and it is not so easy to identify the assumptions which define the model, but they seem to be equivalent to assuming that the fluctuating motion of each solute particle in the solvent is governed by Langevin's equation, the mathematical equation of Brownian motion, while the momentum transfer is estimated using a certain approximation to the hydrodynamic forces operating among macroscopic spheres moving in a viscous medium (Oseen's equation). While these assumptions seem to be adequate for deriving the limiting laws, they may be inadequate for understanding the effects of short-range ion–ion interactions upon the observable transport and relaxation coefficients.

In summary, one cannot take any of the models discussed in Sections VII–IX and calculate any of its nonequilibrium properties with confidence that the effect of the model potential U_N on these properties shown by the calculation is not conditioned by the other assumptions required, namely those concerning the fluctuating motion and the momentum transfer through the solvent.

XII. SUPPLEMENT

Some recent developments in the field of this review are described here.

The Monte Carlo calculations made by Card and Valleau for the primitive model have appeared.[130] They seem more extensive and accurate than those reported earlier and also provide significantly better agreement with the HNC results.

A new method of deriving virial expansions for the thermodynamic properties and distribution functions for ionic systems, described by Kelbg,[153] leads to results in accord with other work.

An equation rather like the PY integral equation, but directly applicable to approximating the behavior of a mixture of charged hard spheres, has been solved *analytically* by Waisman and Lebowitz.[154] The results have been compared with the behavior deduced by the HNC equation for the primitive model for an aqueous 1-1 electrolyte with both ionic radii equal to 2.3 Å. The excess energy calculated from the correlation functions agrees with the HNC result within about 1 % up to 1 M. The osmotic coefficient, calculated from the dependence of E^{ex} on temperature, agrees with the HNC result within 0.006 up to 1 M. It seems likely that this new theory, the *mean spherical model*, will be useful in elucidating a number of difficult questions in ionic solution theory.

The collective coordinate method has been further developed by Anderson and Chandler to produce what they call a *mode expansion*, a sort of cluster expansion of A^{ex} in the collective coordinates.[155] The integrals in the mode expansion converge even for ionic systems, unlike the integrals in equation (19). A mode expansion can also be obtained for $A - A^0$, where A^0 is the free energy of a hypothetical system like the one of interest, but with the ionic charges removed. This aspect is similar to the theory of Stell and Lebowitz.[72] Anderson and Chandler[156] carry the calculation through for a primitive model of an aqueous 1-1 electrolyte, using the known virial coefficients of the hard sphere fluid to estimate A^0. Keeping only as much of the mode expansion as corresponds to the cluster expansion through the third virial coefficient, they find excellent agreement with the HNC results for the same model. If it works this well with other models, the mode expansion method will replace the HNC method in many applications.

The separation of the potential into long- and short-range parts, as in equation (33), plays an essential role in Mayer's derivation of the cluster theory of an ionic system. As emphasized above, the long-range behavior of the Coulomb term must be accurately treated. However this still permits one to replace equation (33) by

$$u(r) = u^*(r) + [1 - S(r)]e^2/\varepsilon r + S(r)e^2/\varepsilon r \qquad (89)$$

where $S(r)$ is some function, chosen at will, such that $1 - S(r)$ approaches zero faster than $1/r^3$ as r increases to infinity. Then the first two terms on the right of equation (89) are still short range while the last still has the proper form as r increases to infinity. Now $S(r)$ may be chosen to eliminate the infinity in the last term at $r = 0$, thereby simplifying the manipulation of the long-range term. If u^* is a hard-sphere potential and we choose $S(r)$ so that it vanishes for r outside the range of u^*, then this procedure offers great computational advantages as found by Anderson and Chandler.[156] Also Allnatt indicates how this procedure in an analytical calculation produces equation (73) from the cluster expansion for an ionic system and transforms the usual cluster expansion of $\ln \gamma_{\pm}$ into a form previously derived by Kirkwood.[157]

Outhwaite has developed[158] his extension of the Debye–Hückel theory based on the Poisson–Boltzmann equation to obtain a refined estimate of the critical concentration (cf. p. 52). For a primitive model electrolyte with ions of the same size and magnitude of charge, the new result is $\kappa^*\hat{a} = 1.2412$.

The association constant problem discussed on p. 58 has been studied by Ebeling by the use of a model system.[159] In this case the statistical mechanical expression for the association constant is obtained simply by comparing the virial expansion with the thermodynamic expression for the pressure of an ideal associating system as a function of concentration. It is well known that association constant K_n for the formation of a complex of n particles, when defined in statistical–mechanical terms in this way, may be negative as well as positive and does not necessarily correspond to the K_n determined in physical chemistry.

ACKNOWLEDGMENT

It is a pleasure to note that this review has benefited from a number of changes suggested by Professors J. C. Rasaiah and G. Stell. The writing has been generously supported, in part, by the Office of Saline Water, U.S. Department of the Interior.

REFERENCES

[1] H. Falkenhagen and G. Kelbg, in *Modern Aspects of Electrochemistry*, Vol. 2, Ed., J. O'M. Bockris, Butterworths Publications, London, 1959.

[2] J. A. Ibers and W. C. Hamilton, *Hydrogen Bonding in Solids*, W. A. Benjamin, New York, 1967.

[3] H. S. Frank and M. W. Evans, *J. Chem. Phys.* **13** (1945) 507.

[4] J. E. Desnoyers and C. Jolicoeur, in *Modern Aspects of Electrochemistry*, Vol. 5, Eds., B. E. Conway and J. O'M. Bockris, Plenum Press, New York, 1969.

[5] R. Zwanzig, *Ann. Rev. Phys. Chem.* **16** (1965) 67.

[6] R. G. Gordon, in *Advances in Magnetic Resonance*, Vol. 3, Ed., J. S. Waugh, Academic Press, New York, 1968, p. 1.

[7] J. M. Deutsch and I. Oppenheim, *ibid.*, p. 43.

[8] H. G. Hertz, in *Progress in NMR Spectroscopy*, Vol. III, Pergamon Press, London, 1967; also H. G. Hertz, G. Staladis, and H. Versmold, *J. Chim. Phys.*, *Numero Special*, October 1969, p. 177.

[9] R. Pottel and U. Kaatze, *Ber. Bunsenges. physik. Chem.* **73** (1969) 437.

[10] K. Giese, U. Kaatze, and R. Pottel, *Ber. Bunsenges. physik. Chem.* **74** (1970).

[11] G. J. Safford and P. S. Leung, in *The Techniques of Electrochemistry*, John Wiley and Sons, New York, 1970, Vol. II.

[12] J. M. J. van Leeuwen and E. G. D. Cohen, *Physics Letters* **A26** (1967) 89.

[13] O. K. Rice, *Phys. Rev. Letters* **19** (1967) 295.

[14a] F. Vaslow, *J. Phys. Chem.* **67** (1963) 2773.

[14b] P. A. H. Wyatt, *Trans. Faraday Soc.* **56** (1960) 490.

[14c] L. E. Orgel and R. S. Mulliken, *J. Amer. Chem. Soc.* **79** (1957) 4839.

[15] M. Klein and H. J. M. Hanley, U.S. Natl. Bur. Std. Tech. Note No. 360 (1967); see also *Trans. Faraday Soc.* **64** (1968) 2927.

[16] S. Strong and R. Kaplow, *J. Chem. Phys.* **45** (1966) 1840.

[17] T. L. Hill, *Statistical Mechanics*, McGraw-Hill Book Co., New York, 1956.

[18] J. S. Rowlinson, *Handbuch der Physik*, Springer-Verlag, Berlin, 1958, Vol. XII.

[19] H. L. Friedman, *Ionic Solution Theory*, Interscience–Wiley, New York, 1962.

[20] J. C. Rasaiah and H. L. Friedman, *J. Chem. Phys.* **50** (1969) 3965.

[21] D. W. Jepsen and H. L. Friedman, *J. Chem. Phys.* **38** (1963) 846; also W. A. Steele, *J. Chem. Phys.* **39** (1963) 3197.

[22] P. Egelstaff, *Disc. Faraday Soc.* **43** (1967) 149.

[23] C. J. Pings and J. Waser, *J. Phys. Chem.* **48** (1968) 3106.

[24] J. C. Rasaiah and H. L. Friedman, *J. Chem. Phys.* **48** (1968) 2742.

[25] D. Henderson, *Ann. Rev. Phys. Chem.* **15** (1964) 31.

[26] J. G. Kirkwood, *J. Chem. Phys.* **3** (1935), 300.

[27] L. Onsager, *Chem. Revs.* **13** (1933) 73.

[28] G. S. Rushbrooke and M. Silbert, *Mol. Phys.* **12** (1967) 505.

[29] J. S. Rowlinson, *Disc. Faraday Soc.* **43** (1967) 55.

[30] J. E. Mayer, *J. Chem. Phys.* **18** (1950) 1426.

[31] J. O. Hirschfelder, C. F. Curtiss, and R. B. Bird, *Molecular Theory of Gases and Liquids*, John Wiley and Sons, New York, 1964.

[32] J. C. Rossi and F. Danon, *Disc. Faraday Soc.* **40** (1965); 97; J. S. Rowlinson, *Disc. Faraday Soc.* **40** (1965) 55.

[33] N. Metropolis, A. W. Rosenbluth, M. N. Rosenbluth, A. H. Teller, and E. Teller, *J. Chem. Phys.* **21** (1953) 1087.

[34] S. G. Brush, H. L. Sahlin, and E. Teller, *J. Chem. Phys.* **45** (1966) 2102.

[35] B. J. Alder and T. E. Wainright, *J. Chem. Phys.* **31** (1958) 459; **33** (1960) 1439.

[36] T. L. Hill, *Thermodynamics of Small Systems*, W. A. Benjamin, New York, 1963.

[37] W. W. Wood and J. D. Jacobson, *J. Chem. Phys.* **27** (1957) 1207.

[38a] S. G. Brush, in *Progress in High Temperature Physics and Chemistry*, Vol. 1, Ed., C. A. Rouse, Pergamon Press, New York, 1967.

[38b] G. Stell, in *The Classical Theory of Fluids*, Eds., H. L. Frisch and J. L. Lebowitz, W. A. Benjamin, New York, 1964.

[39] F. H. Ree, R. N. Keeler, and S. L. McCarthy, *J. Chem. Phys.* **44** (1966) 3407.

[40] G. E. Uhlenbeck and G. W. Ford, in *Studies in Statistical Mechanics*, Vol. 1, Eds., J. De Boer and G. E. Uhlenbeck, North-Holland Publishing Co., Amsterdam, 1962.

[41] F. A. Ree and W. G. Hoover, *J. Chem. Phys.* **46** (1967) 4181.

[42] E. Meeron, *J. Chem. Phys.* **28** (1958) 630.

[43] G. Kelbg, *Annal. Physik* **9** (1962) 159.

[44] W. Ebeling, G. Kelbg, and G. Schmitz, *Annal. Physik* **18** (1966) 29.

[45] W. Ebeling, *Physica* **38** (1968) 378.

[46] J. K. Percus, in *The Classical Theory of Fluids*, Eds., H. L. Frisch and J. L. Lebowitz, W. A. Benjamin, New York, 1964.

[47] V. Volterra, *Theory of Functionals*, Dover Publications, New York, 1959.

[48] R. Fowler and E. A. Guggenheim, *Statistical Thermodynamics*, Cambridge University Press, 1952.

[49] S. Kaneko, Researches of the Electrochemical Laboratory (Tokyo), No. 403 (1937).

[50] H. Falkenhagen and G. Kelbg, *Disc. Faraday Soc.* **24** (1957) 20.

[51] E. A. Guggenheim, *Trans. Faraday Soc.* **55** (1959) 1714; **56** (1960) 1152.

[52] C. W. Outhwaite, *J. Chem. Phys.* **50** (1969) 2277.

[53] M. Wertheim, *Phys. Rev. Letters* **8** (1963) 321.

[54] E. Thiele, *J. Chem. Phys.* **38** (1963) 1959.

[55] J. L. Lebowitz, *Phys. Rev.* **133** (1964) A895.

[56] H. Reiss, *Advan. Chem. Phys.* **9** (1966) 1.

[57] T. W. Leland, J. S. Rowlinson, and G. A. Sather, *Trans. Faraday Soc.* **64** (1968) 1447.

[58] J. M. J. van Leeuwen, J. Groeneveld, and J. de Boer, *Physica* **25** (1959) 792.

[59] A. R. Allnatt, *Molecular Phys.* **8** (1964) 533.

[60] G. Kelbg, *Z. Physik. Chem.* **214** (1960) 8, 26, 141, 153.

[61] J. G. Kirkwood and Z. Salsburg, *Disc. Faraday Soc.* **15** (1953) 28.

[62] N. N. Bogoliubov, *Problems of a Dynamical Theory in Statistical Physics*, Gostekhizdat, Moscow, 1946. English Translation in *Studies in Statistical Mechanics*, Vol. 1, Eds., J. de Boer and G. E. Uhlenbeck, North-Holland Publishing Co., Amsterdam, 1963.

[63a] A. E. Glauberman, *J. Exp. Theor. Phys. (USSR)* **30** (1956) 1092.

[63b] I. R. Juchnowski, *Ber. Akad. Wiss. (USSR)* **126** (1959) 557.

[64] J. G. Kirkwood, *J. Chem. Phys.* **3** (1935) 300.

[65]J. G. Kirkwood and J. C. Poirier, *J. Phys. Chem.* **58** (1954) 591.

[66]G. Stell, *Molecular Phys.* **16** (1968) 209; T. Morita, *Progr. Theor. Phys.* (*Japan*) **41** (1969) 339.

[67]L. Verlet, *Physica* **30** (1964); 95; *Phys. Rev.* **165** (1968) 201.

[68]Y. T. Lee, F. H. Ree, and T. Ree, *J. Chem. Phys.* **48** (1968) 3506; see also D. Henderson, S. Kim, and L. Oden, *Disc. Faraday Soc.* **43** (1967) 26.

[69]S. F. Edwards, *Phil. Mag.* **4** (1959) 1171.

[70]H. A. Kramers, *Proc. Roy. Acad.* (*Amsterdam*) **XXX** (1927) 145.

[71]T. H. Berlin and E. W. Montroll, *J. Chem. Phys.* **20** (1952) 75.

[72]G. Stell and J. L. Lebowitz, *J. Chem. Phys.* **49** (1968) 3706.

[73]G. Kelbg, in *Chemical Physics of Ionic Solutions*, Eds., B. E. Conway and R. G. Barradas, John Wiley and Sons, New York, 1966; see also G. Kelbg, *Annal. Physik* **9** (1962) 159, 168, and I. R. Juchnowski, *J. Exp. Theor. Phys.* **34**, 263 (1958).

[74]K. B. Eisenthal and W. G. McMillan, *J. Chem. Phys.* **42** (1965) 3766.

[75]F. H. Stillinger and R. Lovett, *J. Chem. Phys.* **48** (1968) 3858.

[76]R. Lovett and F. H. Stillinger, *J. Chem. Phys.* **48** (1968) 3869.

[77]F. H. Stillinger and R. Lovett, *J. Chem. Phys.* **48** (1968) 1991.

[78]F. H. Stillinger, Jr., *Proc. Nat. Acad. Sci.* **60** (1968) 1138.

[79]R. M. Fuoss, *Trans. Faraday Soc.* **30** (1934) 967.

[80]J. C. Poirier and J. H. de Lap, *J. Chem. Phys.* **35** (1961) 213.

[81]G. S. Manning and B. H. Zimm, *J. Chem. Phys.* **43** (1965) 4250; G. S. Manning, *J. Chem. Phys.* **43** (1965) 4260, 4268.

[81a]H. L. Friedman, *J. Chem. Phys.* **32** (1960) 1351.

[82]J. G. Kirkwood and J. Mazur, in *Macromolecules*, Ed., P. L. Auer, Gordon and Breach, New York, 1967.

[83]W. G. McMillan and J. E. Mayer, *J. Chem. Phys.* **13** (1945) 276.

[84]T. L. Hill, *Introduction to Statistical Thermodynamics*, Addison-Wesley Publishing Co., Reading, Mass., 1960.

[85]T. L. Hill, *Disc. Faraday Soc.* **21** (1956) 31.

[86]D. W. Jepsen and H. L. Friedman, *J. Chem. Phys.* **38** (1963) 846.

[87]J. B. Hasted, D. M. Riston, and C. H. Collie, *J. Chem. Phys.* **16** (1948) 1, 11.

[88]S. Levine and H. E. Wrigley, *Disc. Faraday Soc.* **24** (1957) 43.

[89]S. Levine and G. M. Bell, in *Electrolytes*, Ed., B. Pesce, Pergamon Press, New York, 1962.

[90]S. Levine and D. K. Rozenthal, in *Chemical Physics of Ionic Solutions*, Eds., B. E. Conway and R. G. Barradas, John Wiley and Sons, New York, 1966, p. 119.

[91]A. Bellemans and J. Stecki, *Bull. de l'Acad. Polonaise des Sciences* **IX** (1961) 339, 343, 349.

[92]J. Stecki, *Bull. de l'Acad. Polonaise des Sciences* **IX** (1961) 429, 435, 483, 489, 663, 669.

[93]J. Stecki, *Advan. Chem. Phys.* **VI**, (1964) 413.

[94]R. A. Marcus, *J. Chem. Phys.* **38** (1963) 1335; **39** (1963) 460.

[95]R. A. Marcus, *J. Chem. Phys.* **43** (1965) 58.

[96]A. Isihara, *J. Chem. Phys.* **36** (1962) 433.

[97]H. L. Friedman, *J. Chem. Phys.* **32** (1960) 1351.

[98]J. C. Poirier, *J. Chem. Phys.* **21** (1953) 965.

[99]P. J. Reilly and R. H. Wood, *J. Phys. Chem.* **73** (1969) 4292.

[100]R. H. Wood and H. L. Anderson, *J. Phys. Chem.* **70** (1966) 992.

[101]T. H. Lilley, *Trans. Faraday Soc.* **64** (1968) 2947.

[102]H. S. Harned and B. B. Owen, *The Physical Chemistry of Electrolyte Solutions*, Reinhold Publishing Corporation, New York, 1958, 3rd ed.

[103]E. A. Guggenheim, *Trans. Faraday Soc.* **62** (1966) 3446.

[104]G. Scatchard, *J. Am. Chem. Soc.* **83** (1961) 2636.

[105]H. S. Harned and R. A. Robinson, *Multicomponent Electrolyte Solutions*, International Encyclopedia of Phys. Chem. and Chem. Phys., Topic 15, Vol. 2, Pergamon Press, London, 1968.

[106]J. Groeneveld, in *Graph Theory and Theoretical Physics*, Ed., F. Harary, Academic Press, New York, 1967.

[107]J. L. Lebowitz, *Ann. Revs. Phys. Chem.* **19** (1968) 389.

[108]J. L. Lebowitz and E. Lieb, *Physical Review Letters* **22** (1969) 631.

[109]T. H. Gronwall, V. K. LaMer, and K. Sandved, *Physik. Z.* **29** (1928) 558.

[110]L. Onsager, *J. Am. Chem. Soc.* **86** (1964) 3421.

[111]R. M. Fuoss and L. Onsager, *Proc. Nat. Acad. Sci.* **47** (1961) 818.

[112]J. F. Skinner and R. M. Fuoss, *J. Am. Chem. Soc.* **86** (1964) 3423.

[113]R. A. Robinson and R. H. Stokes, *Electrolyte Solutions*, Butterworths, London, 1965.

[114]H. L. Friedman, *J. Chem. Phys.* **32** (1960) 1134.

[115]T. L. Hill, *Disc. Faraday Soc.* **21** (1956) 31.

[116]J. G. Kirkwood, *Chem. Revs.* **19** (1936) 275.

[116a]F. Vaslow, *J. Phys. Chem.* **71** (1967) 4585.

[117]J. C. Poirier, *J. Chem. Phys.* **21** (1953) 972; see also E. Meeron, *J. Chem. Phys.* **26** (1957) 804.

[118]L. Pauling, *The Nature of the Chemical Bond*, Cornell University Press, Ithaca, N.Y., 1960.

[119]D. G. Miller and H. L. Friedman, unpublished calculation.

[120]P. G. M. Brown and J. E. Prue, *Proc. Roy. Soc. A* **232** (1955) 320.

[121]C. W. Davies, *Ion Association*, Butterworths, London, 1962.

[122]M. Swarc, *Makromolecular Chemie*, **89** (1965) 44.

[123]R. H. Wood, private communication.

[124]G. Scatchard, *Z. Physik. Chem.* **228** (1965) 354.

[125]A. L. Loeb, J. T. G. Overbeek, and P. H. Wiersma, *The Electrical Double Layer around a Spherical Colloid Particle*, The MIT Press, Cambridge, Mass., 1961.

[126a]W. Ebeling and G. Kelbg, *Z. Physik. Chem.* **233** (1966) 209.

[126b]G. Kelbg, W. Ebeling, and H. Kreinke, *Z. Physik. Chem.* **238** (1967) 76.

[126c]W. Ebeling, *Z. Physik. Chem.* **238** (1968) 400.

[127]D. D. Carley, *J. Chem. Phys.* **46** (1967) 3783.

[128]P. N. Vorontsov-Veliaminov, A. M. Eliashevich, and A. K. Kron, *Elektrokhimiya* **2** (1966) 708.

[29a]P. N. Vorontsov-Veliaminov and A. M. Eliashevich, *Elektrokhimiya* **4** (1969) 1430.

[29b]P. N. Vorontsov-Veliaminov, A. M. Eliashevich, J. C. Rasaiah, and H. L. Friedman, *J. Chem. Phys.* **52** (1970) 1013.

[130]D. N. Card and J. P. Valleau, *J. Chem. Phys.* **52** (1970) 6232.

[131]J. C. Poirier, in *Chemical Physics in Ionic Solutions*, Eds., B. E. Conway and R. G. Barradas, John Wiley and Sons, New York, 1966.

[132]R. H. Stokes and R. A. Robinson, *J. Phys. Chem.* **70** (1969) 2126.

[133]E. Glueckauf, *Trans. Faraday Soc.* **51** (1955) 1235; **53** (1957) 305.

[134]K. B. Eisenthal and W. G. McMillan, *J. Chem. Phys.* **44** (1966) 2542.

[135]R. W. Gurney, *Ionic Processes in Solution*, Dover Publications, New York, 1954.

[136]H. S. Frank, in *Chemical Physics of Ionic Solutions*, Eds., B. E. Conway and R. G. Barradas, John Wiley and Sons, New York, 1966, p. 57.

[137]H. L. Friedman, *ibid.*, p. 101.

[138]J. C. Rasaiah and H. L. Friedman, *J. Phys. Chem.* **72** (1968) 3352.

[139]J. C. Rasaiah, *J. Chem. Phys.* **52** (1970) 704.

[140a]M. Eisenstadt and H. L. Friedman, *J. Chem. Phys.* **44** (1966) 1407; **46** (1967) 2182.

[140b]D. E. Woessner, B. S. Snowden, Jr., and A. G. Ostroff, *J. Chem. Phys.* **49** (1968) 371.

[141]T. Morita, *Prog. Theoret. Phys.* (*Japan*) **22** (1959) 757.

[142]G. Kelb, *Ann. Physik.* **12** (1963) 219, 354.

[143]W. Ebeling, *Ann. Physik.* **21** (1969) 315; *Physica* **38** (1968) 378.

[144]P. S. Ramanathan and H. L. Friedman, *J. Phys. Chem.* **75** (1970) 3756.

[145]S. W. Benson and C. S. Copeland, *J. Phys. Chem.* **67** (1963) 1194.

[146]J. J. Kozak, W. S. Knight, and W. Kauzmann, *J. Chem. Phys.* **48** (1968) 675.

[147]G. R. Haugen and H. L. Friedman, *J. Phys. Chem.* **67** (1963) 1757.

[148]J. Bockris, J. Bowler-Reed, and J. A. Kitchener, *Trans. Faraday Soc.* **47** (1951) 184.

[149]J. H. Stern and J. Nobilione, *J. Phys. Chem.* **72** (1968) 3937.

[150]H. L. Friedman, in *Chemical Physics of Ionic Solutions*, Eds., B. E. Conway and R. G. Barradas, John Wiley and Sons, New York, 1966, p. 487.

[151]H. L. Friedman, *Z. Physik. Chem.* **228** (1965) 318.

[152]H. L. Friedman, *J. Chim. Phys.*, *Numero Special*, October 1969, p. 75.

[153]G. Kelbg, *J. Chim. Phys.*, *Numero Special*, October 1969, p. 83.

[154]E. Waisman and J. L. Lebowitz, *J. Phys. Chem.* **52** (1970) 4307.

[155]H. C. Anderson and D. Chandler, *J. Chem. Phys.* **53** (1970) 547.

[156]H. C. Anderson and D. Chandler, *J. Chem. Phys.* **54** (1971).

[157]A. R. Allnatt, *Molecular Physics* **18** (1970) 409.

[158]C. W. Outhwaite, *Chem. Phys. Letters* **5** (1970) 77.

[159]W. E. Ebeling, *Z. Physik. Chem.* **240** (1969) 265.

2

Surface Potential at Liquid Interfaces

J. Llopis

Instituto de Química Física "Rocasolano"
C.S.I.C. Madrid, Spain

I. DEFINITIONS†

Prior to entering into the main subject of this chapter, it will be convenient first to consider briefly the proposals of different international commissions[1-3] for definitions of the fundamental electrical phenomena to be dealt with in this article. An excellent review of these definitions was given some years ago by Parsons.[5]

A realistic system of definitions was first developed by Lange.[6,7] Later, the CITCE Commission on Electrochemical Nomenclature and Definitions and the IUPAC Subcommission on Electrochemical Symbols and Terminology[3] discussed and revised these definitions. Specialists in surface science have, in general, accepted, the conclusions of the electrochemists, but in some aspects it has not been possible to arrive at a completely satisfactory agreement; for example, the majority of surface scientists found the use of the word *tension* to characterize a potential difference unacceptable. It is not only the possible confusion with surface tension but, more generally, the association of this word with a mechanical rather than an electrical effect which made it objectionable. Therefore, this term has not been used here, since the topic of this review is of interest to both electrochemists and surface scientists.

According to these definitions, the *outer electric potential* ψ of a conducting phase is that defined[5] in the classical theory of

†For notation, see p. 153.

electricity and taken *in vacuo* close to the surface of the phase, immediately beyond the practical range of the image forces. This quantity is measurable.

In order to transfer a charge q from a point at which the electric potential ψ has been defined to a point in the bulk of the phase, an amount of work w must be done and the limit of the quotient w/q, when q tends toward zero, is equal to the *surface electric potential difference* χ.

The *inner electric potential* ϕ at a point in a phase is equal to the sum

$$\phi = \psi + \chi \tag{1}$$

This quantity is definable in principle, but generally is not measurable.[5]

The difference in outer electric potential between two phases in contact (or two which have been in contact) is called the *Voltaic potential difference*,

$$(^{\alpha}\Delta^{\beta})\psi \equiv \psi^{\beta} - \psi^{\alpha} \tag{2}$$

The difference in inner electric potential between two phases is called the *Galvanic potential difference*,

$$(^{\alpha}\Delta^{\beta})\phi \equiv \phi^{\beta} - \phi^{\alpha} \tag{3}$$

and is only measurable if the composition of the two phases is the same.

The Voltaic potential difference between a phase α' covered with an adsorbed (or spread) layer of n molecules per cm^2, and the same phase without such a layer is often indicated by ΔV and should be called the *surface Voltaic potential difference* (frequently called the measured *surface potential*),

$$\Delta V \equiv (^{\alpha}\Delta^{\alpha'})\psi \tag{4}$$

In interpreting ΔV as $4\pi n(p^s)$, this may be considered as an operational definition of the *surface (dipole) moment per adsorbed molecule* (p^s), but it should be realized that p^s is quite a complex function of the structure of the surface layer of the clean phase and of the structure and orientation of the adsorbed molecules.

If α is an electronic conducting phase (the metal), it can be shown that

$$\Delta V \equiv (^{\alpha}\Delta^{\alpha'})\psi = -(\chi^{\alpha'} - \chi^{\alpha}) + [(\mu_e^{\alpha'} - \mu_e^{\alpha})/e] \tag{5}$$

and, unless the adsorbate can penetrate into the bulk of the phases, the chemical potential of the electron μ_e is not affected,

$$\mu_e^{\alpha} = \mu_e^{\alpha'} \tag{6}$$

so that

$$\Delta V = -(\chi^{\alpha'} - \chi^{\alpha}) \equiv -\delta\chi^{\alpha} \tag{7}$$

and the "surface potential" ΔV defined above may be identified with the change in χ of a single metal due to adsorption.

The measurement of the Voltaic potential difference between a metal (reference electrode) and an electrolytic solution δ was shown by Parsons[5] to be

$$(^{\delta}\Delta^{\alpha})\psi e = [(\mu_M^{\gamma} - \alpha_{M^{z+}}^{\delta})/z] - \alpha_e^{\alpha} \tag{8}$$

in which γ is another metal forming an electrode reversible to an ion M^{z+} of valence z present in the solution δ, and α_i^{α} is the *real potential* of the particle i in phase α.

Now, if a layer of molecules is adsorbed (or spread) on the electrolytic solution (air/water interface), the change in $(^{\delta}\Delta^{\alpha})\psi$ will be given by

$$\delta(^{\delta}\Delta^{\alpha})\psi = -\delta\alpha_{M^z}^{\delta}/ze = -\delta(\mu_{M^z}^{\delta} + ze\chi^{\delta})/ze \tag{9}$$

and therefore:

$$\Delta V = \delta(^{\delta}\Delta^{\alpha})\psi = -\delta\chi^{\delta} \tag{10}$$

since the films spread on the solution very often are involatile and changes in the reference electrode α may certainly be neglected. Furthermore, changes in the bulk of the solution must be negligible ($\delta\mu_{M^z} \simeq 0$), for otherwise $\mu_{M^z}^{\delta}$ would be affected.

At an aqueous(α)–liquid(β) interface (oil/water interface), if β is a nonconductor of dielectric constant ε_a, the variation of the potential with distance in β would be similar to that for a perfect conductor surrounded by vacuum. In this case, the ΔV defined above may also be introduced as the difference in the work required to move a unit charge from the bulk of β to a distance $10^{-4}/\varepsilon_a$ cm from the interface before and after spreading (adsorbing) a film at the interface.

II. EXPERIMENTAL METHODS FOR MEASURING SURFACE POTENTIALS AT LIQUID INTERFACES

Theoretical aspects of the experimental methods used for Voltaic potential measurements were reviewed by Möhring[8]; a briefer review of these was given by Parsons.[5] Only methods used for measuring surface potentials at liquid interfaces will be described in the following section.

1. Vibrating Plate Method

The methods to be described here are a development of the Kelvin procedure[9] as modified by Zisman.[10] The electrode in the air is moved with respect to the water surface.

In the case of the air/water interface, the resulting change in the capacity of the air space leads to current flow in the external circuit. The actual measurement is accomplished by a null technique; a potentiometer is incorporated in the circuit to oppose the Voltaic potential and the potentials required to prevent current flow are measured. Yamins and Zisman[11] applied this method to measurements of surface potentials of monomolecular films. As mentioned above, the difference in these null values, with and without the monolayer, are taken as giving ΔV. The method has been used extensively to measure surface potentials of solids,[12] but in this article attention will be restricted to problems concerning liquid interfaces.

An equivalent circuit for surface potential measurements by the vibrating plate method can be simplified as shown in Fig. 1. An approximate analysis was given by Bewig and was briefly summarized later by Gaines.[13]

In making a measurement, what it is observed is a potential difference V between the two electrodes. The measurement is usually accomplished by introducing a potentiometer E to oppose the potential V. Then, if it is assumed that

$$C = C_0 + C_1 \sin \omega t \tag{11}$$

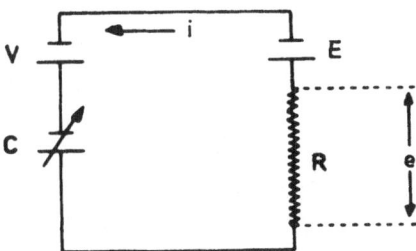

Figure 1. Equivalent circuit for surface potential
measurement by the vibrating plate method.

where ω is the vibration frequency, the current i flowing in the
circuit is given by

$$V - E = iR + (1/C) \int i \, dt \tag{12}$$

If the amplitude of the capacity change $C_1 \ll C_0$, where C_0 is
the static value of the electrode capacity, an approximate solution
of this equation is obtained as

$$e = iR \simeq \frac{(V - E)\omega C_1 R}{[1 + (\omega C_0 R)^2]^{1/2}} \sin(\omega t + \varphi) \tag{13}$$

where φ expresses the fact that the variation of e will not be syn-
chronous with the variation in C.

As expected, the signal e disappears when $V = E$. For maximum
accuracy, e must be as large as possible for small values of $|V - E|$.
This can be accomplished by: (a) making R as large as possible,
(b) making the amplitude of the vibration C_1 larger with respect to
C_0, and (c) increasing the frequency.

For the vibrating electrodes usually employed ($\simeq 1 \text{ cm}^2$ gold
plate and spacing $\simeq 0.1$ mm), the mechanical resonant frequency is
a few hundred cycles (250–300 Hz). The signal e across a 10^9 ohm
resistance will be a few μV (peak-to-peak)/mV off-balance in
$|V - E|$. It is therefore necessary to have a high-gain amplifier to
detect this signal.

A number of circuits have been described in the literature for
reliable measurements at air/water and oil/water interfaces (see,
for example, Davies[14,15]; a complete description of a modified

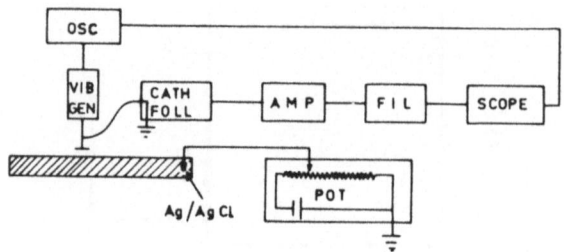

Figure 2. Block schematic diagram of Gaines' apparatus.

apparatus was given by Kinloch and MacMullen[16]. A circuit taking advantage of commercially available instruments was described by Gaines[17] and is shown in the schematic diagram, Fig. 2. By means of this apparatus, using a gold electrode (1 cm ϕ), at a distance 0.3 mm from the water surface, driven at 330 Hz, surface potential variations of ± 0.5 mV are usually detectable.

In the automatic recording system recently described by Suzuki *et al.*,[18] the signal from the electrode is led to a preamplifier and then amplified in two steps within a main amplifier, which is equipped with a twin-T filter circuit and a phase-sensitive detector. The detector indicates both by a difference and a sign whether the given potential is higher or lower than the surface potential. Voltage differences of less than 2 mV can be detected and recorded by this apparatus.

2. Ionizing Electrode Method

This method requires a radioactive source to ionize the air gap between the reference electrode and the water surface and a high-impedance electrometer as the indicating device. The measurement can be accomplished with an electrometer voltmeter with an input resistance r across which the output signal e is measured; in this case, i is the current flowing in the circuit (equivalent circuit a, Fig. 3). V can also be determined by a null method by inserting a potentiometer E, in which case the electrometer M functions only as an indicating meter (equivalent circuit b, Fig. 3).

An analysis of these circuits was given by Bewig[19] and the following conclusions concerning the performance of these measurements were obtained. First of all, only if $r \gg R$ in circuit (a) is M

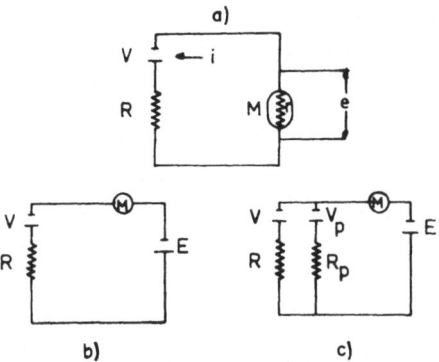

Figure 3. Equivalent circuit for surface potential measurement by the ionizing electrode method.

acting as a true voltmeter and $e = V$; this is an obvious result. However, it has been shown (see Bewig[19]) that true values of V may also be estimated when $r \leq R$ by making two measurements of the apparent output, e_1 and e_2, on two different known impedance values r_1 and r_2, using the electrometer M. The required relations are

$$V = e_1(1 + R/r_1)$$

$$V = e_2(1 + R/r_2) \tag{14}$$

$$R = (e_2 - e_1)[(e_1/r_1) - (e_2/r_2)]$$

and R and V can be calculated from these equations.

With the null method, V can be measured without the use of above equations even though $r \leq R$. The high values of gap resistance R permit only small values of i, so that, for detection of a small unbalance, a highly sensitive instrument M will be required. Under these conditions, problems of noise pickup are common; therefore, adequate electrical shielding must be provided around the ionizing source and the electrometer input.

Bewig[19] has also pointed out that serious errors can arise when the spacing between the air electrode and the water surface is large; thus, scattered ionization can lead to current flow from the air electrode to other nearby surfaces (e.g., of the enclosure) having Voltaic potentials different from that of the water surface. These

parallel current paths (circuit c, Fig. 3) may modify the apparent
value of V as measured by M. His analysis indicates that these errors
cannot be evaluated directly by simple measurements. If a high-
impedance voltmeter (input impedance $> 10^{14}$ ohms) is available,
comparison of values obtained by both the direct voltmeter and the
nulling potentiometer methods can help to detect errors due to
stray conducting paths.[20] The occurrence of such stray conduction
errors depends largely on the geometry of the system and considerable care is needed to eliminate them, especially in the insulation of
the air electrode mounting.

The ionization should be concentrated in the gap, both to
provide sufficient conductivity and to prevent stray currents to
other surfaces. The radiation from an α-particle source is especially
convenient for this purpose and, of the α-emitters, polonium (^{210}Po)
is usually preferred. Within the past few years, other α-emitters have
been used, and a review of the area can be found in the book by
Gaines.[13] At present, these radioactive sources (generally Po sources)
are commercially available.

Although a number of circuits have been described in the
literature for surface potential measurements by the ionizing
electrode method (see, for example, [21]), commercial high-impedance
instruments are now entirely satisfactory and of increasing use in
many laboratories.

3. Liquid Jet Method

Differences of outer potential between two liquids can be
measured by this method, originated by Kenrick.[22] Its principle is
shown in Fig. 4. The method was used by Frumkin[23] with aqueous
solutions of inorganic electrolytes and it has been improved recently
by Randles.[24]

The principle of the method can be summarized as follows.
A jet of liquid is directed down the axis of a tube, the inner surface
of which is covered by a stream of another liquid. If the Voltaic
potentials of the two liquids are different, the jet and the surrounding
liquid will be oppositely charged due to the electric field in the gap
between them. The droplets carry away the charge on the jet, so that
there must be a continuous flow of charge into the jet. If this is
connected to a very high-impedance electrometer M and the liquid
surrounding the jet is connected to a potentiometer E, a condition

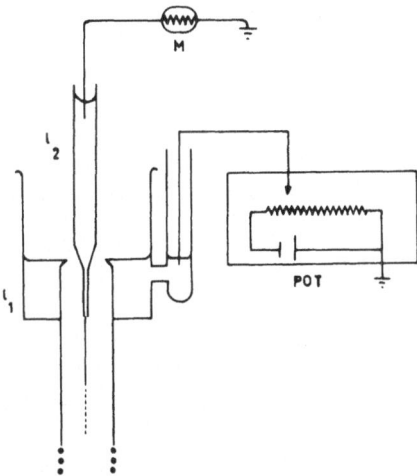

Figure 4. Apparatus for measuring the outer
potential of a liquid jet relative to that of another
liquid phase.

may be established in which the outer potential of the jet equals that
of the surrounding liquid.

Several precautions are necessary with this method to attain
a high degree of reliability. Perfect insulation of the jet is required;
otherwise, leakages would introduce errors. The surface conductivity
of the glass can be reduced by silicone treatment. After such treatments, static charges are left which take a time to leak away, so that
it is convenient to leave the apparatus for a length of time connected
to ground before readings are taken.

As a check on the significance of the measured potential
differences, experiments must be carried out using a jet of electrolyte
solution with the same solution as the external flowing electrolyte.
The measured potential difference, usually less than 1.0 mV, permits
possible corrections to be made for the asymmetry of the system.

An important factor to be considered is that the surface of the
jet is flowing rapidly and therefore the method can only be used
with solutions of substances such as inorganic electrolytes for which
equilibrium at the fresh surface is established instantaneously.
Conversely, with solutions of surface-active substances, establish-

ment of equilibrium requires time, so that care must be taken in the interpretation of the results.

III. SURFACE POTENTIAL OF AQUEOUS SOLUTIONS OF INORGANIC ELECTROLYTES

1. Surface Tension and Adsorption of Ions at Air/Water Interfaces

A large amount of experimental work indicates that addition of an inorganic electrolyte to water increases the surface tension of the solutions over that of the pure solvent. This effect was shown first by Heydweiller[25] (previous measurements were carried out by Quincke[26] and Gradenwitz[27]) and it was also observed by Stocker[28] in experiments carried out using the oscillating jet method, which measures surface tension of a freshly formed surface.

By means of an analysis employing the Gibbs equation, these results indicate that there is a deficiency of the solute in the surface layer (negative adsorption) so that the water is positively adsorbed; that is, the salt solution is covered by a thin film of water. Langmuir[29] discussed the subject briefly and concluded that the surface of an aqueous salt solution probably consists of a single layer of oriented water molecules; this layer is about 4 Å thick and it probably represents the length of the water molecule in the surface. However, the data upon which Langmuir based this generalization were somewhat scanty and the subject was investigated more fully by Goard.[30] The negative adsorption at the surface of aqueous solutions of several inorganic salts was determined over wide ranges of concentration, but the results obtained were not consistent with the supposition that the surface of these solutions consists of a single layer of oriented water molecules, since the mean thickness of the adsorbed layer varies with the concentration of the solution and with the nature of the salt.

In interpretation based on results obtained from this kind of experiment (see Harkins et al.[31,32]), an oversimplified model of the surface layer was assumed, so that the values of the layer thickness obtained represent only apparent quantities. With solutions of NaCl, this apparent thickness is 4.0 Å at a concentration of 0.1 M, and falls to 2.3 Å at 5 M. In the case of $CaCl_2$ solutions, thicknesses of the water film between 3.1 and 2.5 Å were obtained. Thus, it can be seen that these values were of the order of the linear dimension of the water molecules, but they depended on the concentration and nature of the solute. To what extent this is due to a change of the molecules of water or to penetration of the ions into the surface layer is not yet known.

The surface tension of dilute solutions of electrolytes was carefully determined by Schwenker,[33] who improved the Lenard method[34] considerably. Later, Jones and Ray[35–38] studied the relative surface tension (i.e., with respect to that of the solvent) of very dilute salt solutions at concentrations below $10^{-2} M$. This study, which was carried out using the capillary rise method, revealed a noticeable minimum in the relative surface tension at concentrations below $5 \times 10^{-3} M$. After having passed through a minimum, the surface tension increased with concentration.

However, to date, this effect is still under discussion.[39] Thus, Schäffer et al.[40] have measured the surface tension of a number of inorganic electrolyte solutions by means of Lenard's method and the Jones–Ray effect was not confirmed.

Among the explanations which have been offered for the results of Jones and Ray's is the proposal by Coolidge[41] that they are due to imaginary contact angles. Langmuir[42] has pointed out that the anomalous results are due to a change in the effective diameter of the capillary owing to the wetting layer of electrolyte adjacent to the inner glass wall. Therefore, no decision can be made on the possible mechanism for—or even the reality of—the Jones–Ray effect. However, Drost-Hansen[43] believes it is reasonable to accept the existence of such an effect, and its existence would be a manifestation of structural peculiarities of the surface of aqueous solutions.

On the other hand, positive or negative adsorption of ions at the surface of liquids has also been studied from a theoretical standpoint and such work, in turn, has contributed notably to the understanding of the structure of liquid interfaces.

On the basis of interionic attraction theory, Wagner[44] was the first to develop a theory of the surface tension of electrolytic solutions. Owing to the repulsion of the ions from the surface by the electrostatic image surfaces, there will be a deficiency of solute near the surface. The screening effect of the ionic atmospheres limits the depth to which these forces remain operative. However, Wagner's formulation led to equations which were difficult to solve. Onsager and Samaras[45] added further simplifications to Wagner's treatment and were able to obtain an explicit first approximation (a summary of this treatment can be found in Harned and Owen[46]). In the limiting case of dilute solutions of 1–1 electrolytes with water as solvent at 25°C, the limiting law becomes

$$\gamma - \gamma_0 = 1.0124c \log(1.467/c) \qquad (15)$$

The problem was considered by Ssementschenko and co-workers[47,48] and Belton[49] in terms of salting out of the surface region of the solution. The effect of the ions is to produce a reduction in the number of solvent molecules there and so cause a change in the surface tension.

It is to be noted, however, that all these theories (see also Oka[50] and Ariyama[51]) are general for any electrolyte, but insufficient parameters have been considered to take into account ionic specificities.

However, though the increase of surface tension by most inorganic elect lytes is roughly linear with concentration, the effects of different electrolytes are quite different, and many acids even lower the surface tension of water. This means that the individuality of the ions should be considered in the theory. In accord with this line of thought, Lorenz[52] supposed that such individuality was an additive function of the ions present and that the adsorption was the sum of two effects: (a) the electrostatic (negative) adsorption of the ions near the boundary of a dielectric medium; and (b) the specific adsorption of ions as chemical species.

As a result of this preferential adsorption of the ions, their concentrations are, in general, different in the neighborhood of the surface, and a surface charge arises (see Frenkel[53]) which is compensated by the charge of the ions present in excess in the adjacent layer of the solution (ionic atmosphere). In this way, an electrical double layer is formed at the surface. The presence of this electrical double layer must increase the surface free energy and hence the surface tension.

Schäfer and Pérez-Masiá have developed a theory[40,54–56] in which all these factors are considered. The negative adsorption of inorganic electrolytes is due to the higher energy of solvation of ions in the bulk of the liquid phase than in the interface. This energy of solvation is, in general, different for the anions and cations of a given salt and a double layer will hence be formed at the surface.

Let μ_- and μ_+ be the chemical potential of negative and positive ions at a distance δ from the interfaces and let μ_-^∞ and μ_+^∞ be the values of these chemical potentials in the bulk of the liquid phase. Then, for a 1–1 electrolyte,

$$\mu_- - \mu_-^\infty = e\phi$$
$$\mu_+ - \mu_+^\infty = -e\phi \tag{16}$$

We can substitute into these equations the thermodynamic relation for μ,

$$\mu = \mu^\circ + kT \ln c + \mu^{ac} \tag{17}$$

where μ^{ac} represents the part of the chemical potential associated with activity coefficients $\neq 1$; then

$$\Delta\mu_-^\circ + \Delta\mu_-^{ac} + kT \ln(c_-/c_-^\infty) = e\phi$$
$$\Delta\mu_+^\circ + \Delta\mu_+^{ac} + kT \ln(c_+/c_+^\infty) = -e\phi \tag{18}$$

in which the symbol Δ represents the difference between the value at a distance δ from the surface and that for the bulk of the solution. The problem now consists in calculating separately the quantities $\Delta\mu^\circ$, $\Delta\mu^{ac}$, c, and ϕ.

The term $\Delta\mu^\circ$ can be evaluated as the difference between the electrostatic energy of the ion at the surface and that for the ion in the interior of the liquid phase, assuming in both cases that the ion is isolated. The evaluation of the electrostatic energy of the ion situated at a distance δ from the interface was carried out using the method of the mirror-image forces: Figure 5 shows the notation used for this type of evaluation, in which I represents the ion, I' its mirror image, and P the point where the electric field is calculated. The theory allows the following expression to be obtained:

$$\Delta\mu^\circ = \frac{e^2}{4\varepsilon_w}\left(\frac{1-\varepsilon}{1+\varepsilon}\right)\left\{\frac{1}{\delta}\left(\frac{1-\varepsilon}{1+\varepsilon}\right)\left[\frac{1}{4\delta}\left(\ln\frac{2\delta+r_0}{2\delta-r_0}\right) - \frac{r_0}{4\delta^2 - r_0^2}\right]\right\} \tag{19}$$

where ε_w and ε_a are respectively the dielectric constants of water and air, $\varepsilon \equiv \varepsilon_a/\varepsilon_w$, and r_0 is the ionic radius.

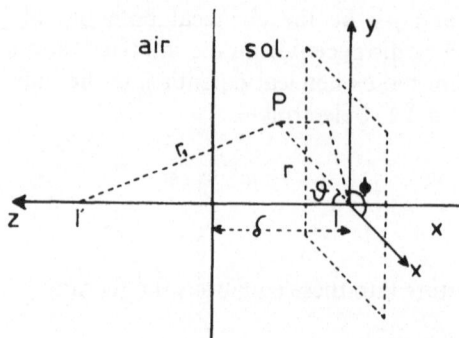

Figure 5. Notation used to calculate the electrical
interactions of an ion situated at a distance δ from the
interface.

This equation holds for distances $\delta \geq r_0$. For the particular
case of $\delta = 0$ (an ion situated exactly in the air/water interface),
it follows from the theory that

$$\Delta\mu^\circ = \frac{e^2}{r_0(\varepsilon_a + \varepsilon_w)} - \frac{e^2}{2\varepsilon_w r_0} \tag{20}$$

The contribution of the Debye–Hückel (electrical) interactions
with other ions to the free energy of a single ion situated at a distance
δ from the interface is given by the approximate expression (see
Pérez-Masiá[54] for a more detailed calculation)

$$\Delta\mu^{ac} = \frac{e^2}{4\varepsilon_w} \left\{ \frac{2\kappa}{(1 + \kappa a)} \exp[-\kappa(\delta - a)] \right.$$

$$- \left(\frac{1 - \varepsilon}{1 + \varepsilon} \right) \frac{1}{\delta} \left[1 + \frac{\kappa\delta - 1}{\kappa\delta + 1} \frac{1}{2} \exp[-\kappa(\delta - \alpha)] \right.$$

$$\left. \left. - \frac{\kappa^2}{2(1 + \kappa a)} \int_\delta^\infty \exp[-\kappa(r - a)] (4\delta^2 + r^2)^{1/2} \, dr \right] \right\} \tag{21}$$

which is valid for values of $\delta \geq a$. If $\delta < a$, the equation to be used is given by

$$\Delta\mu^{ac} = \frac{e^2}{4\varepsilon_w}\left\{\frac{2\kappa}{(1 + \kappa a)} - \left[\frac{\kappa}{1 + \kappa a}\exp[-\kappa(\delta - a)]\right.\right.$$
$$+ \left(\frac{1 - \varepsilon}{1 + \varepsilon}\right)\frac{1}{2\delta}\left\{\left(\frac{1 + 3\kappa\delta}{1 + \kappa a}\right)\exp[-\kappa(\delta - a)]\right.$$
$$\left.\left.\left. - \frac{\kappa^2}{1 + \kappa a}\int_0^\infty \exp[-\kappa(r - a)](4\delta^2 + r^2)^{1/2}\,dr\right\}\right]\right\} \qquad (22)$$

In these equations, κ is the usual Debye–Hückel function and a is the distance of closest approach of oppositely charged ions.

The above equations allow the sum $\Delta\mu^\circ + \Delta\mu^{ac}$ to be obtained for different values of δ. Introducing now the symbols

$$H_+(\delta) = (1/kT)(\Delta\mu_+^\circ + \Delta\mu_+^{ac})$$
$$H_-(\delta) = (1/kT)(\Delta\mu_-^\circ + \Delta\mu_-^{ac}) \qquad (23)$$

it follows from Eq. (18) that the mean ionic concentration $c_\pm$ at a distance δ from the surface is given by

$$c_\pm \equiv (c_-c_+)^{1/2} = c^\infty \exp\{\tfrac{1}{2}[H_+(\delta) - H_-(\delta)]\} \qquad (24)$$

where c^∞ is the concentration of solute far from the surface.

As a consequence of the different ionic distributions of anions $c_-(\delta)$ and cations $c_+(\delta)$, an ionic double layer is built up in the surface layer. The values of potential $\phi(\delta)$ farther out into the solution were obtained by an approximate integration of Poisson's equation (see Pérez-Masiá[54]):

$$d^2\eta/d\delta^2 = \kappa^2 \exp[\tfrac{1}{2}(H_+ - H_-)]\sinh[\tfrac{1}{2}(H_+ - H_-) + \eta] \qquad (25)$$

where $\eta \equiv e\phi/kT$. The boundary conditions are

$$\phi = 0 \quad \text{and} \quad d\phi/d\delta = 0 \quad \text{for} \quad \delta = \infty$$
$$d\phi/d\delta = 0 \quad \text{at} \quad \delta = 0 \qquad (26)$$

Insertion of H and η into Eq. (18) gives

$$c_-(\delta) = c^\infty \exp[-H_-(\delta) + \eta(\delta)]$$
$$c_+(\delta) = c^\infty \exp[-H_+(\delta) - \eta(\delta)] \qquad (27)$$

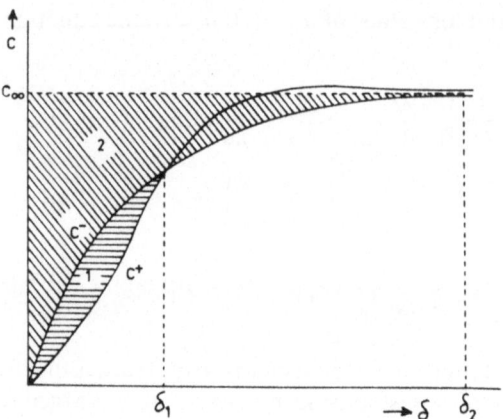

Figure 6. Distribution of ions in the neighborhood of
an air/water interface.

which allow the distribution of anions and cations to be obtained in the interface. Figure 6 shows such a distribution for the case of an anion larger than the cation, so that there will be an excess of negative over positive ion close to the interface and a net negative charge will exist in the region $0 < \delta < \delta_1$. In the region $\delta_1 < \delta < \delta_2$, there will be an opposite charge.

Using Eq. (27), the negative surface concentration, i.e., the "surface excess," Γ in moles/cm^2 can be calculated as a function of c^∞:

$$\Gamma(c^\infty) = c^\infty \left\{ \int_0^{\delta_2} [\exp(-H_- + \eta) + \exp(-H_+ - \eta)] \, d\delta - 2\delta_2 \right\}$$

$$(28)$$

where δ_2 is the thickness of the surface layer, i.e., the value of δ at which $c(\delta) \simeq c^\infty$. Further, $\Gamma(c)$ can be expressed as a function of the activity $\Gamma(a)$; by integrating the Gibbs equation, we get

$$(\Delta\gamma)_G = -RT \int_0^a \Gamma(a) \, d(\ln a) \qquad (29)$$

that is, the part of the surface tension increment for an electrolyte solution arising from the impoverishment of the surface layer is the result of a diminished free-energy of hydration of the ion close to the air/water interface.

In order to calculate the surface charge density σ, area 1 in Fig. 6 must be multiplied by Ne so that

$$\sigma(c^x) = Ne \int_0^{\delta_1} c^x [\exp(-H_- - \eta) - \exp(-H_+ - \eta)] d\delta \qquad (30)$$

Finally, the part of the surface tension increment due, according to Lippmann and Helmholtz, to the charge σ of the surface layer is given by

$$(\Delta\gamma)_{LH} = - \int_0^{c^x} \sigma(c^x) \frac{d\phi_0(c^x)}{dc^x} dc^x \qquad (31)$$

where ϕ_0 is the value of ϕ for $\delta = 0$. This increment $(\Delta\gamma)_{LH}$ is always negative for inorganic electrolytes and it is opposite to the positive effect in $(\Delta\gamma)_G$ [Eq. (29)]. Nevertheless, values of $(\Delta\gamma)_{LH}$ are small and in general they are negligible.

Figure 7 shows some of the experimental results[40] for $\Delta\gamma$ as a function of bulk concentration; it can be seen that they compare very well with the theoretical values, except for HCl. The behavior of aqueous solutions of inorganic acids is not yet clear (for HCl, see Lorenz[52]; for H_2SO_4, see Sabinina and Terpugow[57] and Suggitt et al.[58]).

On the other hand, it can be seen that all the theories mentioned above predict at the lowest concentration an increase of surface

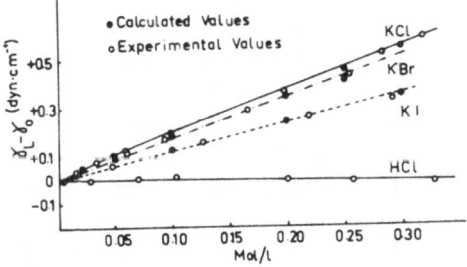

Figure 7. Increments of surface tension for aqueous solution of some inorganic electrolytes (Schäfer, Pérez-Masiá, and Jühtgen[40]).

tension with increase of salt concentration. They therefore contradict the existence of a Jones–Ray effect. However, if the correctness of this effect were accepted, it would be necessary to invent a mechanism by which the positive adsorption ceases at a very low concentration and negative adsorption begins. Dole[59] has proposed a theory to account for this minimum in the relative surface-tension curves. In this theory, it is assumed that a limited number of locations or "active spots" exist on the water surface, in such a configuration that negative ions will be attracted to these spots from the interior of the solutions and be adsorbed until all the active spots are occupied. At higher concentrations, no more negative ions can become adsorbed, but they will be repelled from the surface layer by virtue of repulsive effects such as mirror-image forces or changes of hydration energy, as explained above. The final equation was tested with the data of Jones and Ray, but, overall, this theory seems too phenomenological and its interest depends on the reality of the Jones–Ray effect. However, this effect has acquired a new interest recently[43] because it has been assumed to be related to the more general problem of the surface layer structure of aqueous solutions, to be discussed later.

2. Experimental Results on Surface Potential

Measurements of the surface potential of aqueous electrolyte solutions by Frumkin[23] showed that solutions of most simple salts have surface potentials more positive (inward) than that of pure water. The magnitude of this effect depends on the nature of the anion; in the case of monovalent anions, the order of the effect follows the lyotropic series. In general, the effects of ions are roughly in the order of decreasing ionic size and, probably, of increasing hydration energy (see also the results by Garrison[60]).

More accurate measurements were carried out later by Randles[61] for solutions of low concentration and with well-defined liquid junctions. In Fig. 8, the surface potential $\Delta\chi$ is plotted against $c^{1/2}$ for a number of salts. It can be seen that the surface-potential change due to KPF_6 is extraordinarily large, being greater than that for $NaClO_4$ or $KCNS$ at the same concentration. In many cases and for concentrations less than 0.1 M, the values of $\Delta\chi$ amount to no more than a few mV, so that it is understandable that Guyot[62]

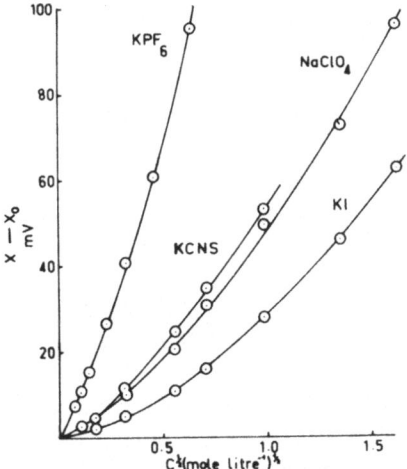

Figure 8. Surface potentials plotted against
$c^{1/2}$ for the salts indicated (Randles[61]).

considered the Voltaic potential of very dilute aqueous solutions of electrolytes to be independent of concentration.

According to the theoretical work referred to above, these facts can be interpreted in the sense that all the ions are repelled from the surface but that most of the anions are repelled less strongly than are the cations, and some anions very much less strongly.

In order to calculate these surface potentials, Pérez-Masiá[56] has applied Eq. (12) assuming that a layer of thickness τ ($\simeq 2$ Å) is inaccessible to both anions and cations. With this in mind, the boundary conditions are

$$\phi = 0 \quad \text{and} \quad d\phi/d\delta = 0 \quad \text{for} \quad \delta = \infty$$
$$d\phi/d\delta = 0 \quad \text{for} \quad \delta = \tau \tag{32}$$

The value ϕ_τ is the potential at a plane τ relative to that in the bulk of the solution, and it is the negative of the contribution of the ionic double layer to the surface potential of the solution. However, this is too simple a model and gives values much less than those found experimentally.

Randles[61] has proposed a more detailed model in which he assumes that a surface layer of about 4–5 Å is inaccessible to the

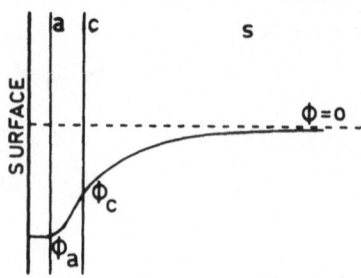

Figure 9. Variation of ϕ close to the surface of an aqueous solution of inorganic electrolyte according to Randles.[61]

cations, but anions can penetrate more closely to the surface. In Fig. 9, plane c is an infinite potential barrier for cations and plane a is the corresponding plane for anions. The region between both planes acquires a negative charge and equivalent positive charge remains in the region c-s.

To calculate the distribution of ions, the theory of the diffuse double layer is applied and the expression for the potential gradient is integrated and solved for the boundary conditions

$$\phi = 0 \quad \text{and} \quad d\phi/d\delta = 0 \quad \text{for} \quad \delta = \infty$$
$$d\phi/d\delta = 0 \quad \text{for} \quad \delta = \delta_a \tag{33}$$

In this way, the distance between planes c and a is a parameter because the position of plane c is fixed. Furthermore, he introduced another parameter as an adsorptive potential by assuming that, within the layer from which cations are excluded, the chemical potential of anions is reduced by a fixed amount.

However, it follows from the experimental results that this theory is still inadequate, not only because it is unable to account for large values of surface potential, but also because the experimental and theoretical graphs of $\Delta\chi$ against $c^{1/2}$ have opposite curvature. It has been suggested[63] that this discrepancy between the experimental results and the theoretical prediction is due to neglect of the discrete nature of ions and it would be more reasonable to assume a linear potential gradient between planes a and c.

The concentration dependence of the surface potential of electrolyte solutions was first studied by Andauer and Lange[64] using the ionizing electrode method for aqueous solutions of Ag^+ and Cu^{2+} at very low concentration $(10^{-9}-10^{-1} N)$. The variation observed was very small, $\partial \chi_{sol}/\partial(\log a_i) \simeq -0.01$ V or less.

Higher-precision experiments were carried out later by Klein and Lange[65] in which they showed that, for the systems Ag/Ag_{aq}^+ and Hg/Hg_{aq}^+,

$$\partial(\Delta\phi)/\partial(\log a_i) = \partial(\Delta\psi)/\partial(\log a_i) \qquad (34)$$

and therefore

$$\partial \chi_{sol}/\partial(\log a_i) = 0 \qquad (35)$$

where a_i is the activity of the potential-determining ion. These facts led to the conclusion that for low concentrations ($<0.1 N$) the surface potentials of aqueous solutions could be assumed to be equal to that of pure water. However, this conclusion was not confirmed by Randles' experiments when large anions such as I^-, ClO_4^-, or PF_6^- were present in the solution.

Measurements of the surface potentials of concentrated solutions of inorganic acids (HCl, HBr, and H_2SO_4) by Gurenkov[66] were carried out by the Kenrick method using the following cell without liquid junctions:

$$Pt, H_2|0.01 \ N \ HX|air|M \ HX|H_2, Pt \qquad (36)$$

In this way, assuming that $\Delta\chi_{0.01} \simeq 0$ and that the activity a_{H^+} can be replaced by the mean activity $a_\pm$, we obtain

$$\chi_{sol} = E + (RT/F)\ln(a_\pm)_{sol} + \text{const} \qquad (37)$$

where E is the measured electromotive force of cell (36) and F the Helmholtz free energy. The experimental values of the variation $\partial \chi_{sol}/\partial(\ln a_\pm)$ exceeded RT/F in all cases, but they were $<2.8RT/F$ in the case of air/HCl solution and $<2.5RT/F$ for the air/HBr solution interfaces.

The results were accounted for in terms of the Ershler's theory.[67] Clear evidence was provided that the effect of discreteness of charges at the interface offered not only qualitative but satisfactory quantitative correspondence with experimental values. According to this theory,

$$\frac{\partial \phi_0}{\partial(\ln a_\pm)} = \frac{1}{(RT/F\phi_0) - 2.22\lfloor|\delta_c - \delta_a|/r\rfloor} \frac{RT}{F} \qquad (38)$$

in which $|\delta_c - \delta_a| \simeq 0.75$ Å is the distance between the planes for adsorption of cations and anions, respectively, and r is the mean distance between the adsorbed anions. For the cases considered here involving high electrolyte concentrations, the effective thickness of the diffuse layer may be neglected in comparison with that of the compact layer.

Krylov[68] has set forth recently a general theory in which he takes into consideration both the discreteness of the adsorbed charges and the statistical distribution of ions in the diffuse layer. The theory was applied to the air/water interface and the results obtained with the aid of the "hexagonal lattice" and "cutoff-disk" models were compared with Gurenkov's experimental results. In this way, the following theoretical expressions were obtained:

$$\partial\phi_0/\partial(\ln a_\pm) \geq 3.9RT/F\,; \qquad \partial\phi_0/\partial(\ln a_\pm) \geq 2.7RT/F \qquad (39)$$

depending on the model used for the calculations, namely the "hexagonal lattice" model and the "cutoff-disk" model, respectively.

As will be seen later, the thermal variation of χ_{sol} allows us to draw conclusions about the sign of χ_{H_2O} (or that for other solvents) and in general on the structure of any liquid surface layer. Unfortunately, values of $d\chi_{sol}/dT$ are not measurable, but they can be estimated from measurements of emf E between the terminals of cells such as

Hg,Hg$_2$Cl$_2$, N KCl	0.01 N KCl	0.01 N KCl	air	0.01 N KCl	Hg,Hg$_2$Cl$_2$, N KCl	
at T_1	at T_1	at T_2		at T_1	at T_1	
(1)	(2)	(3)	(4)	(5)	(6)	(40)

in which T_1 is constant and T_2 is variable (both in °C).

The measured emf E includes, besides the thermal coefficient of χ_{sol}, the difference of potential between solutions 2 and 3, which are of identical composition but are at different temperatures. However, values of such thermal potentials cannot be determined experimentally nor evaluated from thermal coefficients of ionic mobilities (for a discussion of this problem, see Conway and Bockris[83]); consequently, interpretation of such results will always be uncertain.

Employing this method, Jofa (see Frumkin et al.[69]) has measured values of dE/dT for 0.01 N KCl in H_2O; 0.01 N NH$_4$NO$_3$ in ethyl alcohol; 0.015 N (C$_2$H$_5$)$_4$Cl in chloroform; and 0.01 N LiCl in acetone. If the thermal effect at the 2|3 interface is negligible, it might be assumed that $dE/dT \simeq d\chi_{sol}/dT$.

For the range of temperature studied, it was found in all cases that

$$E = \text{const} + (B/T) \qquad (41)$$

Bordi and Vannel[70] have also studied the surface potential of aqueous solutions of $CdCl_2$ and Na_2SO_4 as a function of temperature; their results indicate the presence of thermal anomalies near 30–35°C. A distinct minimum at about the same interval of temperature has also been found for the surface potential of water.[71,72]

Thus, it is evident that reliable experimental data are lacking, especially concerning the variation of surface potential with concentration and the nature of the electrolyte; it is also clear that the theoretical models proposed hitherto to account for the adsorption of ions at the free surface of their solutions are quite inadequate.

3. Values of the Surface Potential χ_{H_2O}

As it has been indicated above, the quantity measured in practice with aqueous solutions of electrolytes is $\Delta\chi = \chi_{sol} - \chi_{H_2O}$, which represents the *difference* between the potential drop across the surface layer of a given solution and the potential drop across the surface layer of pure water; the values we ascribe to this latter quantity are of great importance in the interpretation of the experimental results. Unfortunately, our "molecular insight" into the structure of the surface layer of water is far from satisfactory and there is even no agreement with regard to the sign of χ_{H_2O} at the air/water interface.

A direct measurement of χ_{H_2O} was attempted by Chalmers and Pasquill[73] using the cell

ref. elect.|dil. aq. sol.|air|filter paper|dil. aq. sol.|ref. elect.   (42)

in which, on the assumption that the water molecules were randomly orientated at the surface of the filter paper, so that the χ potential was zero at this surface, the emf of the cell was taken to be equal to the χ value of the other undisturbed surface. In this way, a value of $\chi_{H_2O} = +0.26$ V was obtained. However, it is questionable whether the assumption that water molecules adsorbed upon the surface of a porous material are disoriented, and there is also the objection of a possible contribution to the potential arising from the filter paper itself.[74]

According to Frumkin,[69,75] it seems most likely that the value of χ_{H_2O} must be a small, positive quantity of the order of 0.1–0.2 V. This conclusion was argued indirectly from measurements of $\Delta\chi$ made following the addition of surface-active substances to dilute aqueous solutions of electrolytes. In these cases, the smallness of χ_{H_2O} would be in agreement with the similar maximum shifts of χ toward more positive or negative values, depending on the nature of the added surface-active substances. The positive sign of χ_{H_2O} indicates that the water molecules are oriented in the surface with their negative poles toward the air phase.

To justify this value of χ_{H_2O}, Frumkin et al.[69] gave the following arguments. Let $(^{Hg}\Delta^{H_2O})\psi$ denote the outer (Voltaic) potential of mercury relative to water (with a very dilute solution of inorganic electrolyte) at the zero-charge potential; ϕ_0 the potential difference (inner potential) at the mercury/solution interface in the absence of electrochemical double layers; and ϕ_e the potential difference at the mercury/vacuum interface. Then, we may write

$$(^{Hg}\Delta^{H_2O})\psi - \chi_{H_2O} - \phi_0 + \phi_e = 0 \qquad (43)$$

The value ϕ_0 may be subdivided into a first term ϕ'_e, which, like ϕ_e, depends on the distribution of the "electron cloud" in the surface layer of the metal, and a second term, $(^{Hg}\Delta^{H_2O})\chi$, due to the orientation of dipoles of water at the mercury/water interface. Hence,

$$\phi_0 = \phi'_e - (^{Hg}\Delta^{H_2O})\chi \qquad (44)$$

The quantity $(^{Hg}\Delta^{H_2O})\chi$ is experimentally measurable[65] and a value of -0.33 V was given by Frumkin.[76] Recent measurements by Randles[24] of the Voltaic potential difference between Hg and an aqueous solution of Hg_2^{2+} ion of activity 1 gave the value $\Delta\psi°(Hg, Hg_2^{2+}) = 0.729$ (for this quantity, Klein and Lange found 0.69 V). On the other hand, the determination of the potential of zero charge at Hg by Grahame et al.[77] gave a value of -0.47 V, referred to the normal calomel electrode, and therefore -0.99 V referred to the normal Hg_2^{2+} ion electrode. The value of $(^{Hg}\Delta^{H_2O})\psi$

so calculated is hence -0.26 V. Thus,

$$\chi_{H_2O} = ({}^{Hg}\Delta^{H_2O})\chi + \phi_e - \phi'_e - 0.26 \qquad (45)$$

Therefore, assuming that $\phi_e \simeq \phi'_e$ and that $\chi_{H_2O} \simeq 0.1$ V, a positive value is obtained for $({}^{Hg}\Delta^{H_2O})\chi$, of the order of 0.36 V, which is close to the value 0.2–0.3 V obtained from electrocapillary data.

Another series of arguments was based on the temperature coefficient of χ. Since this quantity depends on the orientation of the molecules in the surface layer, an increase of temperature will decrease the degree of such orientation and must therefore lead to a decrease in the absolute value of χ. In other words, the sign of $d\chi/dT$ must be opposite to that of χ. In practice, the only thing that can be done is to measure electromotive forces E of cells like (40). It was observed from such a measurement that, in the case of alcohols as solvents, for which a negative value of χ was expected, the values of dE/dT were positive. In the case of solutions in water and chloroform, where positive values of χ were most probable, negative values of dE/dT were obtained. However, for solutions in acetone, the results were not so conclusive.

On the other hand, the values of the constant B in expression (41) were respectively 24, -89, and -67 V/°C for H_2O, C_2H_5OH, and $C_5H_{11}OH$. Now, as B/T represents the absolute value of χ for the temperature T, it is to be expected that the 18°C, $\chi_{H_2O} = 0.082$ V and $\chi_{C_2H_5OH} = -0.306$ V. From this, it follows that $\chi_{H_2O} - \chi_{C_2H_5OH} = 0.388$ V, which is close to the value 0.378 found by measuring the emf of cells such as

Hg, Hg$_2$Cl$_2$, N KCl in H$_2$O	sol. in H$_2$O 0.01 N HCl	air	sol. in C$_2$H$_5$OH 0.01 N HCl	Hg, Hg$_2$Cl$_2$, N KCl in H$_2$O
(1)	(2)	(3)	(4)	(5)

$$(46)$$

for which the difference of potential at the liquid junction 4/5 (between the aqueous and alcoholic solutions) is assumed to be negligible.

However, an alternative point of view is held by Kamienski[78] based on experimental results. According to this author, the orientation of about one-tenth of the water molecules at the free surface is sufficient to establish a potential drop of $\chi_{H_2O} = 1$ V.[79] This surface potential is positive, so that the aqueous phase is on the positive side of the oriented dipoles, and the air is on the negative side.[80] The shifts of χ toward negative values, observed in the presence of organic substances, are explained by Kamienski not by orientation of polar groups, but by a decrease of the initial value of χ_{H_2O}. The contribution from the dipolar character of the solute molecules is only considered in the case of substances shifting χ_{H_2O} toward more positive values.[80,81]

This point of view was criticized by Frumkin[75,82] in the following terms: (a) the surface potential χ_{H_2O} can be shifted in both directions by dissolved surface-active substances and by insoluble monolayers; this fact makes a high value for χ_{H_2O} improbable; (b) a value of $\chi_{H_2O} \simeq 1$ V raises objections when the relationship between this quantity and the hydration energies of ions are considered.

Besides this experimental research on the value of χ_{H_2O}, there also exist the results of theoretical work on the comparison of the real energies of hydration of ions $\alpha_{M^{z+}}^{H_2O}$ and the so-called "chemical" energies of hydration. According to the terminology of Lange and Miscenko, the real (free or total) energy is described as the work done in transferring the ion M^{z+} through the gas/solution interfaces. This value may be further subdivided into a part, the "chemical" energy of hydration $\mu_{M^{z+}}^{H_2O}$, due to the interaction between the ions and the surrounding water molecules and a part $(ze\chi_{H_2O})$ due to the electrical energy change in passing the ions through the surface. Then,

$$\alpha_{M^{z+}}^{H_2O} = \mu_{M^{z+}}^{H_2O} + ze\chi_{H_2O} \qquad (47)$$

On the other hand, the real potential of ions $\alpha_{M^{z+}}^{H_2O}$ can be calculated from the Voltaic potential difference between a metal electrode M and a solution (aq. soln. S in the present case) containing its ions in equilibrium with the metal; then,

$$ze(^M\Delta^S)\psi = \alpha_{M^{z+}}^M - \alpha_{M^{z+}}^S = -ze\alpha_e^M + S_M + I_z - \alpha_{M^{z+}}^S \qquad (48)$$

since the electrochemical potential of "ions" in the metal is equal to that of the ions in the solution. In this expression, S_M is the (free or total) energy of sublimation of the metal, I_z the energy of ionization of the isolated atoms, and $-\alpha_e^M$ is the electronic work function of the metal.

Therefore, if values of $\alpha_{M^{z+}}^{H_2O}$ and $\mu_{M^{z+}}^{H_2O}$ where known, the value of χ_{H_2O} could be calculated. The problem now is how to calculate $\mu_{M^{z+}}^{H_2O}$, which is not the object of this chapter (for a general review of this problem, see Conway and Bockris[83]). Table 1 shows the values

Table 1

Method	χ_{H_2O}, V	Reference
Porous material	+0.26	Chalmers and Pasquill[73]
Jet electrode	+0.1	Frumkin et al.[69]
Jet electrode	+1	Kamienski[81]
Theoretical	+0.4	Bernal and Fowler[84]
		Eley and Evans[85]
Theoretical	−0.48	Verwey[86]
Theoretical	−0.30	Hush[87]
Theoretical	−0.36	Strehlow[88]
Theoretical	+0.28	Passoth[89]
Theoretical	−0.3	Miscenko and Kwait[90]

of χ_{H_2O} obtained from both theoretical and experimental methods. It is of interest to note the negative sign for χ_{H_2O} obtained in some of the theoretical work. These results indicate that the water molecules are oriented with their positive poles toward the gas phase. To account for a negative value of χ_{H_2O}, Verwey[86] considered the question of how the liquid structure of water is altered at its surface. The arguments, however, seem rather sophisticated and in the opinion of Passoth[89] the positive sign is probable for the following reasons. Owing to the symmetry of the H_2O molecule, the two "centers of gravity" of the charge distribution are in a straight line with the center of mass which nearly coincides with the nucleus of the oxygen atom. Then, when a molecule of H_2O approaches the liquid phase (dielectric constant ε_w) from the gaseous phase, it will turn itself under the action of the image forces (< 10 Å) so that the center of gravity of electric charges in the dipole can be as near as possible to the liquid phase. In this way, the molecules will tend to orient at the surface with the negative poles toward the gas phase. However, specific directional H-bond effects may outweigh this tendency.

It is evident that, despite the large amount of work that has been devoted to this problem, little can be said conclusively about the value or the sign of χ_{H_2O}. However, it seems most likely that the value of χ_{H_2O} must be a small, positive quantity (∼0.1–0.2 V).*

*Recent evidence, however, seems to point to a small negative value for χ_{H_2O}. Case and Parsons[87] critically discussed the question and concluded that there were insufficient grounds to choose either χ_{H_2O} = +0.1 or χ_{H_2O} = −0.1 V. More recently, de Ligny et al.[188] have derived a negative value, χ_{H_2O} = −0.3 ± 0.1 V.

4. Surface Structure of Water

As Drost-Hansen[43] has pointed out, when considering the structure of liquid interfaces (the surface of water is a particular case) two problems are involved: What is the bulk structure of the liquid (or liquids in contact), and how is the surface (or interface) layer to be defined?

With regard to the first problem, the X-ray diffraction pattern, together with the known properties of free water molecules and those in the ice lattice, led Bernal and Fowler[84] to propose their well-known model of water structure. An analysis by Morgan and Warren[91] led to the abandonment of this model, but its main features, namely the existence of extensively hydrogen-bonded regions and the gradual breakdown of hydrogen bonding with increasing temperature, have been incorporated into most of the more recently proposed models. Since then, a number of crystal structures have been proposed as a basis for a structural model for liquid water.

Another important type of approach involves the so-called mixture model. According to this model, liquid water consists of discrete polymolecular units in equilibrium with "monomeric" molecules. For example, Grojtheim and Krogh-Moe[92] assumed liquid water to be a mixture of icelike species and non-hydrogen-bonded, close-packed species. These forms are in equilibrium, which may be altered by changes in temperature and pressure and by the presence of ionic solutes.

Very recently, Eyring et al. have assumed that two icelike states of water are in equilibrium with each other in liquid water. One of these is a cagelike cluster with a density close to that of ice-I. These clusters in turn are dispersed in an ice-III-like structure, which consists of smaller clusters and, at the same time, fluidized vacancies occur. With this model, calculated and observed values for many thermodynamic parameters are in excellent agreement with the experimental values, but the calculated critical temperature and specific-heat values are in poor agreement with observed values.

A second category of mixture theories envisages polymolecular clusters in which the number of molecules in a cluster is more important than its structure. Frank and Wen[93] postulated that the formation of hydrogen bonds in water has a predominantly cooperative character, so that when one bond breaks, a whole cluster will "dissolve." The ions alter this structure and they may be ordered according to their net structure-promoting or structure-breaking influence. This "flickering cluster" model was used by Nemethy and Scheraga[94] to calculate thermodynamic parameters of liquid water. In this way, it has been estimated, for example, that the average cluster size ranges from 91 to 25 water molecules over a temperature range from 0 to 70°C, with the mole fraction of non-hydrogen-bonded molecules increasing from 0.24 to 0.39 over the same range of temperature.

In recent work by Gurikov,[95,96] a theory was given in which he assumed the existence in liquid water of a framework of hydrogen bonds as a legacy of the structure of ice after melting. To determine the average number of broken bonds as a result of melting, the conditions of minimum free energy are written down. The resulting equation has two real roots. One corresponds to ordinary water and the other (anomalous water) gives, near 0°C, a vapor pressure lower than that of ordinary water. However, the conformity of this theory with the whole range of experimental facts concerning water is still an open question.

On the other hand, Pauling called attention to the clathrate, cagelike structures. In this model, the basic structural unit is the pentagonal dodecahedron. At each corner, there is a water molecule and the structure is sustained by hydrogen bonds among these molecules. Such dodecahedra leave cavities between them that can accommodate or trap single water molecules. By placing trapped water molecules within some of the dodecahedra, Pauling was able to calculate a value for the density of water in fair agreement with that observed. However, there are several difficulties and objections to this model, mainly because it does not account for the fluidity of liquid water. Frank and Quist have proposed the concept of a "flickering" cluster, and in this way, the necessary fluidity can be introduced into the Pauling model.

The models of α-quartz structure, cubic ice structure, clathrate structure, and ice-I structure have been examined very recently by Narten et al.[97] to explain the X-ray diffraction patterns of liquid water in the temperature range 4–200°C. The clathrate type of model proposed by Pauling[98] was investigated in detail, but it does not seem adequate to explain the X-ray data. Of the crystal structures, examined the ice-I lattice was the most promising. Each oxygen atom is tetrahedrally surrounded by other oxygen atoms, forming layers in six-membered, puckered rings. Two adjacent layers, related by mirror symmetry, form dodecahedral cavities. Vacant lattice sites and occupancy by water molecules of the large cavities typical of this structure was permitted. With proper adjustment of parameters, this model explains the X-ray data over the whole range of temperature.

The cell theory of fluids has been applied[99] to liquid water, using the expanded ice lattice model. A "free volume" and the distance between nearest neighbors were computed for several values of the temperature. It was found that the experimental density of liquid water could only be obtained if approximately 20 % of the molecules were placed in interstitial sites.

Unfortunately, however, the "molecular insight" into the structure of water at interfaces is much less satisfactory. It was previously believed that the ordered region defining the surface layer of water was quite thin, about 2–3 Å.[100] However, in addition to this layer of very strongly bonded water, the perturbation of the water structure may extend into the depth of the liquid to a distance of more than several molecular diameters.[101] In the case of the solid/water interface, Derjaguin[102] presented further evidence, now confirmed by various workers, for a thick layer of "anomalous" water.

For a number of years, it has been suggested anomalies exist in the properties of water at a number of different discrete temperatures (a general review is given by Drost-Hansen[43]). These anomalies have become known as "kinks" in the properties of water and they owe their existence to structural transitions of the nature of higher-order phase transitions. Thus, liquid water appears to undergo a higher-order transition somewhere around 40°C,[103] although Falk and Kell[104] have questioned the reality of this higher thermal transition and the matter appears to be unsettled.

An analysis of the most reliable surface tension measurements of liquid water indicate the existence of at least three minima in the values of $d\gamma/dT$ at 13, 35, and 37°C. These anomalies, which have been observed also in the surface potential of aqueous solutions of electrolytes, reflect structural changes in the surface layer. Thus, Drost-Hansen[43] has suggested that the thermal anomaly near 30°C was due to the "melting" of structural clusters. The minimum in the entropy at slightly higher temperature in turn corresponds to a state of greater ordering such as found in a close packing of spherical molecules. Horne et al.[105,106] have shown recently that the interfacial water structure is more stable with respect to pressure than the bulk structure; it is also less susceptible to thermal destruction than the bulk form. This interfacial structure gives rise to anomalies in water properties near 40°C, but although its structure is unknown, it is possibly the Pauling clathrate type with the cages occupied by monomeric water molecules. However, at the present time, this is no more than a suggestion.

IV. IONIZED MONOLAYERS

1. Surface Potential of Charged Monolayers

If a monolayer of neutral molecules is spread or adsorbed on a clean water surface, all dipoles at the interfaces will be rearranged.[107] The total surface dipole moment per adsorbed molecule will be given by

$$\Delta V \equiv (4\pi/A)p^s \qquad (49)$$

Consider now the potential due to a monolayer of long-chain ions spread on an aqueous solution of an electrolyte. An "ionic atmosphere" (or ionic double layer) will be set up in the aqueous phase, underneath the film. According to the operational definition of surface moment p^s given above, the values of p^s will include in this case the electrostatic contribution of this ionic double layer to the measured surface potential ΔV. To distinguish this electrostatic distribution from the ordinary dipole term, Davies[108] assumed that

$$\Delta V = (4\pi/A)p_u + \phi_0 \qquad (50)$$

where ϕ_0 is the electrical potential drop between the plane of the charged groups at the interface and in the bulk phase.

Furthermore, he assumed that, on the basis of the Gouy–Chapman theory of the electric double layer, ϕ_0 (called ϕ_G) can be calculated for a 1–1 electrolyte,

$$\sigma = (2\varepsilon_w kTn/\pi)^{1/2} \sinh(e\phi_0/2kT) \qquad (51)$$

where n is the number of ions of one sign per cm^3 in the bulk of the solution, ε_w is the dielectric constant of water, and σ is the surface charge density.

At 20°C, Eq. (51) simplifies to

$$\phi_G \equiv \phi_0 = (2kT/e)\sinh^{-1}(134/Ac_i^{1/2}) \tag{52}$$

with ϕ_G expressed in mV, A in $Å^2$ per charged group, and c_i is the total electrolyte concentration (in moles per liter).

If p_u is unaffected by the electric field and ϕ_G may be identified with ϕ_0, then

$$\Delta V = (4\pi/A)p_u + \phi_G \tag{53}$$

Davies has found that indeed for films of $C_{18}H_{36}N(CH_3)^+$ on salt solutions and $A = 85\ Å^2$/molecule, p_u is constant (450 mD).

However, the following assumptions are implicit in the use of the Gouy–Chapman equation to calculate potentials: (a) the charge is uniformly spread over the surface ("smeared out"); (b) the possibility of penetration of some counter-ions between the film-forming ionic group is neglected; (c) the counter-ions do not have appreciable size; (d) the dielectric constant in the vicinity of the charged film is the same as that in the bulk phase.

Concerning the first two assumptions, several authors have suggested that in practice the monolayers of long-chain ions are neither uniformly charged nor impenetrable; there is a pronounced penetration by the counter-ions into and above the plane of the charged groups.[109-112] Certainly, this plane is unlikely to coincide with the phase boundary and it has been suggested, e.g., that an adsorbed long-chain sulfate ion has over half its chain immersed in the aqueous phase. There is, then, a region above the head groups which can accommodate counter-ions and the electrical potential will fall off on each side of the charged groups.

In the theory developed by Haydon and Taylor,[113] allowance is made for the head groups being situated at an equilibrium distance d from the interface and for the fact that counter-ions will be present above them. Figure 10 gives a qualitative picture of the distribution of electrical potential in the region of an ionized monolayer; the dashed line indicates the plane of the ionized head groups.

Allowance is also made for a limitation of space above the plane of the head groups because of the presence of the hydrocarbon

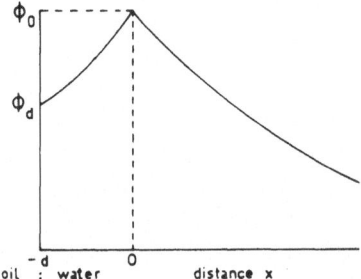

Figure 10. Distribution of electrical potential in the region of an ionized monolayer according to the theory of Haydon and Taylor.[113]

chains so that, in this region, a modified Boltzmann equation is used

$$n_i = n_0[\exp(-z_i e\phi/kT)][1 - (A_0^1/A)] \qquad (54)$$

where A_0^1 is the cross-sectional area of the hydrocarbon chain.

For this equation, the following expressions are derived for a uni-univalent electrolyte:

$$\frac{d\phi}{dx} = -\left[\frac{32\pi n_0 kT}{\varepsilon_w}\left(1 - \frac{A_0^1}{A}\right)\right]^{1/2} \sinh\frac{e\phi}{2kT} \qquad (55)$$

and

$$\kappa d\left(1 - \frac{A_0^1}{A}\right)^{1/2} = \ln\left[\frac{\tanh(e\phi_0/4kT)}{\tanh(e\phi_d/4kT)}\right] \qquad (56)$$

The total diffuse charge is given by

$$\sigma = \sigma_1 + \sigma_2 = -\int_0^\infty \rho\, dx - \int_0^d \rho\, dx \qquad (57)$$

where ρ is the bulk charge density at a distance x from the plane zero.

Using the Poisson equation and the simple Gouy–Chapman theory for the first term (concerned with the aqueous side of the film) and the expression (55) for the second one (concerned with the penetration of the film by counter-ions), the following relation is obtained:

$$\sigma = \frac{10^{16}e}{A} = \left(\frac{2n_0\varepsilon_w kT}{\pi}\right)^{1/2}\left[1 + \left(1 - \frac{A_0^1}{A}\right)^{1/2}\right]\sinh\frac{e\phi_0}{2kT}$$
$$- \left[\frac{2n_0\varepsilon_w kT}{\pi}\left(1 - \frac{A_0^1}{A}\right)\right]^{1/2}\sinh\frac{e\phi_d}{2kT} \tag{58}$$

As $A \to \infty$, we have

$$\sigma = \left(\frac{2n_0\varepsilon_w kT}{\pi}\right)^{1/2}\left[2\sinh\frac{e\phi_0}{2kT} - \sinh\frac{e\phi_d}{2kT}\right] \tag{59}$$

As $d \to 0$ and $\phi_d \to \phi_0$, Eq. (59) reduces to the simple Gouy equation (51).

For experiments carried out, for example at the oil/water interface, ϕ_0 and d are related directly by

$$e\phi_0/kT = (\Delta G_{o/w}^{CH_2}/NkT)\,d/l \tag{60}$$

where $\Delta G_{o/w}^{CH_2}$ is the nonelectrical component of the free energy of adsorption per CH_2 and l is the effective carbon–carbon bond length. Using Eqs. (56), (58), and (60), the quantities ϕ_d, ϕ_0, and d can be calculated from σ (or A) and $\Delta G_{o/w}^{CH_2}$.

The measured surface potential ΔV may be related to ϕ_d and the Davies expression (53) rewritten

$$\Delta V = (4\pi/A)p_u + \phi_d \tag{61}$$

For large areas, ϕ_d is either constant or varies by no more than 5 mV; then, if p_u is constant, a plot of ΔV against $1/A$ should be linear with an intercept of ϕ_d. This is tested in Fig. 11 and found to be correct for experiments carried out with monolayers of $C_{12}H_{25}N(CH_3)_3^+$ spread at the oil/water interface (the oil was petroleum ether and the aqueous phase NaCl solution). Close agreement between experimental and theoretical values of ϕ_d has been claimed.

These results demonstrate that the inner potential ϕ_d close to the phase boundary differs from that in the plane of the charged groups (ϕ_0) and that the true ϕ_0 is lower than that calculated from the simple Gouy equation (ϕ_G). Davies' assumption that $\phi_0 = \phi_G$ is only valid for large areas ($A \to \infty$).

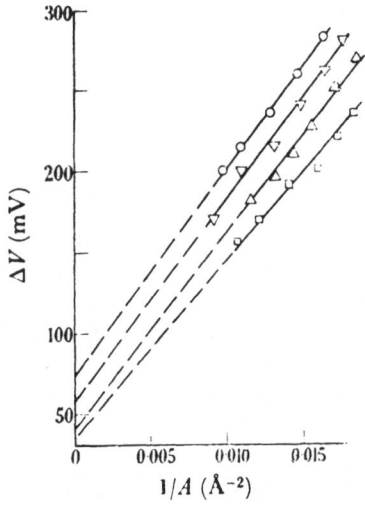

Figure 11. Plot of ΔV versus $1/A$ for monolayers of $C_{12}H_{25}N(CH_3)_3^+$ spread at the
oil/water interface (Haydon and Taylor[113]).

However, this theory predicts potentials which are too low
under conditions where either the area A is low (high values of σ) or
the total electrolyte concentration exceeds a certain value. The
direction and magnitude of the discrepancies suggest that they
arise from a neglect of the finite sizes of the ions present. Previous
attempts to introduce an ionic size term into the Gouy–Chapman
equation are mainly concerned with the theory of the electrochemical
double layer at electrode interfaces. For liquid interfaces, however
the most interesting work is that of Ohlenbusch's[114] whose approach
is slightly simpler but essentially similar to that developed by
Haydon and Taylor[115,116]; in this case, the finite sizes of both the
adsorbed and counter-ions were taken into account.

In general, for systems where the plane of the charged groups in
the film is situated a finite distance d from the phase boundary, the
distribution of potential (see Fig. 12) and of ions normal to the
interface is considered in three regions:

(a) The region where $\tau < x < \infty$, τ being defined as the distance
of closest approach of the centers of the counter-ions to the plane of

the head groups. This is a "diffuse region" in which only the size of the counter-ions needs to be considered.

(b) The region $\tau > x > 0$, in which the potential drops from ϕ_0 to ϕ_τ. This region is treated as a condenser in which

$$\phi_0 = \phi_\tau + (4\pi\sigma\tau/\varepsilon_w) \tag{62}$$

(c) The region $-\tau > x > d$ above the plane of the charged head groups in which the counter-ions can penetrate. In this region, allowance is made in addition for the restriction in the space available caused by the presence of the long hydrocarbon chains.

To calculate electrical potentials in the "diffuse" region, a modified Boltzmann equation is used to allow for the space occupied by the ions present in the system[117,118]

$$n_i^s = n_{i0}[(1 - \sum v_i n_i^s)/(1 - \sum v_i n_{i0})] \exp(-z_i e\phi/kT) \tag{63}$$

where n_{i0} are bulk phase concentrations, n_i^s are concentrations of adsorbed ions, and v_i are "space factors," which in general may be taken as the ionic volumes of species i.

Figure 12 gives the potential distribution calculated by Haydon and Taylor according to this refined theory. In this particular system, the electrolyte concentration was $0.07\,M$, $A = 48.2\,\text{Å}^2$, $d = 5.28\,\text{Å}$, $\phi_0 = 144\,\text{mV}$, and $\tau = 0.19\,\text{Å}$.

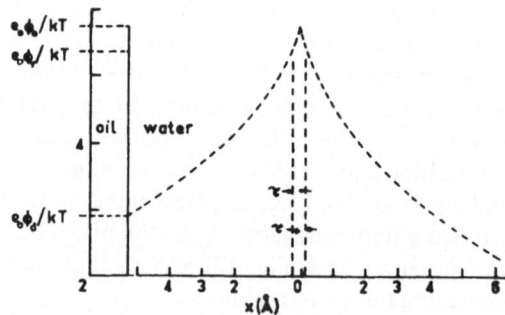

Figure 12. Distribution of electrical potential according to Haydon and Taylor's refined theory.[115]

For systems such as those under consideration, Eq. (60) must be replaced by

$$\frac{e\phi_0}{kT} = \left[\frac{\Delta G_{o/w}^{CH_2}}{NkT}\right]\frac{d}{l} = \frac{e\phi_\tau}{kT} + \frac{4\pi 10^{13} e\tau}{A\varepsilon_w} \tag{64}$$

Comparisons of the values of ϕ_d and ϕ_τ calculated from experiments carried out with monolayers of $C_{12}H_{25}SO_4^-$, $C_{12}H_{25}N(CH_3)_3^+$, and $C_8H_{17}SO_4^-$ at the oil/water (or NaCl solution) interface reveal that an important potential drop occurs between the plane of the head groups and the phase boundary. The number of ions per cm^2 of interface present above the head groups can be calculated and it can be seen that this number increases sharply as A decreases, and it decreases as the ionic strength increases (A being kept constant) mainly because d also decreases.

From these results, Haydon and Taylor have stated that, in the systems they have studied, ions such as Na^+ and Cl^- have no specific binding to the interface of the plane of the head groups. The "penetration" of the interface by small ions in the sense of the above theory seems sufficient to provide an interpretation of the results.[119]

However, this theory has been criticized by Levine et al.[120] on the ground that Eq. (64) is not correct. In addition, the solutions of the Poisson equation for diffuse layers do not satisfy the condition $d\phi/dx = 0$ at the plane of nearest approach of diffuse layers. So that, while Haydon and Taylor concluded that, for $C_{12}H_{25}SO_4^-$ and $C_{12}H_{25}N(CH_3)_3^+$, d varies between 3 and 6 Å for NaCl concentrations 0.05–0.1 M and areas $A = 100$ Å^2, Levine et al. showed that d is less than 1.9 Å at $A = 50$ Å^2 and it diminishes to a limiting value of 1.3 Å as A increases.

Furthermore, there are certain features which seem to be inconsistent with the Gouy theory of the electrical double layer at the film. Thus, maxima in the magnitude (irrespective of the sign) of the surface potential ΔV against A have been reported by Philips and Rideal[111] for monolayers of $C_{18}H_{37}SO_4^-$ on 10^{-2} and 10^{-1} M HCl and NaCl at air/water and oil/water interfaces. The maximum is relatively sharp in 10^{-2} M solutions and it occurs at $A = 75$ Å^2. However, these results were criticized by Haydon and Taylor[121] on the grounds that they were obtained with unpurified NaCl. The presence of alkaline earth ions in the subsolution could account for the discrepancies, in view of the known preferential adsorption of Ca^{++} ions at anionic monolayers.[122,123]

Pethica and Few[124] have found maxima in $|\Delta V|$ for $C_{18}H_{37}SO_4^-$ at the air/water interface at somewhat smaller areas (65 $Å^2$) and also for adsorbed films of $C_{12}H_{25}SO_4^-$ at the air/water interface in the presence and absence of NaCl at $A = 60$–70 $Å^2$. At the oil/water interface, Philips and Rideal[111] found no maxima with $C_{22}H_{45}SO_4^-$ spread on aq. HCl as substrate, and Stenhagen[125] observed a similar behavior for the same film spread at the air/water interface. Haydon and Philips[126] observed no maximum in ΔV for $C_{12}H_{25}SO_4^-$ adsorbed at the petroleum ether/water interface, but more recent experiments carried out by Haydon[127] at the n-heptane/NaCl solution interface indicate that $|\Delta V|$ has a maximum when $A > 80$ $Å^2$.

It is important to note, however, that the above data on maxima in $|\Delta V|$ have been restricted to anionic surfactants, while no maxima have been found for cationic films. This can be attributed to differences in the magnitude of the dipole contribution to ΔV, as discussed by Levine et al.[120]

On the other hand, careful measurements of surface pressure and surface potential of spread monolayers of $C_{18}H_{36}SO_4^-$ at the air/water interface were carried out recently by Mingins and Pethica[128] for surface areas from 80 to 5000 $Å^2$ molecule^{-1} on NaCl solution (0.1, 0.01, and 0.001 M). These results were analyzed in terms of the above theories of the ionic double layer and it was found that, in this case, the Haydon–Taylor model gives poor agreement with experimental data except for large areas.

On the basis of experimental results like these, several authors[124,129] have claimed that specific interactions occur between long-chain ions and counter-ions at monolayers. Davies[130] weighed these rival claims and concluded by adopting the intermediate view that there are small specific interactions. It is well known, for example, that counter-ion Ag^+ is also held on a monolayer of $C_{22}H_{45}SO_4^-$ by the energy of polarization λ, while $C_2H_5NH_3^+$ ions are adsorbed partly due to the van der Waals energy W associated with ethyl groups. In general, the total specific interaction energy Φ of the counter-ions with the film-forming

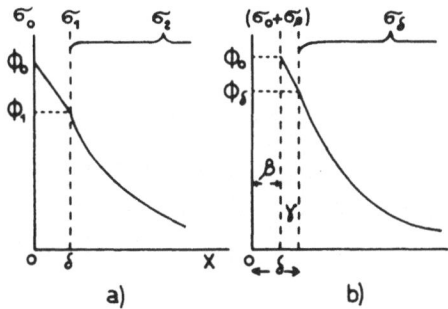

Figure 13. Stern models of the electrical double
layer.

long-chain ions is given by

$$\Phi = \lambda + W \tag{65}$$

In order to allow for this specific interaction, the Stern theory of
the ionic double layer has been employed.[124] Figure 13(a) represents
one form of the Stern model in which: (a) the primary surface charge
(ionic monolayer) is not penetrated by the counter charge; (b) the
counter-ions may approach the charged surface to within a distance
($\simeq$ ionic radius) and may be "specifically" adsorbed; (c) the adsorbed
or compact layer (charge σ_1) is taken to be concentrated at the plane
of δ; and (d) the rest of the counter charge (σ_2) is a diffuse Gouy layer.
As is known, for a uni-univalent electrolyte, the equations are

$$\sigma_2 = \left(\frac{2\varepsilon_w kTn}{\pi}\right)^{1/2} \sinh \frac{e\phi_1}{2kT} \tag{66}$$

$$\sigma_1 = \frac{N_1 e}{1 + (N/Mn) \exp(e\phi_1/kT) \exp(\Phi/kT)} \tag{67}$$

$$\sigma_0 = \sigma_1 + \sigma_2 = \frac{\varepsilon'}{4\pi\delta}(\phi_0 - \phi_1) \tag{68}$$

where N is the number of adsorption sites per unit area for the
counter-ions making up σ_1, and may be taken as equal to $|\sigma_0/e|$; M
is the molecular weight of solvent (water); Φ was defined above; and
ε' is the dielectric constant of the region 0 to δ.

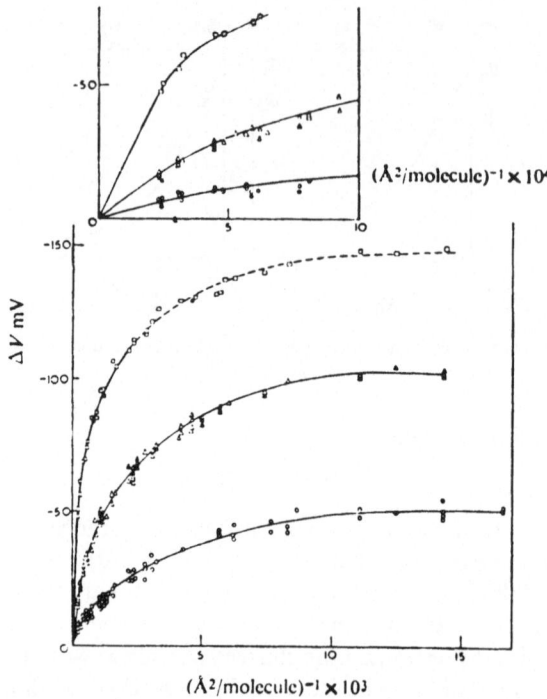

Figure 14. Plot of ΔV versus $1/A$ for monolayers of $C_{18}H_{37}SO_4^-$ spread on NaCl solutions (Mingins and Pethica[128]).

Figure 14 shows the results of Mingins and Pethica,[128] expressed as ΔV versus $1/A$, for monolayers of $C_{18}H_{37}SO_4^-$ spread on NaCl solution. The curves tend to be linear as $A \to \infty$ and the plot of the initial slopes of these ΔV versus $1/A$ curves against $(1/n)^{1/2}$ yields a straight line. The analysis of these results showed, as mentioned above, the inadequacy of the Gouy model, except at the largest area. The Stern model seems to be more appropriate, but the values of Φ, which at this stage are simply "fitting parameters," showed a variation with area A per film molecule.

It must be pointed out that the high values of σ_0 attainable with fully ionized monolayers would lead to improbably high potential drops between the plane of the primary ions and that of the counterions in the above model.[124] A penetration of the plane of the head groups by substrate ions (in particular, counter-ions) would reduce this potential drop in the Stern region. In this sense, van Voorst Vader[122] assumed that a part of the counter-ions are adsorbed (interpenetrate) between the head groups of the surfactant ions over a thin layer, giving an overall charge density of $\sigma_0 + \sigma_1$.

This model was simplified by Levine et al.[120] in order to make it more amenable to mathematical analysis for calculating the discrete-ion effect. Figure 13(b) represents this model, in which it is assumed that the primary and adsorbed charges are in the same plane (adsorption plane) at a depth β from the interface. The limit of the diffuse layer is situated at a distance δ from the interface, forming one boundary of the inner region. This "inner region" is assumed to be a homogeneous medium of dielectric constant ε_1 and thickness $\gamma = \delta - \beta$.

If the electrolyte is uni-univalent and the Gouy–Chapman theory governs the ion distribution in the diffuse layer, σ_δ is related to the potential ϕ_δ by

$$\sigma_\delta = (2\varepsilon_w kTn/\pi)^{1/2} \sinh(e\phi_\delta/2kT) \qquad (69)$$

Since it is assumed that the counter-charge either resides in the diffuse layer or on the adsorption plane, the potential ϕ_0 at this plane is given by the electrostatic relation

$$\sigma_\delta = (\varepsilon_1/4\pi\gamma)(\phi_0 - \phi_\delta) \qquad (70)$$

Also, the condition of electroneutrality is

$$\sigma_0 + \sigma_\beta = \sigma_\delta \qquad (71)$$

Finally, it is assumed that the specific binding is small and that the counter-ions and ionized monolayer molecules are highly mobile as separate entities in the common adsorption plane. A Langmuir-type model of adsorption sites for the counter ions is chosen. Then, the adsorption isotherm for counter-ions may be written as

$$kT \ln[\sigma_\beta/(N_1 e - \sigma_\beta)] = -\Phi - e\phi_0 - \varphi(\sigma_0, \sigma_\beta) + kT \ln(n/n_0) \qquad (72)$$

where n_0 is the volume density of water molecules in the electrolyte and $\varphi(\sigma_0, \sigma_\beta)$ is the fluctuation (discrete-ion effect) potential at the center of an adsorbed counter-ion. From relations (69)–(72), it can be shown that the condition for a maximum in $|\phi_d|$ with variation in

σ_0 at fixed ionic strength in the substrate requires that

$$\frac{\partial \varphi}{\partial \sigma_0} - \frac{\partial \varphi}{\partial \sigma_\beta} = kT \left[\frac{1}{\sigma_\beta} + \frac{1}{N_1 e - \sigma_\beta} \right] \qquad (73)$$

and the potential $|\phi_0|$ at the adsorption plane likewise shows a maximum.

In determining φ, Levine and coworkers[120,131] applied the so-called "cutoff-disk" approximation, in which only the discrete charge in the vicinity of a given adsorbed ion is taken into account, whereas the charges of all other ions are assumed to be continuously distributed in the adsorption plane. Consequently, this model allows to some extent for the thermal motion of ions smoothing the discreteness effect. On the other hand, the Soviet authors who originated the notion of the charge-discreteness effect (see Krylov[68]) were mainly concerned with the so-called "hexagonal-lattice" model in which no allowance was made for the thermal motion of specifically adsorbed ions, so that attention is restricted to the maximum influence of the discreteness effect.

Since an accurate determination of $\varphi(\sigma_0, \sigma_\beta)$ is difficult, a simple "cutoff" approximation was used by Levine for the problem of ionized monolayers at liquid interfaces. It is assumed in this approximation that the electrolyte concentration is small and that the fluctuation potential φ may be expressed as a sum of separate functions of σ_0 and σ_β, namely

$$\varphi(\sigma_0, \sigma_\beta) = \varphi^{\mathrm{I}}(\sigma_\beta) + \varphi^{\mathrm{II}}(\sigma_0) \qquad (74)$$

Then, it can be shown that the contribution to the fluctuation (self-atmosphere) potential from the charge distribution $\sigma_\beta(r)$ is

$$\varphi^{\mathrm{I}}(\sigma_\beta) = - \frac{4\pi^2 \sigma_\beta^2 r_\beta^2}{3\varepsilon_1}(1 + P) - \frac{\pi^2 \sigma_\beta^2 a^2}{6\varepsilon_1}(1 - 6L) \qquad (75)$$

where $r_\beta = (e/\pi\sigma_\beta)^{1/2}$; a is the distance of nearest approach (exclusion diameter) of two counter-ions; and P and L are functions plotted by Levine et al.[120]

With regard to the contribution $\varphi^{\mathrm{II}}(\sigma_0)$ from $\sigma_0(r)$, only the gaseous model of the monolayers was considered. It was found as a lower estimate that

$$\varphi^{\mathrm{II}}(\sigma_0) = -(4be\sigma_0/\varepsilon_1)(1 + H) \qquad (76)$$

where b is the "mutual exclusion" diameter between primary charges and counter-ions and H is a function also plotted by Levine et al.[120]

This theory of the charge-discreteness effect has hitherto, only been applied quantitatively to ionized monolayers for the one series of experiments carried out by Mingins and Pethica,[128] namely $C_{18}H_{37}SO_4^-$ spread at the air/0.01 M NaCl solution interface at $9.5°C$. The method outlined below illustrates the procedure required to apply the above theory to this system.[132]

The results of Mingins and Pethica, namely that ΔV is a linear function of $1/A$ for large A and that a plot of the initial slopes of these ΔV versus $1/A$ curves against $(1/c)^{1/2}$ yields a straight line parallel to the one given by the Gouy–Chapman model, show that under these conditions the equation

$$\Delta V = (4\pi p_u/A) - (\pi/2\varepsilon_w kTn)^{1/2}(2kT/A) \qquad (77)$$

is obeyed. This implies that $\sigma_\beta \ll \sigma_0$ for such large areas. The dipole term p_u is assumed to be independent of A and c and it can be estimated from the initial slopes of ΔV versus $1/A$ curves; a value of $p_u = 120$ mD was calculated.

The value of ϕ_0 at any given A is now calculated from (50) with the assumption that p_u is independent of A. A maximum in $|\phi_0|$ is seen to occur at $A_{max} = 180$ Å^2 molecule^{-1}. By assuming $\gamma = 0$ (i.e., $\phi_0 = \phi_\delta$) and making use of (69), σ_β and therefore r_β can be calculated from (71). The value of r_β at the maximum is found to be 8.8 Å. Values of r_β at the maximum in $|\phi_0|$ as a function of ε_1 for different β at a fixed γ can be determined from a Grahame disk model in the absence of diffuse-layer screening. If $\beta = 3$ Å, ε_1 at the maximum is 28 when $\gamma = 1$ Å. Substitution of this value into (70) shows that at the maximum the potential drop $(\phi_0 - \phi_\delta)$ is small compared with ϕ_δ and therefore the assumption $\gamma = 0$ employed to obtain r_β at the maximum was justified.

Assuming ε_1 to be independent of A, Eqs. (69) and (70) allow both ϕ_δ and σ_δ to be determined. Then, values of σ_β are calculated from (71), since σ_0 represents experimental data. For $A < 500$ Å^2 molecule^{-1}, the calculated values of σ_β with $\gamma = 1$ Å are only slightly higher than those with $\gamma = 0$.

Finally, the fluctuation potentials $\varphi(\sigma_0, \sigma_\beta)$, which is the sum of (75) and (76), was calculated for a value of the dielectric constant of the nonaqueous phase taken as 1, $a = 5$ Å, $b = 5$ Å, $\gamma = 1$ Å, $N_1 = 10^{15}$ cm^{-2}, several values of β, $T = 9.5°C$, and $\varepsilon_w = 84$. In this way, all terms in (72) are known except Φ, which is easily calculated. For a value of $\beta = 3$ Å, it can be seen that Φ is small $(-\Phi/kT = 0.8\text{--}0.9$ for $A = 100\text{--}1000$ Å^2 molecule^{-1}) and is almost independent of A. For a value of $\beta = 5$ Å at $\gamma = 1$ Å, ε_1 at the maximum is 30. Under these conditions, the constancy of Φ with a change in A is still maintained and this is consistent with Stern's idea that it is a purely chemical term.

In the above calculations of fluctuation potential $\varphi(\sigma_0, \sigma_\beta)$, the screening influence of the diffuse layer and the presence, in the inner region, of ions of opposite sign to σ_β were ignored. This theory was improved[133] by allowing for such a diffuse layer screening effect, but up to now, no applications have been made to ionized monolayers. However, using a much simpler method of calculating this screening effect, Bell et al.[134] showed that, for a 0.1 M uni-univalent electrolyte solution, this screening term amounted to a significant factor whose magnitude depended on the value of A, but for a 0.001 M solution, the effect was negligible except for large areas per long-chain ion. The diffuse-layer correction tended to reduce the magnitude of φ by about 15 % at the maximum in $|\phi_0|$ and became a greater proportion of this potential with increase in area A.

2. Equation of State for Charged Monolayers

The surface pressure Π is defined as the difference in the surface tension of a clean surface and a film-covered surface. For an uncharged film, both the kinetic pressure Π_k and cohesive pressure Π_s contribute to the total pressure Π, but if the film is charged (ionized monolayer), an electrostatic interaction of charged head groups Π_e must also be allowed for. If these terms are separable, the total pressure of a charged film will be made up of three independent components,

$$\Pi = \Pi_k + \Pi_s + \Pi_e \tag{78}$$

All these terms are usually investigated by means of a surface equation of state. The determination of the electrostatic contribution Π_e from experimental data is difficult, but its examination is worthwhile because of its relation to surface potential in the case of ionized monolayers.

The equations of state for charged monolayers spread at the air/water and oil/water interfaces have been of interest since Davies[108] applied the theory of the double layer to calculate the repulsive energy between long-chain ions at the interface. According to Verwey and Overbeek,[135] the total free-energy decrease per unit area resulting from an ionic double layer formed by localized adsorption at an interface is given by

$$\Delta F = -\sigma_0 \phi_0 + \int_0^{\sigma_0} \phi \, d\sigma = -\int_0^{\phi_0} \sigma \, d\phi \qquad (79)$$

where ϕ_0 and σ_0 are the final equilibrium values of the "wall" potential and the charge, respectively. Davies[108] supposed that the increase Π_e in surface pressure (in erg cm^{-2}) due to the electrical double layer could be equated with $-\Delta F$ to give

$$\Pi_e = \int_0^{\phi_0} \sigma \, d\phi \qquad (80)$$

and, by applying the Gouy–Chapman model of the double layer to a uni-univalent electrolyte, the Davies equation is derived,

$$\Pi_e = \left(\frac{8\varepsilon_w c_i}{\pi}\right)^{1/2} \frac{(kT)^{3/2}}{e}\left[\cosh\left(\frac{e\phi_0}{2kT}\right) - 1\right] \qquad (81)$$

For large areas, the cohesive pressure Π_s is shown to vary in accord with the empirical equation first suggested by Guastalla[136]

$$\Pi_s = -Km/A^g \qquad (82)$$

where m is the number of CH_2 groups in each surface-active molecule or ion and g is a parameter having the value 3/2 or 2.[137]

In a recent paper by Chattoraj and Chatterjee[138] it was found that $g = 2$ in the range of low surface concentration. This value was explained on the basis of the van der Waals equation applied to the surface phase. At the region of higher surface concentration, the long-chain ions form surface micelles and Π_s then tends to be independent of A.

This leads to the equation of state for coherent charged films,[139,140]

$$\sinh^{-1}\left(\frac{\theta}{Ac_i^{1/2}}\right) = \cosh^{-1}\left[1 + \frac{\theta}{2kTc_i^{1/2}}\left(\Pi \quad \Pi_s - \frac{kT}{A - A_0}\right)\right] \qquad (83)$$

where $\theta = e(\pi/2\varepsilon_w kT)^{1/2}$ ($\theta = 136$ for $t = 20°C$ and A expressed in Å^2 per long chain); A_0 corrects for the area actually occupied by the film-forming molecules.

At the oil/water interface, the cohesive term is very small and Eq. (83) can be expressed as a virial expansion of ΠA versus $1/A$,[139]

$$\Pi A = kT[1 + (4\xi + A_0)(1/A) + A_0^2(1/A^2)$$

$$+ (A_0^2 - 16\xi^3)(1/A^3) + \cdots] \qquad (84)$$

in which $\xi = \theta/4c_i^{1/2}$.

If the ionic concentration increases, $\xi \to 0$ and Eq. (84) reduces to

$$\Pi(A - A_0) = kT \qquad (85)$$

On the other hand, if the salt concentration c_i is small, Eq. (84) reduces to the approximate form

$$\Pi(A - A_0) = 3kT - (2kTA_0/A) - (2kTc_i^{1/2}/\theta)(A - A_0) \qquad (86)$$

which cannot be extrapolated to $A \to \infty$ since the condition under which (86) holds is that $Ac_i^{1/2}$ is small. Nevertheless, Eq. (84) allows an extrapolation of ΠA versus $1/A$ curves to be made for a given value of c_i, namely

$$\Pi A \to kT \qquad \text{if} \quad A \to \infty \quad \text{and } c_i \text{ is kept constant} \qquad (87)$$

Llopis *et al.*[139] employed such an extrapolation in experiments carried out with monolayers of $C_{16}H_{33}(C_6H_5CH_2)(CH_3)_2N^+$ and $C_{20}H_{41}(CH_3)_3N^+$, spread at benzene/NaCl solution (0.01, 0.1, and 1 M) interfaces, and areas between 10^3 and 10^2 Å^2 per long-chain ion were found. Figure 15 shows the results for $C_{20}H_{41}(CH_3)_3N^+$; also plotted (broken lines) are the theoretical curves calculated with $\Pi_s = 0$ and $A_0 = 30 \text{Å}^2$ from Eq. (83). The agreement between theory and experimental results is only qualitative. This may be attributed either to a breakdown in the assumption of the Gouy–Chapman model or to the cohesive term, which for these long-chain ions at the oil/water interface, is small but not quite negligible.

At the air/water interface, the cohesive energy between the long hydrocarbon chains is no longer negligible. Figure 16 shows the curves of Π_s versus A calculated from Eq. (83) for a series of long-chain alkyl quaternary alkylammonium salts spread or adsorbed at the air/1 N HCl solution interface. These results demonstrate how

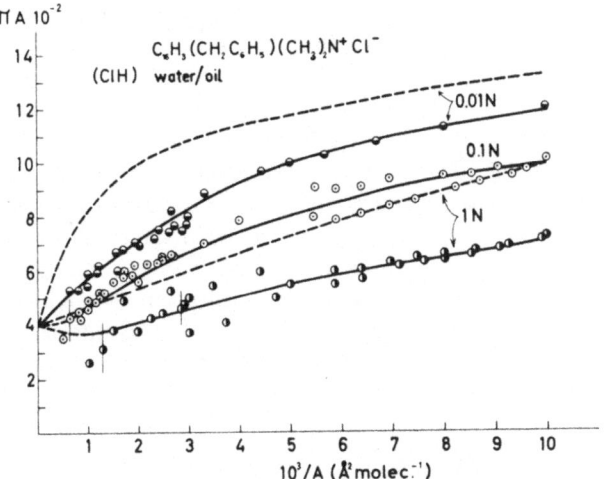

Figure 15. Plot of ΠA versus $1/A$ for monolayers of $C_{20}H_{41}(CH_3)_3N$ spread at the benzene/water interface; broken lines are theoretical curves calculated from Eq. (83) assuming $\Pi_s = 0$ and $A_0 = 30 \text{ Å}^2$ (Llopis et al.[139]).

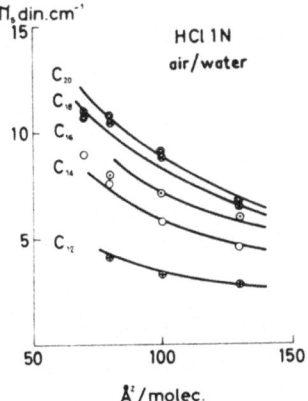

Figure 16. Curves of Π_s versus A for a series of alkyl quaternary ammonium salts (Llopis et al.[139]).

Π_s increases with the length m of the hydrocarbon chain. However, it must be taken into account that to evaluate this cohesive pressure it has been necessary to choose a model for the charge–potential relationship. As a first approximation, the simplest model of the Gouy–Chapman theory was chosen.

Studies of adsorption of surface-active materials at liquid interfaces (oil/water and air/water interfaces) provide a convenient method for testing double electrochemical layer theory and for obtaining information about the structure of the interface. In the theoretical treatment of this type of adsorption, it is reasonable to assume that adsorbed ions or molecules at the interface will be completely mobile, in which case[137] the equation of the adsorption isotherm becomes

$$\frac{A_0}{A - A_0} \exp\left(\frac{A_0}{A - A_0}\right) = \tau A_0 a_d \exp\left[\frac{\Delta G_{a/w}^{CH_2} m v(\Pi^\Delta)}{kT}\right] \quad (88)$$

where a_d is the bulk activity of the surface-active substance in the aqueous phase; τ is a parameter that can be considered as the thickness of the interface for $c_d \to 0$; $\Delta G_{a/w}^{CH_2}$ is the free energy of adsorption per CH_2 group; $\Pi^\Delta = \Pi - \Pi_k$; and

$$v(\Pi^\Delta) = \int_\infty^A A(\partial \Pi^\Delta / \partial A)\, dA \quad (89)$$

If it is assumed now that at the oil/water interface $\Pi_s = 0$ and the Davies equation (81) holds, equation (89) yields

$$v(\Pi_e) = e\phi_0 \quad (90)$$

and equation (88) reduces to

$$\frac{A_0}{A - A_0} \exp\left(\frac{A_0}{A - A_0}\right) \exp\left(\frac{e\phi_0}{kT}\right) = \tau A_0 a_d \exp\left(\frac{\Delta G_{a/w}^{CH_2} m}{kT}\right) \quad (91)$$

Knowledge of the activity coefficients of the surfactants is necessary for this type of calculation. In assessing the validity of this equation, a plot of the left-hand side versus the bulk activity should be linear. Deviations with respect to this behavior are found.

These deviations from the Davies equation were analyzed by Haydon and Taylor[113,115] in terms of a penetration of counter-ions behind the layer of adsorbed long-chain ionic head groups, taking into account the ion-size correction; thus, these authors claimed that their theory for obtaining ϕ_0 fits the experimental results if the finite dimensions of ions are considered.

However, the objection has recently been made[134] that the increase of surface pressure resulting from ionization of an uncharged film is, in general, not given by expression (80). The deviations from the Davies equation may therefore be due as much to the failure of equation (80) as to the inadequacy of the model used to evaluate the charge–potential relationship.

Let it now be assumed that a monolayer of area S spread at an air/water or oil/water interface becomes charged by dissociation of N_i molecules to give surface active ions of charge $-e$ and counter-ions of charge $+e$. The number of surface-active ions per cm^2 is denoted by n_i, the total number of surface-active ions and molecules per unit area by n, and the charged density by $\sigma = +en_i$. We denote by $S\,\Delta F$ the difference between the Helmholtz free energy of the actual equilibrium state of the system and the free energy of the state when none of the monolayer molecules are dissociated (i.e., $n_i = 0$). The total increase in surface pressure resulting from ionization of the monolayer will then be given by

$$-[\partial(S\,\Delta F)/\partial S]_{T,V,N,\alpha} \qquad (92)$$

where $\alpha\,(= n_i/n)$ is the degree of dissociation of the monolayer.

Let it now be assumed that a monolayer of area S spread at an undissociated molecules, the surface-active ions in the monolayer, and the other species of ions released by dissociation. Then,

$$[\partial(\Delta F)/\partial n_i]_{T,V,N} = \bar{\mu}_i + \bar{\mu}_e - \mu_u = \Delta\mu \qquad (93)$$

If ϕ is the average electrostatic potential of the monolayer with respect to that of the bulk electrolyte, then for the particular case considered here

$$\bar{\mu}_i = \mu_{ic} - e\phi \qquad (94)$$

where the term μ_{ic} may be termed the chemical part of $\bar{\mu}_i$ but will also contain a "fluctuation" term due to the fact that the actual potential at an ion is not equal to the average potential. For the common case, $\bar{\mu}_e$ is both constant and independent of ϕ; therefore,

$$\Delta\mu = \mu_{ic} + \bar{\mu}_e + \mu_u - e\phi = \Delta\mu_c - e\phi \qquad (95)$$

where $\Delta\mu_c$ denotes all components of $\Delta\mu$ other than $e\phi$. At equilibrium,

$$\Delta\mu = \Delta\mu_c - e\phi = 0 \tag{96}$$

and

$$\Delta F = \int_0^{n_i} \Delta\mu \, dn_i = n \int_0^\alpha \Delta\mu_c \, d\alpha + \int_0^{\sigma_0} \phi \, d\sigma \tag{97}$$

Thus, ΔF may be regarded as derived from a charging process in which α increases from zero to its equilibrium value as n remains constant. The two terms of Eq. (97) may be denoted by

$$\Delta F_c = n \int_0^\alpha \Delta\mu_c \, d\alpha; \qquad \Delta F_e = \int_0^{\sigma_0} \phi \, d\sigma \tag{98}$$

Then, it is easy to show that

$$\Delta\Pi = \Pi_c + \Pi_e \tag{99}$$

where

$$\Pi_c = n^2 \int_0^\alpha [\partial(\Delta\mu_c)/\partial n] \, d\alpha \tag{100}$$

and

$$\Pi_e = \int_0^{\phi_0} \sigma \, d\phi \tag{101}$$

so that

$$\Delta\Pi = \int_0^{\phi_0} \sigma \, d\phi + n^2 \int_0^\alpha [\partial(\Delta\mu_c)/\partial n]_\alpha \, d\alpha \tag{102}$$

Thus, for equation (80) to be true, it is necessary that

$$n^2 \int_0^\alpha [\partial(\Delta\mu_c)/\partial n]_\alpha \, d\alpha = 0 \tag{103}$$

Payens[112,141] has argued that (in the present notation)

$$\Delta\mu_c = \text{const} + kT \ln [\alpha/(1 - \alpha)] \tag{104}$$

so that, if $\Delta\mu_c$ is a function of α but not of n, Eq. (102) gives equation (80). For equation (104) to hold, Payens assumes both that the entropy of mixing is ideal and that $\mu_u^0 - \mu_{ic}^0$ must be independent of n. This latter assumption is not thermodynamically necessary and in general it may be expected that $\mu_u^0 - \mu_{ic}^0$ varies with n. There will, for example, be a fluctuation term in μ_{ic}^0 due to electrostatic interactions in the monolayers which is not present in μ_u^0, and which must depend on n.

The whole surface pressure of the charged monolayer is therefore given by

$$\Pi = \Pi_k + \Pi_s + \Pi_c + \Pi_e = \Pi_u + \Pi_c + \Pi_e \qquad (105)$$

where Π_u is the surface pressure of the uncharged film. Unfortunately, most of the experimental data refer to compounds for which it is impossible to measure Π_u, the pressure of the hypothetical uncharged film. Only for octaionic acid adsorbed from solutions in 0.1 M HCl and 0.1 M NaOH has it been possible to establish a comparison between the behavior of the uncharged and charged films.[142] These results supported the use of equation (82) to determine Π_s, so that it is reasonable to put

$$[\Pi_u + (Km/A^g)](A - A_0) = kT \qquad (106)$$

At oil/water interfaces, the cohesive pressure $\Pi_s \simeq 0$.

The evaluation of Π_e requires a model to be chosen for the electrical double layer. Bell et al.[134] used the simplest model of the Gouy–Chapman theory. The deviations of the actual experimental surface pressure from the sum $(\Pi_u + \Pi_e)$ calculated from Eqs. (106) and (81) give the values of Π_c. Therefore, the experimental route to the evaluation of Π_c depends on assessing Π_e separately (by surface potential studies) and thereafter obtaining Π_c from the difference, in the case of an oil/water interface, from the surface pressure measurements.

An important contribution to $\Delta\mu_c$ will come from the "discreteion" potential term. If ϕ_0 is the average potential at the adsorption plane, the electrical potential at an adsorbed ion will be $\phi_0 + \varphi$, where the extra term φ is the fluctuation or "self-atmosphere" potential. By applying the "cutoff" model and making the assumptions: (a) that the primary charges are situated in a plane parallel to and a depth β from the interface (β is taken as 5 Å); (b) that the dielectric constant of the aqueous phase is 80 right up to the interface (ε_2 is taken as 1); and (c) that the electrolyte solution is dilute enough for us to ignore diffuse-layer screening, Bell *et al.* were able to calculate the contribution $\Pi_c^{(f)}$ of the fluctuation potential to the surface pressure Π_c.

It is seen that the calculated $\Pi_c^{(f)}$ is in fair agreement with Π_c derived from the air/water data for octainoic acid[142] and lauryl sulfate.[143] However, it is still far from clear whether the $\Pi_c^{(f)}$ term could explain all the differences with respect to the experimental data.

It must be emphasized that the fluctuation contribution to Π_c is only one of the extra contributions required for a full equation of state. Ordinary "chemical" terms will be required in Π_c arising from different effects.

Bell and Levine[144] have shown recently that expression (80) is admissible when all the surface species are soluble. They have also considered the case of an insoluble monolayer of molecules dissociating into ions and also adsorbing counter-ions from the solution. In this case, $\Pi_c^{(f)}$, the part of the surface pressure due to surface self-atmosphere (fluctuation) potentials, can be appreciable. As mentioned above, Levine *et al.*[120] have discussed fluctuation potentials in a mixed film (long-chain ions interpenetrated by counter-ions) in connection with surface potential measurement, but no attempts have been made to study surface pressures. The calculations by Bell *et al.*[134] refer only to a simple model with no adsorption of the counter-ions in the film.

On the other hand, the Π versus A relationship for monolayers composed of an equimolecular mixture of anionic and cationic surface-active substances was first studied by Philips and Rideal.[111] Haydon and Taylor[113,115] regarded the behavior of these equimolecular films as uncharged gaseous monolayers, but Brooks and Pethica[145] have shown recently that such behavior is more complicated. With monolayers composed of an equimolecular mixture of $C_{18}H_{37}SO_4^-$ and $C_{18}H_{37}(CH_3)_3N^+$, they demonstrated a phase transition at low areas ($55\,\text{Å}^2$/ion) which was attributed to the attractive interaction of the discrete charges of the mixed films. These results were confirmed recently by Corkill *et al.*[146] The size of this discrete-ion effect was estimated by supposing that a two-dimensional lattice of cations and anions is formed at the transition. For larger areas compared to this point, presumably the lattice assumption breaks down. The important fact is that this attraction between positive and negative ions in the monolayer gives a negative electrostatic contribution to the surface pressure Π. This attractive term increases with compression of the monolayer and decreases with increasing electrolyte concentration in the substrate.

In a recent paper, Levine *et al.* have attempted to analyze the behavior of these mixed monolayers on the basis of a more realistic model which takes account of diffuse-layer screening and thermal motion of ions at interfaces. The discrete-ion theory hitherto has only been developed for large areas and it predicts that, for an uncharged (equimolecular) mixed film, Π should increase with salt concentration, in agreement with the data of Brooks and Pethica,[145] who showed that an increase in NaCl concentration from 0.001 M to 0.1 M in the substrate produced a very small increase in the pressure of the equimolecular film.

3. Study of Charged Insoluble Monolayers by Using the Donnan Equations

An alternative approach to considering the properties of insoluble charged monolayers consists in introducing the concept of the "surface phase." This is a mathematical fiction, but its application to ionized monolayers leads in some cases to interesting results. In the model proposed by Davies and Rideal,[147] it is assumed that the ionized groups of the long-chain ions restrained at the surface are dissolved uniformly in a "surface phase" with a definite thickness τ, and ions of the aqueous substrate are distributed between the surface and bulk phase. Counter-ions will tend to concentrate in the surface phase and the distribution of ions between the two phases will contribute to both surface pressure and surface potential.

In this model, the boundary between the surface and bulk phases is assumed to behave like a membrane permeable only to small inorganic ions and the distribution of these ions is governed by a Donnan membrane equilibrium. If the thickness of the "surface phase" is taken to be the Debye–Hückel term $1/x$, Davies and Rideal[148] showed that the potential ϕ_D of the surface phase at 20°C could be calculated from the relation

$$\phi_D = (kT/e)\sinh^{-1}(2 \times 134/Ac^{1/2}) \tag{107}$$

which may be compared with equation (52). The potential ϕ_D is less than ϕ_G, since the former is calculated assuming a thickness $1/x$ for the "surface phase" while for the latter it is assumed that all primary charges are concentrated in a plane. The authors claim that this model does not require a separate consideration of penetration or of finite ionic size. However, the thickness τ is certainly a fitting parameter and sometimes it is preferable to take for it a value other than $1/\kappa$.

This problem was studied again by Isemura and coworkers,[149,150] who showed that, if the surface phase is assumed to be electrically neutral, the distribution of ions between the surface and bulk phases is in complete agreement with that derived from the Gouy–Chapman theory of the diffuse layer if the thickness of surface phase is chosen as $\tau = 2/\kappa$ and the potential $\phi_D = \phi_G/2$.

In an improved approximation, they assumed that the surface phase deviates slightly from electroneutrality to an extent defined by a quantity λ given by

$$(1/\tau A) + n_+ - n_- = (1/\tau A)\lambda \tag{108}$$

with

$$\lambda \ll 1 \tag{109}$$

Evidently, the problem is reduced to that of a membrane equilibrium if $\lambda = 0$. Now, it is easy to show that

$$\sinh(e\phi/kT) = (1 - \lambda)/2n\tau A \tag{110}$$

Expanding $e\phi/kT$ as a series in λ and ignoring the terms higher in order than 1,

$$\frac{e\phi}{kT} = \ln\left[\frac{1 + [1 + (2n\tau A)^2]^{1/2}}{2n\tau A}\right] - \frac{\lambda}{[1 + (2n\tau A)^2]^{1/2}} \tag{111}$$

is obtained, and therefore

$$\lambda = [1 + (2n\tau A)^2]^{1/2}(e/kT)(\phi_D - \phi) \tag{112}$$

Solution of the Poisson equation then allows the mean electric potential in the surface phase to be obtained, which is given by

$$\langle \phi \rangle = \phi_D\{1 - [(\tanh \rho\tau)/\rho\tau]\} \tag{113}$$

where

$$\rho^2 = (4\pi e^2/\varepsilon_w kTA\tau)[1 + (2n\tau A)^2]^{1/2} \tag{114}$$

The surface pressure Π_e due to the ionization of the monolayer can be expressed by

$$\Pi_e = 2kT\left\{-2n\tau + \frac{[1 + (2n\tau A)^2]^{1/2}}{A} - \frac{\langle\lambda\rangle}{A[1 + (2n\tau A)^2]^{1/2}}\right\} \quad (115)$$

In the approximation for the membrane equilibrium, $\langle\lambda\rangle = 0$ and $\tau = 2/\kappa$, so that equation (115) reduces to

$$\Pi_e = 2kT\{-(4n/\kappa) + [(1/A^2) + (4n/\kappa)^2\}^{1/2}\} \quad (116)$$

However, the main purpose of this theory is to produce a basis for the experimental fact that an ionized monolayer tends to be relatively more expanded on dilute salt solutions, but the treatment cannot explain completely the effect of such an ionization. Evidently, in these cases, theories based on the structure of the double layer give a clear insight into the surface phenomena.

V. INCOMPLETELY AND UN-IONIZED MONOLAYERS

1. Monolayers of Un-ionized Compounds

(i) Insoluble Monolayers

The simplest way to relate the observed surface potential ΔV to molecular properties of monolayers is to consider the value of $\Delta V/n$.[151] It is found that for many films the value of ΔV changes with n so that $\Delta V/n$ remains nearly constant. This has been interpreted in the sense that

$$\Delta V = 4\pi n p_\perp$$

where $p_\perp$ is the effective perpendicular component of the dipole moment of the molecules in the film. This expression appears to have been introduced by Schulman and Rideal,[152] who assumed also that $p_\perp$ is due to an intrinsic moment $\bar{p}$ making some angle $\frac{1}{2}\pi - \theta$ with respect to the surface,

$$p_\perp = \bar{p}\cos\theta \quad (118)$$

so that $\Delta V/n$ will be constant when the orientation of molecules at the interface is nearly independent of the surface concentration.

If the monolayer concentration is expressed in terms of the molecular area A in Å^2 molecule^{-1} and ΔV is in mV, the operational surface moment per adsorbed molecule will be

$$p^s \equiv (A\Delta V)/12\pi \equiv p_\perp \qquad (119)$$

The value of $p_\perp$ then obtained is in mD, the Debye unit being taken, of course as 10^{-18} cgs unit.

As usual, the measured surface potential ΔV is taken as positive if the air electrode becomes more positive with respect to the trough electrode when the film is spread. When ΔV is positive, the value of $p_\perp$ is also positive, so that the positive end of the molecular dipole is oriented upward. Note that this sign convention is opposite to that used in the study of the surface potential of aqueous solutions of inorganic electrolytes considered earlier.

However, it must be recognized that the reorientation of subphase molecules contributes significantly to ΔV. This factor was introduced by Guastalla,[153,154] who considered, in the case of aqueous substrates, that the surface potential would be the sum of an effect due to the dipoles of the spread molecules and another due to reorientation of the water molecules near the interface. The equation proposed for very dilute monolayers may be written (in volts) as

$$\Delta V = 12\pi p_\perp [(1/A) - (K/A^{3/2})] \qquad (120)$$

where K is an empirical constant. This equation predicts a change in sign of surface potential for large areas. With myristic acid films on distilled water at high surface dilution (molecular area 1.000–30.000 Å^2), the surface potential is negative, according to the above sign convention. Furthermore, if $A\Delta V$ is plotted against $A^{1/2}$, a straight line is obtained, and the author claims that extrapolation to infinite dilution gives $p_\perp$, the vertical component of the dipole moment of the film molecules. Values of $p_\perp$ close to 1.8 D were obtained in these measurements. As predicted, for concentrated films, the surface potentials were of opposite sign. However, with stearylamide monolayers spread on distilled water, the surface potentials were positive at high dilutions and no change in sign was observed as the films were concentrated.

It is evident that equation (120) is an oversimplification and does not prove at all satisfactory when applied to condensed monolayers. For such films, effects due to reorientation of the water dipoles become predominant and may depend on the group dipole moment of the film-forming molecules.

The surface potential ΔV is thus related in a simple way to the concentration of molecules in the film, but in a complicated and unknown way to the standard dipole moments. Hence, it has been proposed recently that equation (119) be regarded as an operational definition of the surface moment p^s.

Many authors, however, have considered that the constituent dipoles in the polar parts of the film-forming molecules can be summed vectorially to give the resultant vertical components of the effective dipole moment.[155] A comparison of the calculated and experimental moments allows conclusions to be drawn concerning the orientation of the polar groups in the monolayer. For example, large, negative values of ΔV have been observed for monolayers of fatty acids in which a terminal hydrogen atom has been substituted by a halogen atom;[156–159] these have been interpreted as resulting from the carbon–halogen bond. Since, at present, it is impossible to separate the contributions of the hydrophilic polar group, the orientation of the water molecules, and the adjacent ionic double layer, it is usual to combine the three terms into a single term p_0. Consequently, p^s for the unhalogenated adsorbed molecule can be expressed as

$$p^s = p_0 + p_{CH_3} \qquad (121)$$

and when a terminal hydrogen atom is substituted by a halogen atom (X),

$$p^s = p_0 + p_{CH_2X} \qquad (122)$$

If p_0 is assumed to be the same in both cases, comparison of values of ΔV for monolayers of both halogenated and unhalogenated compounds allows the moments of the methylene–halogen group p_{CH_2X} to be calculated.[159] These values range from -0.68 to -0.97 in going through the series of halogens from fluorine to iodine. The negative sign originates from the orientation of the carbon–halogen bond with the negative end outwards.

Finally, no attempts appear to have been made to allow for the effect of induced polarization within the film itself.[160] However, this effect is important for a quantitative interpretation of ΔV. For example, if the above-mentioned values of p_{CH_3} and p_{CH_2X} are plotted against the covalent radii of the halogen atom, a straight line results. This would be expected, since the polarizability of an atom increases with the radius of its outer electron shells. However, the standard dipole moment of the CH_2X groups when obtained for a paraffinic compound are nearly the same. That the values of ΔV are not the same for each ω-monohalogenated fatty acid may be attributed to the mutual lateral polarization in a closely packed monolayer, an effect which modifies to a different extent the dipole moment of each carbon–halogen bond.[159]

The intermolecular interactions are also apparent in the general form of the ΔV versus n curves, which in many cases are Langmuirian in shape, i.e., at low values of n, they are linear and ΔV tends to a constant value at higher surface concentrations.[161] This means that p^s is only constant at low values of n when intermolecular interactions are small.

It is certain, therefore, that the quantitative interpretation of ΔV is still far from satisfactory and more experimental and theoretical work is desirable if the surface potential data are to be analyzed more quantitatively.

(ii) Soluble Substances

The surface potential shift at the air/water interface has been determined recently for aqueous solutions of soluble substances by using the Kenrick method. A review is given by Frumkin and Damaskin.[162] For solutions of saturated aliphatic compounds, the surface potentials were measured several years ago by Frumkin,[163] but at that time, the concentration dependence was not yet investigated. Recent results for butylamine[164] and propyl alcohol[165] show that the dependence of ΔV upon the amount adsorbed Γ is, to a first approximation, linear at the air/water interface.

The compounds considered above are oriented at this interface with the polar groups turned towards the water and the hydrocarbon chain towards the air. Judging by the minimal value of the area, 29 Å^2 molecule^{-1}. this orientation is not perfect, but it should be better defined the larger the amount of substance adsorbed. The adsorption is mainly due to the expulsion of the hydrocarbon chain from the bulk of the solution.

Like the saturated aliphatic compounds, the perfluorinated compounds are known to display a high surface activity at the air/water interface.[157] Values of the energy of adsorption of 1.5 kcal/mole for the CF_2 group were obtained, in comparison with a value of 0.73 kcal/mole for the CH_2 group. With perfluorodecanoic acid, a value of -1.00 V has been obtained from surface potential measurements. For β-iodopropionic acid in 0.3 M solution, ΔV is only -0.13 V with a limiting value of -0.15 V.[166]

The adsorption behavior of some perfluorinated aromatic compounds has also been investigated.[167,168] An increase in adsorptivity is observed when passing from aniline to perfluoroaniline. In the case of perfluorinated aniline and phenol, the molecules adsorbed at the air/water interface are oriented "edgewise," and some of the fluorine atoms prove to be turned outward. This results in the appearance of high, negative surface potentials.

Other compounds, such as p-fluorobenzoic acid, p-nitrobenzoic acid, and p-chlorocresol, also increase the negative potential of the air/water interface as a result of the halogen atom being directed toward the gaseous phase.[80] The *meta* compounds act in a less pronounced way than the p-isomers. However, the o-isomers of chlorophenol decrease the negative potential owing to the orientation of the carbon–halogen bond at the interface.[169] Aqueous solutions of trimethylacetic acid show positive surface potentials, but in the case of trichloroacetic acid, high, negative values were obtained.[170] Solutions of hydroxybenzoic acid, hydroquinone, resorcinol, pyrogallol, phloroglucinol, fumaric acid, maleic acid, chloracetic acid, chloropropionic acid, and α-bromopropionic acid behave in a similar way, giving negative surface potentials.[80]

However, it is interesting to note that the surface potentials of aqueous solutions of many long-chain compounds change throughout the adsorption process (surface ageing effect). Thus, Posner and Alexander[171] studied the variation of surface potential of aqueous solutions of isoamylalcohol and *sec*-octylalcohol as a function of the age of the surface. For these experiments they used a channel method suitable for times between 10^{-2} and 1 sec, and a jet suitable for times down to 10^{-3} sec. It was apparent from these results that surface potentials of freshly formed liquid surfaces may differ in some cases from those of corresponding aged surfaces.

Addison and Litherland[172,173] observed at stationary surfaces of long-chain compound solutions that the surface potentials at the instant when adsorption is complete differed widely from the ultimate equilibrium value. In interpreting these results, they considered that adsorbing molecules arrive at the surface in a disoriented state and that this disorientation is maintained throughout the adsorption process. Nevertheless, the experimental techniques used seem unreliable, so that these results might be artifacts.

2. Monolayers of Weak Acids and Bases

(i) Surface Potential Measurements

The surface potentials of monolayers of weak acids and bases (amines) change sharply as the subphase pH is altered. The relation between ΔV and bulk pH resembles a titration curve.[174,175] At a constant area of about $20 \, \text{Å}^2$ molecule^{-1}, the myristic acid, e.g., exhibits a negative potential of $\Delta V = 50 \, mV$ at pH 12 and beyond, which is that of the long-chain ion ($C_{13}H_{26}COO^-$); between pH 11 and 4, it rises to 400 mV, where it remains nearly constant until pH 1. This is the potential characteristic of the undissociated acid. The change in surface potential shown by nonadecylamine, on passing from the alkaline pH 12 to a more acid pH of 8, is a little smaller than that shown by the stearic acid.[176]

Betts and Pethica[177] have given the following expression for the surface potential of a partially charged weak acid or base monolayer:

$$\Delta V = (12\pi/A)[(1 - \alpha)p_u^s + \alpha p_i^s] + \phi_0 \qquad (123)$$

where ΔV is given in mV, A is in Å^2, α is the degree of dissociation, and p_u^s and p_i^s are the surface dipole moments (in mD) for the uncharged and charged forms of the ionizing molecule, respectively.

Now, it must be remembered that, as a result of the potential ϕ_0, once the film bears net charge, the pH at the surface is no longer equal to that in the bulk.[178,179] If pH^s is the "effective surface pH" in the region of ϕ_0, the ionization of the acid or base is determined by

$$pH^s = pK - z \log[\alpha/(1 - \alpha)] \qquad (124)$$

where z is the sign (± 1) of the ions in the film and K is the intrinsic dissociation constant of the ions.

Further, the surface and bulk pH values must be related by

$$pH^s = pH^b + (e\phi_0/2.3 \, kT) \qquad (125)$$

The surface density for an incompletely ionized film is given by

$$\sigma = \alpha e/A \qquad (126)$$

so that from the Gouy–Chapman equation it follows that

$$\phi_0 = (2kT/ze)\sinh^{-1}(134\alpha/Ac^{1/2}) \tag{127}$$

which applies for monolayers spread at 20°C on solutions of univalent electrolytes.

In general, p_u^s and p_i^s will probably not be constant with varying α, but in the region of small α values, the assumption that they are constant is plausible. Then, using equations (124), (125), and (127), it follows that

$$\left[\frac{\partial(\Delta V)}{\partial(pH^b)}\right]_{A,c,T} = -z\frac{12\pi}{A}(p_i^s - p_u^s)\alpha(1 - \alpha)$$

$$\times\left\{1 + \frac{e}{kT}\left[\frac{\partial\phi_0}{\partial(pH^b)}\right]_{A,c,T}\right\} + \left[\frac{\partial\phi_0}{\partial(pH^b)}\right]_{A,c,T} \tag{128}$$

$$\left[\frac{\partial\phi_0}{\partial(pH^b)}\right]_{A,c,T} = -\frac{2kT}{e}\frac{1 - \alpha}{2(1 - \alpha) + \coth(ze_0\phi_0/2kT)} \tag{129}$$

For stearic acid, the dipole correction has been calculated as $(p_i^s - p_u^s) = -264$ mD, and extrapolation to $\alpha = 0$ gives $pK = 5.55$. For nonadecylamine, no dipole correction has been offered, and extrapolation to $\alpha = 0$ gives an uncorrected $pK = 10.1$. Comparison with the bulk pK values for soluble fatty acids and amines (4.9 and 10.6, respectively) shows that, in monolayers, there is a shift of pK of about 0.7 units for both amines and acids towards the neutral pH.

Previous work carried out by Glazer and Dogan[180] led to a pK value of 8.0 for monolayers of stearic acid spread at the air/water interface. This high value of the pK probably was due to Ca^{++} impurities in the substrate that produce ΔV results too low at higher pH. The sensitivity of charged fatty acid monolayers to the presence of alkaline earth metal ion has been a subject of investigation by a number of authors.[181–184] This effect is governed by two factors, an electrical effect due to the double charge of the metal ions, and the association constant of the metal soap. The addition of either Ca^{++} or Mg^{++} ions to the substrate underneath fatty acid monolayers produces a lowering of the ΔV values.

A careful study of the ionization of monolayers of fatty acids from C_{14} to C_{18} spread at the air/water interface was undertaken by

Spink.[185] Both force–area and surface potential–area characteristics were studied as a function of pH. It was found that the C_{14}–C_{17} acids begin to dissolve at pH 10 and only the C_{18} acid could be studied above this value. The surface potential measurements are more appropriate for the study of ionization phenomena than are the surface pressure–area curves. Application of the above theory based on the Gouy–Chapman equation gave a pK value of 5.1 for stearic acid monolayers, which is only slightly higher than the usual pK of 4.9 for soluble fatty acids in solution. Sears and Schulman[186] claimed recently that for stearic acid monolayers spread on substrates of pH 12 the cation size is an important factor, an effect which is due to hydration of the cation, and which leads to the cation sequence Li < Na < K.

For long-chain alkyl amine monolayers, Glazer and Dogan[180] obtained, in the case of octanodecylamine, a suface pK value of 8.5, which seems too low. Betts and Pethica[177] indicate that on acid solutions the octadecylamine dissolves slowly, and it was for this reason that these authors synthesized the nonodecylamine.

On the other hand, from surface pressure measurements, Payens[141] calculated a value of $pK = 3.8$ for stearyl phosphonic acid spread at the oil/water interface. This value seems rather high in comparison with that of shorter aliphatic phosphonic acids ($pK \simeq 2.5$) and, as in the case of stearic acid, the pK shift is toward neutral pH. Payens showed that the most probable cause of such a pK shift is the lower free energy of hydration of the long-chain phosphonic ion at the interface than in bulk solution.

(ii) Equation of State for Incompletely Ionized Monolayers

When equation (126) is employed in the derivation of the Davies equation from the Gouy–Chapman theory, the following surface equation of state for an incompletely ionized film is obtained:

$$\sinh^{-1}\left(\frac{\theta\alpha}{Ac_i^{1/2}}\right) = \cosh^{-1}\left[1 + \frac{\theta}{2kTc_i^{1/2}}\left(\Pi - \Pi_s - \frac{kT}{A - A_0}\right)\right] \quad (130)$$

where $\theta = e(\Pi/2\varepsilon_w kT)^{1/2}$, Π_s is the cohesive pressure, and $z = 1$.

The use of Π versus A curves at varying pH and c to obtain the ionization constants of spread weak acid monolayers has been described by Payens.[141] However, Betts and Pethica[177] have claimed for monolayers of weak acids and bases the surface potential method is more sensitive than the surface pressure method, because it is applicable in regions of low α where the Gouy–Chapman theory is most valid.

NOTATION

a	mean diameter of ions; activity of solute; mean activity of an electrolyte
b	mutual exclusion diameter
c	concentration of solute
d	equilibrium distance from the interface of the head groups of a monolayer
e	elementary charge; signal from a vibrating plate instrument
g	a parameter (expression (82))
i	electric current
k	Boltzmann constant
l	effective carbon–carbon bond length
m	number of CH_2 groups in a molecule
n	number of adsorbed (spread) molecules per cm^2 of interface
p^s	surface (dipole) moment per adsorbed molecule
q	electric charge
r	electric resistance; radial distance
r_o	ionic radius
t	time
u	un-ionized species
v_i	ionic volumes of species i
w	work
x	variable (distance from a reference)
z	valence of an ion
A	area per molecule
B	constant
C	capacity of a condenser
E	reading on a potentiometer
F	Helmholtz free energy
G	Gibbs free energy
H	symbol introduced in expression (23); function plotted in ref. 120
I_z	energy of ionization of an isolated atom
K	a constant
L	function plotted in ref. 120
N	Avogadro number
P	function plotted in ref. 120
R	electrical resistance
S	total surface area
S_M	(free or total) energy of sublimation
T	absolute temperature
ΔV	measured surface potential
V	potential difference
W	van der Waals energy
α_i^α	real potential of particle i in the phase α
α	degree of dissociation

β depth of the adsorption plane from the interface
γ thickness of the "inner region"; surface tension
Γ surface concentration
δ distance from the surface
Δ difference between values
ε_w dielectric constant of water
ε_a or ε_0 dielectric constant of a nonconducting phase (air or oil phase)
ε ratio $\varepsilon_a/\varepsilon_w$
η $= e/kT$
θ defined in expression (83)
κ Debye–Hückel function
λ parameter introduced in expression (108); energy of polarization
μ_i^α chemical potential of particle i in the phase α
$\bar{\mu}_i^\alpha$ electrochemical potential of particle i in the phase α
v defined in expression (89)
ζ defined in expression (84)
Π surface pressure
ρ bulk charge density; defined in expression (114)
σ surface charge density
τ thickness
φ phase angle; fluctuation potential (discrete-ion effect)
ϕ^α inner electric potential of phase α
$^\alpha\Delta^\beta\phi$ Galvanic potential difference
Φ specific interaction energy in the Stern theory
χ^α surface electric potential difference ($\chi^\alpha = \phi^\alpha - \psi^\alpha$)
ψ^α Outer electric potential of phase α
$(^\alpha\Delta^\beta)\psi$ Voltaic potential difference
ω vibration frequency

REFERENCES

[1] P. van Rysselberghe, *Electrochim. Acta* **5**, (1961) 28.
[2] R. Defay, N. Ibl, E. Levart, G. Milazzo, G. Y. Valensi, and P. van Rysselberghe, *Electroanal. Chem.* **7** (1964) 417.
[3] P. van Rysselberghe, *Electrochim. Acta* **9** (1964) 1343.
[4] Commission on Colloid and Surface Chemistry, "Manual of Terminology and Symbols in Colloid and Surface Chemistry," IUPAC Secreteriat, January 1970.
[5] R. Parsons, in *Modern Aspects of Electrochemistry*, Vol. 1 Ed., J. O'M. Bockris, Butterworths, London, 1954, p. 103.
[6] E. Lange and K. P. Miscenko, *Z. Phys. Chem.* **149** (1930) 1.
[7] E. Lange, *Z. Elektrochem.* **55** (1951) 76.
[8] K. Möhring, *Z. Elektrochem.* **59** (1955) 102.
[9] Lord Kelvin, *Phil. Mag.* **46** (1898) 82.
[10] W. A. Zisman, *Rev. Sci. Instr.* **3** (1932) 367.
[11] H. G. Yamins and W. A. Zisman, *J. Chem. Phys.* **1** (1933) 656.
[12] K. A. MacFayden and T. A. Holbech, *J. Sci. Instr.* **34** (1957) 101.
[13] G. L. Gaines, *Insoluble Monolayers in Liquid–Gas Interfaces*, John Wiley and Sons, New York, 1966, p. 79.
[14] J. T. Davies, *Z. Elektrochem.* **55** (1951) 559.
[15] J. T. Davies, *Nature* **167** (1951) 193.

[16] C. D. Kinloch and A. I. MacMullen, *J. Sci. Instr.* **36** (1959) 347.

[17] G. L. Gaines, *Insoluble Monolayers in Liquid-Gas Interfaces,* John Wiley and Sons, New York, 1966, p. 81.

[18] A. Suzuki, S. Ikeda, and T. Isemura, *Ann. Rep. Biol. Works, Faculty of Science, Osaka Univ.* **15** (1967) 83.

[19] K. W. Bewig, *Rev. Sci. Instr.* **35** (1954) 1160.

[20] J. A. Bergeron and G. L. Gaines, *J. Colloid. Interface Sci.* **23** (1967) 292.

[21] A. V. Few and B. A. Pethica, *Research (London)* **5** (1952) 290.

[22] F. B. Kenrick, *Z. phys. Chem.* **19** (1896) 625.

[23] A. N. Frumkin, *Z. phys. Chem.* **109** (1924) 34.

[24] J. E. B. Randles, *Trans. Faraday Soc.* **52** (1956) 1573.

[25] A. Heydweiller, *Ann. Physik* **33**(4) (1910) 145.

[26] G. Quincke, *Ber. Münch. Akad. Wiss.* **1** (1876) 3.

[27] A. Gradenwitz, *Physik. Z.* **3** (1902) 329.

[28] H. Stocker, *Z. phys. Chem.* **94** (1920) 149.

[29] I. Langmuir, *J. Am. Chem. Soc.* **39** (1917) 1848.

[30] A. K. Goard, *J. Chem. Soc.* (1925) 2451.

[31] W. D. Harkins and H. M. McLaughlin, *J. Am. Chem. Soc.* **47** (1925) 2083.

[32] W. D. Harkins and E. C. Gilbert, *J. Am. Chem. Soc.* **48** (1926) 604.

[33] G. Schwenker, *Ann. Physik* **11** (1931) 525.

[34] P. Lenard, *Ann. Phys. Chem.* **46** (1892) 584.

[35] G. Jones and W. A. Ray, *J. Am. Chem. Soc.* **59** (1937) 187.

[36] G. Jones and W. A. Ray, *J. Am. Chem. Soc.* **63** (1941) 288.

[37] G. Jones and W. A. Ray, *J. Am. Chem. Soc.* **63** (1941) 3262.

[38] G. Jones and W. A. Ray, *J. Am. Chem. Soc.* **64** (1942) 2744.

[39] M. Dole and J. A. Swartout, *J. Am. Chem. Soc.* **62** (1940) 3039.

[40] K. Schäfer, A. Pérez-Masiá, and H. Jüntgen, *Z. Electrochem.* **59** (1955) 425.

[41] A. S. Coolidge, *J. Am. Chem. Soc.* **71** (1949) 2153.

[42] I. Langmuir, *Science* **88** (1930) 430.

[43] W. Drost-Hansen, *Ind. Eng. Chem.* **57**(3) (1965) 40; **57**(4) (1965) 18.

[44] C. Wagner, *Physik. Z.* **25** (1924) 474.

[45] L. Onsager and N. N. T. Samaras, *J. Chem. Phys.* **2** (1934) 528.

[46] H. S. Harned and B. B. Owen, *The Physical Chemistry of Electrolytic Solutions,* Reinhold, New York, 1943, pp. 59–62.

[47] W. K. Ssementschenko and E. A. Dawidowskaja, *Kolloid Z.* **73** (1935) 24.

[48] W. K. Ssementschenko and A. F. Gratshewa, *Kolloid Z.* **73** (1935) 30.

[49] J. W. Belton, *Trans. Faraday Soc.* **33** (1937) 1449.

[50] S. Oka, *Proc. Phys. Math. Soc. Japan* **14** (1932) 649.

[51] K. Ariyama, *Bull. Chem. Soc. Japan* **11** (1936) 687.

[52] P. B. Lorenz, *J. Phys. Chem.* **54** (1950) 685.

[53] J. Frenkel, *Kinetic Theory of Liquids,* Dover Publications, New York, 1955, p. 358.

[54] A. Pérez-Masiá, *Anal. real. soc. españ. fís. quím.* **50A** (1954) 5.

[55] K. Schäfer, *Z. Electrochem.* **59** (1955) 233.

[56] A. Pérez-Masiá, *Anal. real. soc. españ. fís. quím.* **54B** (1958) 629.

[57] L. Sabinina and L. Terpugow, *Z. phys. Chem.* **A173** (1935) 237.

[58] R. M. Sugitt, P. M. Aziz, and F. E. W. Wetmore, *J. Am. Chem. Soc.* **71** (1949) 676.

[59] M. Dole, *J. Am. Chem. Soc.* **60** (1938) 904.

[60] A. Garrison, *J. Phys. Chem.* **89** (1925) 1517.

[61] J. E. B. Randles, *Disc. Faraday Soc.* **24** (1957) 194.

[62] J. Guyot, *Ann. Physique* **2**(10) (1924) 501.

[63] R. Parsons, *Disc. Faraday Soc.* **24** (1957) 233.

[64] M. Andauer and E. Lange, *Z. Phys. Chem.* **166** (1933) 219.

[65] O. Klein and E. Lange, *Z. Elektrochem.* **43** (1937) 570.

[66] B. S. Gurenkov, *Zh. Fiz. Khim.* **30** (1966) 1830.

[67] B. V. Ershler, *Zh. Fiz. Khim.* **20** (1946) 679.

[68] V. S. Krylov, *Electrochim. Acta* **9** (1964) 1247.

[69] A. N. Frumkin, Z. A. Jofa, and M. A. Gerovich, *Zh. Fiz. Khim.* **30** (1956) 1455.

[70] S. Bordi and F. Vannel, *Gaz. Chim. Ital.* **92** (1962) 82.

[71] S. Bordi and F. Vannel, in *Electrolytes,* Ed., B. Pesce, Pergamon, New York, 1962, p. 196.

[72] S. Bordi and F. Vannel, *Ann. Chimica* **52** (1962) 80.

[73] J. A. Chalmers and F. Pasquill, *Phil. Mag.* **23** (1937) 88.

[74] G. Passoth, *Z. Elektrochem.* **60** (1956) 420.

[75] A. N. Frumkin, *Electrochim. Acta* **2** (1960) 351.

[76] A. N. Frumkin, *J. Chem. Phys.* **7** (1939) 552.

[77] D. C. Grahame, E. Coffin, J. Cummings, and M. A. Poth, *J. Am. Chem. Soc.* **74** (1952) 1207.

[78] B. Kamienski, in *Proc. 2nd Intern. Congr. Surface Activity,* Vol. III, Butterworths, London, 1957, p. 103.

[79] B. Kamienski, *Electrochim. Acta* **1** (1959) 272.

[80] B. Kamienski, *Proc. 3rd Intern. Congr. Surface Activity,* Vol. II, Cologne, Universitätdruckerei Mainz, 1960, p. 296.

[81] B. Kamienski, *Electrochim. Acta* **3** (1960) 208.

[82] A. N. Frumkin, in *Proc. 2nd Intern. Congr. Surface Activity,* Vol. III, Butterworths, London, 1957, p. 121.

[83] B. E. Conway and J. O'M. Bockris, *Modern Aspects of Electrochemistry,* Vol. I, Ed., J. O'M. Bockris, Butterworths, London, 1954, p. 47.

[84] J. D. Bernal and R. H. Fowler, *J. Chem. Phys.* **1** (1933) 515.

[85] D. D. Eley and M. G. Evans, *Trans. Faraday Soc.* **34** (1938) 1093.

[86] E. J. W. Verwey, *Rec. trav. Chim. Pays Bas* **61** (1942) 564.

[87] N. S. Hush, *Austr. J. Sci. Res.* **A1** (1948) 482.

[88] M. Strehlow, *Z. Elektrochem.* **56** (1952) 119.

[89] G. Passoth, *Z. Phys. Chem.* **203** (1954) 275.

[90] K. P. Miscenko and E. I. Kwait, *Zh. Fiz. Khim.* **28** (1954) 1451.

[91] J. Morgan and B. E. Warren, *J. Chem. Phys.* **6** (1938) 666.

[92] K. Grjotheim and J. Krogh-Moe, *Acta Chem. Scand.* **8** (1954) 1193.

[93] H. S. Frank and W. Y. Wen, *Disc. Faraday Soc.* **24** (1957) 133.

[94] G. Nemethy and H. A. Scheraga, *J. Chem. Phys.* **36** (1962) 3382.

[95] Yu. V. Gurikov, *Zh. Strukt. Khim.* **6** (1965) 817.

[96] Yu. V. Gurikov, *Zh. Strukt. Khim.* **7** (1966) 8.

[97] A. H. Narten, M. D. Danford, and H. A. Levy, *Disc. Faraday Soc.* **43** (1967) 97.

[98] L. Pauling, in *Hydrogen Bonding,* Eds., D. Hadzi and H. W. Thompson, Pergamon Press, London, 1959.

[99] M. Weissman and L. Blum, *Trans. Faraday Soc.* **64** (1968) 2605.

[100] J. W. McBain, R. C. Bacon, and H. D. Bruce, *J. Chem. Phys.* **7** (1939) 818.

[101] W. A. Weyl, *J. Colloid Sci.* **6** (1951) 389.

[102] B. V. Derjaguin, *Disc. Faraday Soc.* **42** (1966) 109.

[103] A. V. Panfilov and O. M. Dolgaya, *Zh. Fiz. Khim.* **37** (1963) 1800.

[104] M. Falk and G. S. Kell, *Science* **154** (1963) 1013.

[105] R. A. Horne and J. D. Birkett, *Electrochim. Acta* **12** (1967) 1153.

[106] R. A. Horne, A. F. Day, R. P. Young, and N. T. Yu, *Electrochim. Acta* **13** (1968) 397.

[107] J. T. Davies and E. K. Rideal, *Interfacial Phenomena*, Academic Press, New York, 1961, p. 70.

[108] J. T. Davies, *Proc. Roy. Soc.* **208A** (1951) 224.

[109] R. G. Aickin and R. C. Palmer, *Trans. Faraday Soc.* **40** (1944) 116.

[110] J. T. Davies and E. K. Rideal, *J. Colloid Sci. Supplement* **I** (1954) 1.

[111] J. N. Philips and E. K. Rideal, *Proc. Roy. Soc.* **A232** (1955) 149, 159.

[112] T. A. J. Payens, *Philips Res. Rep.* **10** (1955) 425.

[113] D. A. Haydon and F. H. Taylor, *Phil. Trans. Royal Soc. London* **A252** (1960) 225.

[114] H. D. Ohlenbusch, *Z. Elektrochem.* **60** (1956) 607.

[115] D. A. Haydon and F. H. Taylor, *Phil. Trans. Royal Soc. London* **A253** (1960) 255.

[116] D. A. Haydon and F. R. Taylor, *Proc. 3rd Intern. Congr. Surface Activity*, Vol. II, Cologne, Universitätdruckerei Mainz, 1960, p. 266.

[117] J. J. Bikerman, *Phil. Mag.* **33** (1942) 384.

[118] R. Schlögl, *Z. phys. Chem.* **202** (1953) 379.

[119] D. A. Haydon, *Proc. Roy. Soc. London* **A258** (1960) 319.

[120] S. Levine, J. Mingins, and G. M. Bell, *J. Phys. Chem.* **67** (1963) 2095.

[121] D. A. Haydon and F. H. Taylor, *Trans. Faraday Soc.* **58** (1962) 1233.

[122] F. van Voorst Vader, *Proc. 3rd Intern. Congr. Surface Activity*, Vol. II, Cologne, Universitätdruckerei Mainz, 1960, p. 276.

[123] C. V. C. Pak and N. L. Gershfeld, *J. Colloid Sci.* **19** (1964) 831.

[124] B. A. Pethica and A. V. Few, *Disc. Faraday Soc.* **18** (1954) 258.

[125] E. Stenhagen, *Trans. Faraday Soc.* **36** (1940) 496.

[126] D. A. Haydon and J. N. Philips, *Trans. Faraday Soc.* **54** (1958) 698.

[127] D. A. Haydon, *Kolloid Z.* **185** (1962) 148.

[128] J. Mingins and B. A. Pethica, *Trans. Faraday Soc.* **59** (1963) 1892.

[129] P. J. Anderson, *Trans. Faraday Soc.* **55** (1959) 1421.

[130] J. T. Davies, *Proc. 3rd Intern. Congr. Surface Activity*, Vol. II, Cologne, Universitätdruckerei Mainz, 1960, p. 309.

[131] S. Levine, G. M. Bell, and D. Calvert, *Can. J. Chem.* **40** (1962) 518.

[132] S. Levine, J. Mingins, and G. M. Bell, *J. Electroanal. Chem.* **13** (1967) 280.

[133] S. Levine, J. Mingins, and G. M. Bell, *Can. J. Chem.* **43** (1965) 2834.

[134] G. M. Bell, S. Levine, and B. A. Pethica, *Trans. Faraday Soc.* **58** (1962) 904.

[135] E. J. W. Verwey and J. T. G. Overbeek, *Theory of the Stability of Lyophobic Colloids*, Elsevier, Amsterdam, 1948, p. 52.

[136] J. Guastalla, *J. Chim. Phys.* **43** (1946) 184.

[137] J. Llopis and P. Artalejo, *An. real. soc. esp. f is. quim.* **58B** (1962) 367.

[138] D. K. Chattoraj and A. K. Chatterjee, *J. Colloid and Interface Sci.* **21** (1966) 159.

[139] J. Llopis, A. Albert, J. A. Subirana, and M. D. Conde, *An. real. soc. esp. fis. quim.* **58B** (1962). 379.

[140] J. T. Davies, *J. Colloid Sci.* **11** (1956) 377.

[141] T. A. J. Payens, *Proc. 2nd Intern. Congr. Surface Activity*, Vol. I, Butterworths, London, 1957, p. 64.

[142] E. Matijevic and B. A. Pethica, *Trans. Faraday Soc.* **54** (1958) 1400.

[143] E. Matijevic and B. A. Pethica, *Trans. Faraday Soc.* **54** (1958) 1382.

[144] G. M. Bell and S. Levine, *Z. phys. Chem. (Leipzig)* **231** (1966) 289.

[145] J. H. Brooks and B. A. Pethica, *Trans. Faraday Soc.* **60** (1964) 208.

[146] J. M. Corkill, J. F. Goodman, S. P. Harrold, and J. R. Tate, *Trans. Faraday Soc.* **63** (1967) 247.

[147] J. T. Davies and E. K. Rideal, *J. Colloid Sci.* **3** (1948) 313.

[148] J. T. Davies and E. K. Rideal, *Interfacial Phenomena*, Academic Press, New York, 1961, p. 83.

[149] S. Ikeda and T. Isemura, *Bull. Chem. Soc. Japan* **34** (1960) 131.

[150]K. Shinoda, T. Nakagawa, B. I. Tamamushi, and T. Isemura, *Colloidal Surfactants*, Academic Press, New York-London, 1963, p. 286.

[151]W. D. Harkins and E. K. Fisher, *J. Chem. Phys.* **1** (1933) 852.

[152]J. H. Schulman and E. K. Rideal, *Proc. Roy. Soc. London* **A130** (1931) 259.

[153]J. Guastalla, *Compt. Rend.* **224** (1947) 1498.

[154]J. Guastalla and J. Michel, *Surface Chemistry*, Research Suppl., Butterworths, London, 1949, p. 127.

[155]A. E. Alexander and J. H. Schulman, *Proc. Roy. Soc. London* **A161** (1937) 115.

[156]M. Gerovich and A. N. Frumkin, *J. Chem. Phys.* **4** (1936) 624.

[157]H. B. Klevens and J. T. Davies, *Proc. 2nd Intern. Congr. on Surface Activity*, Vol. I, Butterworths, London, 1957, p. 31.

[158]M. K. Bernett and W. A. Zisman, *J. Phys. Chem.* **67** (1963) 1534.

[159]M. K. Bernett, N. L. Jarvis, and W. A. Zisman, *J. Phys. Chem.* **68** (1964) 3, 520.

[160]G. L. Gaines, *Insoluble Monolayers in Liquid–Gas Interfaces*, John Wiley and Sons, New York, 1966, p. 192.

[161]B. D. Powell and A. E. Alexander, *J. Colloid Sci.* **7** (1952) 482.

[162]A. N. Frumkin and B. Damaskin, *Pure and Applied Chem.* **15** (1967) 263.

[163]A. N. Frumkin, *Ergebnis, exact. Naturwiss.* **7** (1928) 235.

[164]R. I. Kangovich and V. M. Gerovich, *Elektrokhimiya* **2** (1966) 977.

[165]A. N. Frumkin, B. B. Damaskin, and A. Survila, *Elektrokhimiya* **1** (1965) 738.

[166]A. N. Frumkin, Z. A. Jofa, and P. Tshugunoff, *Acta physicochimica URSS* **1** (1935) 883.

[167]V. Kusnezov and B. B. Damaskin, *Elektrokhimiya* **1** (1965) 1153.

[168]B. B. Damaskin, M. M. Andresev, V. M. Gerovich, and R. I. Kaganovich, *Elektrokhimiya* **3** (1967) 667.

[169]B. Kamienski, I. Kulawik, and J. Kulawik, *Electrochim. Acta* **12** (1967) 219.

[170]G. Pytasz, *Proc. 3rd Intern. Congr. on Surface Activity*, Vol. II, Cologne, Universität-druckerei, Mainz, 1960, p. 303.

[171]A. M. Posner and K. E. Alexander, *Trans. Faraday Soc.* **45** (1949) 651.

[172]C. C. Addison and D. Litherland, *Nature* **171** (1953) 393.

[173]C. C. Addison and D. Litherland, *J. Chem. Soc.* (1953) 1143, 1150, 1159.

[174]J. R. Schulman and E. K. Rideal, *Proc. Roy. Soc. London* **A130** (1931) 284.

[175]J. H. Schulman and A. H. Hughes, *Proc. Roy. Soc. London* **A138** (1932) 430.

[176]G. L. Gaines, *Insoluble Monolayers in Liquid–Gas Interfaces*, John Wiley and Sons, New York, 1966, p. 229.

[177]J. J. Betts and B. A. Pethica, *Trans. Faraday Soc.* **52** (1956) 1581.

[178]J. T. Davies and E. K. Rideal, *Interfacial Phenomena*, Academic Press, New York, 1961, p. 95.

[179]E. Havinga and M. den Hertog Polak, *Rec. trav. chim.* **71** (1952) 64.

[180]J. Glazer and M. Z. Dogan, *Trans. Faraday Soc.* **49** (1953) 448.

[181]J. A. Spink and J. V. Sanders, *Trans. Faraday Soc.* **51** (1955) 1154.

[182]J. V. Sanders and J. A. Spink, *Nature* **175** (1955) 644.

[183]E. Havinga, *Rec. trans. chim.* **71** (1952) 72.

[184]E. D. Goddard and J. A. Ackilli, *J. Colloid Sci.* **18** (1963) 585.

[185]J. A. Spink, *J. Colloid Sci.* **18** (1963) 512.

[186]D. F. Sears and J. H. Schulman, *J. Phys. Chem.* **68** (1964) 3529.

[187]B. Case and R. Parsons, *Trans. Faraday Soc.* **63** (1967) 1224.

[188]C. L. de Ligny, M. Aljenaar, and N. G. van der Ween, *Rec. Trav. Chim.* **87** (1968) 585.

3

Transport Phenomena in Electrochemical Kinetics

A. J. Arvia and S. L. Marchiano

Instituto Superior de Investigaciones, Facultad de Ciencias Exactas
Universidad Nacional de La Plata, La Plata, Argentina

I. INTRODUCTION

In the kinetic study of heterogeneous reactions, which comprises, among others, electrode reactions, it is necessary to take into account the rates of the reacting species in reaching the reaction interface. In this sense, particularly for electrochemical reactions, much work has been done with the double purpose of establishing the phenomenological laws linked to the transport phenomena, and of elucidating the inner mechanism of this type of process.

This chapter is concerned with the first objective, namely a macroscopic treatment of transport phenomena in electrochemistry; the second objective mainly refers to the microscopic interpretation of the problem.

Transport phenomena in the kinetics of electrode processes have been considered on a more rigorous mathematical basis during the last decades as a consequence of the work of Levich on the rotating disk electrode[1-6]. This development is based on the solution of the mass transport differential equation taking into account concepts from hydrodynamic boundary-layer theory.[7] The rate equations obtained are of importance in theoretical electrochemistry, for instance, in hydrodynamic voltammetry, and of interest for the technology of electrochemical processes.

A heterogeneous reaction may be considered, in a simplified scheme, as a sequence of steps each taken independently and each occurring at a definite rate. There are at least three steps:

1. Transport of the reacting species from the bulk of one phase to the reacting interface, characterized by rate v_1.
2. Chemical or electrochemical reaction at the interface taking place at rate v_2, forming the reaction products.
3. Transport of the reaction products from the interface into the bulk of the phase at rate v_3.

Steps 1–3 correspond to the total process and, as they are consecutive, the slowest one will be the rate-determining step of the total process in the steady state. Based on this scheme, heterogeneous reactions, in general, may involve three cases. In the first, when v_1 or v_3 tends to be much slower than v_2, the process is controlled by mass transport; in the second case, when v_1 or v_3 tends to be much faster than v_2, the process is controlled by the chemical or electrochemical reaction taking place at the interface. Quite often, too, v_1 and/or v_3 is of the same order as v_2, and therefore the heterogeneous process, from the kinetic standpoint, is under intermediate control.†

As the equations governing mass transport phenomena are well established, the mass transport rate can be conveniently modified in a rather simple way by changing the hydrodynamic conditions prevailing in the system, such as the fluid velocity, the geometry of the system, and the physicochemical properties of the electrolytic solution. Therefore, it is possible, in principle, to change an electrochemical process under convective-diffusion control to a process under activation control or vice versa. This fact is relevant in electrochemical kinetics and, together with knowledge of the current distribution, is required for the proper design of electrolytic cells.

1. Basic Equations Related to Mass Transport

To obtain the rate of mass transport in a heterogeneous system, it is necessary to solve a mass-balance differential equation. In general, the concentrations of the species participating in the transport process are functions of position coordinates and time. Concentration changes are due to combined mechanisms which, depend-

†Strictly, in the steady state, the actual velocities of consecutive processes becomes equal.

ing on the operational conditions, are molecular mass transport (diffusion), macroscopic mass transport (convection), and all those particular movements induced by the strength of an applied field, as is the case of migration in electrochemical systems. If, in addition to the above-mentioned rate processes, there is also a homogeneous reaction occurring in the bulk of the electrolytic solution which either generates or consumes some reactant, the rate of this reaction appears as an additional term in the mass-balance equation. Thus, for an electrochemical system, the general equation is

$$\partial C_i/\partial t = \nabla \cdot (D_i \nabla C_i) - \mathbf{V} \cdot \nabla C_i + z_i F \nabla \cdot (u_i C_i \nabla \phi) + R_i \qquad (1)$$

where D_i is the diffusion coefficient of species i, whose concentration is C_i; $\mathbf{V}$, the flow velocity vector; z_i, the number of charges transported by i; u_i, the ionic mobility; and ϕ, the strength of the electric field. The first term depends on the concentration gradient of species i; the second, on the macroscopic velocity of the fluid and on the concentration of i; and the third term depends on the mobility of i, on its concentration, and on the potential gradient. The fourth term, R_i, is the rate of the homogeneous chemical reaction.

Besides the mass-balance equation, when electrolytic solutions are involved, the macroscopic treatment of the transport phenomena requires the electroneutrality condition, given by

$$\sum_i z_i C_i = 0 \qquad (2)$$

It is known that, for electrode reactions, the condition given by (2) is not valid in the electrochemical double-layer region. Phenomena taking place in the latter, represented by step 2, are generally involved in the boundary conditions emerging from each particular problem.

Equation (1), as it is a partial differential equation with variable coefficients, can be accurately solved only in a few cases of definite geometry. However, it is possible, for certain particular systems, to make reasonable assumptions which lead to a simpler equation. Thus, for a system formed by an incompressible fluid where only the electrode reaction is assumed, the term R_i is zero. Furthermore, if no migration contribution exists and the diffusion coefficient is supposed to be concentration-independent, equation (1) becomes even simpler and the result is

$$(\partial C_i/\partial t) + \mathbf{V} \cdot \nabla C_i = D_i \nabla^2 C_i \qquad (3)$$

This equation is valid only when a large excess of supporting electrolyte exists and, therefore, the contribution from the reacting ions to the migration is practically zero, as has already been demonstrated by Levich[6] and Newman,[8] and the diffusion coefficient D_i is taken as concentration independent. This situation arises in many electrochemical problems. For steady-state processes, which are the only type of processes to be considered in this chapter, equation (2) becomes even simpler, as $\partial C_i/\partial t = 0$. Then, the steady convective-diffusion equation is

$$\mathbf{V} \cdot \nabla C_i = D_i \cdot \nabla^2 C_i \tag{4}$$

Another simplification arises when a system consists of a binary electrolyte, a case often found in electrochemistry. Here, equation (4) must be solved for each ionic species. The resulting equation for a binary electrolyte is formally equal to equation (4) after replacing C_i by C, the molecular concentration, and D_i by the effective diffusion coefficient D_{eff}, which are defined respectively by

$$C = C_1/z_2 = C_2/z_1 \tag{5}$$

and

$$D_{eff} = D_1 D_2(z_1 + z_2)/(z_1 D_1 + z_2 D_2) \tag{6}$$

where the subscripts refer to the ionic species of the binary electrolyte.

The resolution of these equations requires knowledge of the velocity distribution functions. They are obtained by solving the Navier–Stokes and the continuity equations, which, for an incompressible fluid, are

$$\rho[(\partial \mathbf{V}/\partial t) + \mathbf{V} \cdot \nabla \mathbf{V}] = -\nabla p + \mu \nabla^2 \mathbf{V} + \rho g \tag{7}$$

and

$$\nabla \mathbf{V} = 0 \tag{8}$$

where ρ, μ, and g, respectively, stand for density, viscosity coefficient, and the gravitational constant, and p represents the local pressure.

The left-hand side of equation (7) corresponds to the momentum change of a unit volume of fluid upon which the forces represented in the right-hand side are acting, that is, pressure, viscosity, and gravitational forces.

When solving the Navier–Stokes and Fick equations, two definite cases appear. If the system is under forced convection, the resolution of the Navier–Stokes equation is independent of the mass transport processes. Equation (7) has been solved for fixed boundary conditions, and these solutions may be found in the hydrodynamics literature,[9–11] generally expressed, for laminar flow, as serial expansions.

The other case arises when the forces producing the motion of the fluid are originated by a concentration gradient in a gravitational field. Then, a set of coupled differential equations must be solved. This is the case of natural or free convection, which is similar to that of heat transfer.[6,9–14]

Finally, another simplification of equation (3) is obtained when a system involving a static fluid is considered ($\mathbf{V} = 0$). Thus, the only transport process is molecular diffusion. Here, equation (3) becomes

$$\partial C_i / \partial t = D_i \nabla^2 C_i \qquad (9)$$

The solutions of this equation for definite boundary conditions are well known and are related to chronopotentiometry, polarography, and other related electrochemical techniques.[15–17]

The validity of the transport relations, particularly the mass transport rate equations as applied to electrochemical problems, has been recently discussed by Newman[3] and repetition here is not justified. It is enough to say that, in fact, equation (7) is strictly valid for dilute solutions. However, its successful application to more complex electrochemical systems imposes the use of refinements which introduce too many complications.

2. Systematic Resolution of the Steady-State Mass Transfer Problem

The systematic solution of the problems of mass transport in steady state, such as those which may appear, for instance, in hydrodynamic voltammetry, comprises the following steps:

1. Choose the proper coordinate system, for each particular geometry, to express the momentum and mass transfer differential equations.

2. Replace in the convective-diffusion differential equation the components of the flow velocity by the distribution functions result-

ing from the resolution of the Navier–Stokes equation. These equations are dealt with in detail in the hydrodynamics literature.

3. Set the boundary conditions related to the problem.

4. Transform the partial differential equations into a total differential equation.

5. Solve the total differential equation in order to find the concentration distribution of the ith species.

6. On the basis of the equation of concentration distribution and Fick's law, find the local flow rate j.

7. Find the average mass transport rate $\bar{j}_i$ at the reaction surface. The rate of mass transport at the reaction surface, in the case of an electrochemical reaction, can be expressed in terms of current density i as follows:

$$i = z_i F \bar{j}_i \tag{10}$$

When solving equation (4), it is useful to bear in mind the already known solutions for mass transport with nonelectrolytes[4,9] and for heat transfer.[7–9] When the theoretical rate equations are applied to real systems, these must approach the ideal conditions assumed in the mathematical derivations.

In the following, the basic mass transport rate equations that are of importance for hydrodynamic voltammetry, including different types of electrodes under a given laminar flow, will be considered.

II. FLOWS CONTROLLED BY THE MOTION OF THE WORKING ELECTRODE

1. The Rotating Disk Electrode

(*i*) *Hydrodynamic Equations*

The rotating disk electrode is a useful device for studying the kinetics of heterogeneous reactions controlled by mass transport or under intermediate control.

The hydrodynamics of the rotating disk was studied by Cochran[18] and von Kárman,[19] by solving the corresponding Navier–Stokes and continuity equations. For this system, at distances far from the disk, a motion of fluid toward its surface occurs. Within the film immediately adjacent to the disk, as a consequence of the nonslip condition at the wall, the liquid acquires a rotating motion whose angular velocity increases as the fluid

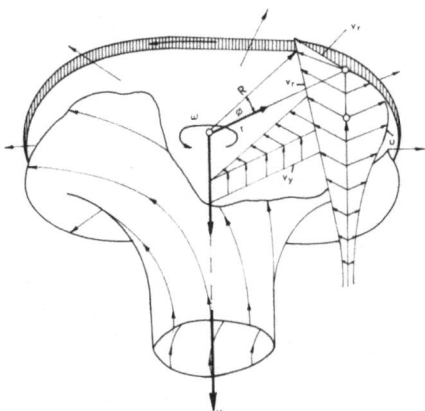

Figure 1. Diagram of a rotating disk (from
Ref. 7).

approaches the disk surface, finally reaching the disk angular
velocity at the surface of the disk. The centrifugal force related to the
rotational motion also gives rise to a radial component of the
velocity within the boundary layer, which is zero at the disk surface,
as is the axial velocity. Figure 1 shows a diagram of the rotating
disk, and gives the meaning of the coordinates and the flow pattern.

Taking into account the symmetry of the system, it is convenient
to express the Navier–Stokes and the continuity equations in cylindrical coordinates. Therefore, for a rotating disk the following is
the system of hydrodynamic equations to be solved:

$$v_r \frac{\partial v_r}{\partial r} - \frac{v_\varphi^2}{r} + v_y \frac{\partial v_r}{\partial y} = v \left[\frac{\partial^2 v_r}{\partial y^2} + \frac{\partial^2 v_r}{\partial r^2} + \frac{1}{r} \frac{\partial v_r}{\partial r} - \frac{v_r}{r^2} \right] \quad (11)$$

$$v_r \frac{\partial v_\varphi}{\partial r} + \frac{v_r v_\varphi}{r} + v_y \frac{\partial v_\varphi}{\partial y} = v \left[\frac{\partial^2 v_\varphi}{\partial y^2} + \frac{\partial^2 v_\varphi}{\partial r^2} + \frac{1}{r} \frac{\partial v_\varphi}{\partial r} - \frac{v_\varphi}{r^2} \right] \quad (12)$$

$$v_r \frac{\partial v_y}{\partial r} + v_y \frac{\partial v_y}{\partial y} = -\frac{1}{\rho} \frac{\partial p}{\partial y} + v \left[\frac{\partial^2 v_y}{\partial y^2} + \frac{\partial^2 v_y}{\partial r^2} + \frac{1}{r} \frac{\partial v_y}{\partial r} \right] (13)$$

and

$$\frac{\partial v_r}{\partial r} + \frac{\partial v_y}{\partial y} + \frac{v_r}{r} = 0 \quad (14)$$

with the boundary conditions:

$$y = 0: \qquad v_r = 0, \quad v_\varphi = \omega r, \quad v_y = 0$$
$$y \to \infty: \qquad v_r = 0, \quad v_\varphi = 0, \qquad v_y = -U_0 \tag{15}$$

where ω is the angular velocity, v is the kinematic viscosity, and v_i the velocity component along the i-coordinate. These conditions emerge from the qualitative flow pattern described above. The boundary condition at $y = 0$ imposed on tangential velocity ($v_\varphi = \omega r$) implies that the fluid on the disk turns with it. The boundary condition at $y = 0$ imposed on tangential velocity the fact that the liquid flow toward the disk must be kept constant. Here, U_0 represents a velocity at infinite determined by the resolution of the problem itself and its negative sign is given by the choice of coordinates.

To transform equations (11)–(14) into total differential equations, v_r, v_φ, v_y, and p must be expressed in terms of the following dimensionless variable ξ:

$$\xi = (\omega/v)^{1/2} y \tag{16}$$

Thus, the velocity components and pressure, given in terms of ξ, assume the form:

$$v_r = r\omega F(\xi)$$
$$v_\varphi = r\omega G(\xi)$$
$$v_y = (v\omega)^{1/2} H(\xi) \tag{17}$$
$$p = -\rho v\omega P(\xi)$$

The functions to be determined, $F(\xi)$, $G(\xi)$, $H(\xi)$, and $P(\xi)$, satisfy the equations

$$2F(\xi) + H'(\xi) = 0 \tag{18}$$

$$F(\xi)^2 - G(\xi)^2 + F'(\xi)H(\xi) = F''(\xi) \tag{19}$$

$$2F(\xi)G(\xi) + G'(\xi)H(\xi) = G''(\xi) \tag{20}$$

$$H(\xi)H'(\xi) = P'(\xi) + H''(\xi) \tag{21}$$

with the following boundary conditions:

$$\xi = 0: \qquad F(\xi) = 0, \quad G(\xi) = 1, \quad H(\xi) = 0$$
$$\xi \to \infty: \qquad F(\xi) \to 0, \quad G(\xi) \to 0, \quad H(\xi) \to -\alpha \tag{22}$$

where the primes correspond to the order of the derivatives of the functions. Equations (18)–(21) with boundary conditions (22), as solved by Cochran[18] and by von Kármán,[19] yield the following asymptotic expressions for the functions, in terms of ξ, when $\xi \to \infty$:

$$F = Ae^{-\alpha\xi} - \frac{A^2 + B^2}{2\alpha^2}e^{-2\alpha\xi} + \frac{A(A^2 + B^2)}{4\alpha^4}e^{-3\alpha\xi} + \cdots \tag{23}$$

$$G = Be^{-\alpha\xi} - \frac{B(A^2 + B^2)}{12\alpha^4}e^{-3\alpha\xi} + \cdots \tag{24}$$

$$H = -\alpha + \frac{2A}{\alpha}e^{-\alpha\xi} - \frac{A^2 + B^2}{2\alpha^3}e^{-2\alpha\xi} + \frac{A(A^2 + B^2)}{6\alpha^5}e^{-3\alpha\xi} + \cdots \tag{25}$$

The solutions derived for $\xi \to 0$ are

$$F = a\xi - \tfrac{1}{2}\xi^2 - \tfrac{1}{3}b\xi^3 + \cdots \tag{26}$$

$$G = 1 + b\xi + \tfrac{1}{3}a\xi^3 + \cdots \tag{27}$$

$$H = -a\xi^2 + \tfrac{1}{3}\xi^3 + \cdots \tag{28}$$

Constants A, B, a, b, and α, obtained through numerical integration, should be chosen in such a way that the functions and their derivatives be continuous when both series expansions are matched. Thus, according to the motion equations, the higher-order derivatives are also continuous. The constants, as given by Cochran[18] and Sparrow and Gregg,[20] are: $A = 0.934$, $B = 1.208$, $a = 0.51023$, $b = -0.616$, $\alpha = 0.88447$.

Figure 2 shows the dependence of the functions $F(\xi)$, $G(\xi)$ and $H(\xi)$ on ξ. These plots contain the velocity distributions and allow the evaluation of the thickness of the hydrodynamic boundary layer δ_h. The latter is defined as the distance from the solid surface where the greatest change in velocity occurs. Figure 2 shows that these velocity changes occur within the range $0 < \xi \leq \xi_0 = 3.6$, from the definition of ξ,

$$\delta_h = 3.6(v/\omega)^{1/2} \tag{29}$$

According to this expression the hydrodynamic-boundary-layer thickness, in this case, is independent of the radius of the disk. As is seen in Figure 2, appreciable components of the radial, axial, and angular velocities exist within the boundary layer ($y < \delta_h$), while outside it ($y > \delta_h$), only the axial contribution can be appreciable.

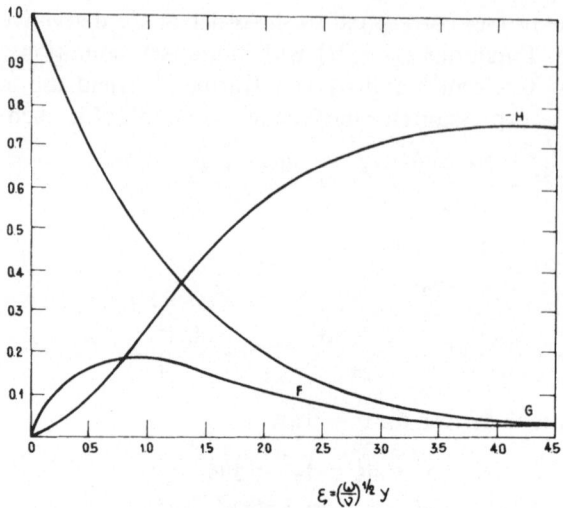

$$\xi \cdot \left(\frac{\omega}{v}\right)^{1/2} y$$

Figure 2. Velocity distributions in the rotating disk, according to Cochran.[18]

The velocity component in the direction normal to the disk, when $\xi \to 0$, is expressed by

$$H(\xi) = v_y/(v\omega)^{1/2} = -(0.510\xi^2 - 0.333\xi^3 + 0.103\xi^4 - \cdots) \quad (30)$$

The flow-line pattern on the disk surface is represented in Figure 1 considering the thickness of the hydrodynamic boundary layer to be much smaller than the disk radius.

(ii) Solutions of the Convective-Diffusion Equation

(a) Solution by Levich.[1,4,6] After the resolution of the hydrodynamic problem, the corresponding differential equation for convective diffusion can be solved. Considering the symmetry of the system and neglecting edge effects, the equation is, for cylindrical coordinates,

$$v_y(y)\, dC_i/dy = D_i\, d^2 C_i/dy^2 \quad (31)$$

and the boundary conditions are

$$\begin{aligned} y = 0, & \quad C_i = 0 \\ y \to \infty, & \quad C_i = C_i^\circ \end{aligned} \quad (32)$$

These boundary conditions are satisfied for any electrochemical system when the overpotential is high enough to maintain the surface concentration of the reacting species equal to zero. This implies that the rate of the interfacial reaction is very much larger than that of the transport process.

By integrating equation (31) twice,

$$C_i = a_1 \int_0^y \exp\left\{(1/D_i) \int_0^t v_y(z)\, dz \right\} dt + a_2 \tag{33}$$

From the boundary conditions, it is deduced that the integration constant a_2 is equal to zero, and a_1 is obtained after solving the integral of the equation

$$C_i^0 = a_1 \int_0^\infty \exp\left\{(1/D_i) \int_0^t v_y(z)\, dz \right\} dt \tag{34}$$

To solve the integral of equation (34), the velocity distribution equations for $y \to 0$ and $y \to \infty$ must be considered. Therefore, the integral becomes the sum of two integrals. The first one covers the integration interval between 0 and δ_h, where the velocity distribution equation for $y \to 0$ is valid. The second integral covers the integration between δ_h and infinity with the velocity distribution equation for $y \to \infty$. As Levich has demonstrated, for liquids the second integral is much smaller than the first one and thus, no appreciable error is introduced if integral (34) is solved for the integration interval between 0 and δ_h with the velocity distribution function given by (30). For liquids, where $v \gg D_i$ this implies, as will be demonstrated below, that the distance along which the largest concentration variation occurs is much smaller than the thickness of the hydrodynamic boundary layer. Consequently, only the first terms of the serial expansion of the velocity function can be taken, without appreciable error:

$$v_y \simeq -0.51(\omega^3/\nu)^{1/2} y^2 \tag{35}$$

the result then is

$$C_i^0 = a_1 \int_0^{\delta_h} \exp[-\omega^{3/2} t^3 / 5.88 D_i \nu^{1/2}]\, dt \tag{36}$$

In terms of the variable u, defined by

$$u = \omega^{1/2} t / (5.88)^{1/3} D_i^{1/3} v^{1/6} \tag{37}$$

the equation can be approximated by the expression

$$C_i^\circ \approx a_1 (1.81 D_i^{1/3} v^{1/6}/\omega^{1/2}) \int_0^\infty \exp(-\mu^3)\, du$$

$$= 1.6166 (D_i^{1/3} v^{1/6}/\omega^{1/2}) a_1 \tag{38}$$

where the change of the upper limit of the integral from δ_h to infinity is a consequence of $(v/D_i) \gg 1$ and of the rapid convergence of the argument of the integral for values of u greater than unity.

Finding the value of a_1 from equation (38) and replacing into (33), the concentration distribution equation satisfying equation (31) and boundary conditions (32) is:

$$C_i = \frac{C_i^\circ}{1.61(D_i/v)^{1/3}(v/\omega)^{1/2}} \int_0^y \exp\left\{ \frac{1}{D_i} \int_0^t v_y(z)\, dz \right\} dt \tag{39}$$

The maximum mass transport rate toward the disk, j_{max}, is obtained by differentiating equation (39) according to Fick's law:

$$j_{max} = D_i(\partial C_i/\partial y)_{y=0} = 0.62 D_i^{2/3} v^{-1/6} \omega^{1/2} C_i^\circ \tag{40}$$

From this equation, the thickness of the diffusion boundary layer, δ_d, may be defined as:

$$\delta_d = D_i C_i^\circ / j_{max} = 1.61(D_i/v)^{1/3}(v/\omega)^{1/2} = 0.5 \delta_h (D_i/v)^{1/3} \tag{41}$$

This equation states that the diffusion boundary layer is independent of r. Taking the usual values of D_i and v for liquids at room temperature, δ_d is much smaller than δ_h, a fact which justifies the approximation employed by Levich.[6]

The maximum rate of mass transport toward the disk is then proportional to the concentration of the diffusing species, to the square root of the electrode angular velocity, to the inverse of the 1/6-power of the kinematic viscosity, and to the 2/3-power of the diffusion coefficient. Furthermore, δ_d does not depend on the distance to the rotating axis, at least within those regions distant from the disk edge. Therefore, it is essentially uniform over the whole disk surface.

(b) *Solution by Gregory and Riddiford*.[21,22] The problem of the rotating disk electrode has been also approached by Gregory and Riddiford. These authors too, considered the solutions proposed by Cochran[18] and von Kármán[19] for the velocity distribution, but took higher-order terms in the velocity distribution function.

The convective-diffusion equation is solved for any flux, and v_y is substituted in equation (31) by its expression (30) taking the first and second terms of the serial expansion with the following boundary conditions:

$$y = 0: \quad C_i = C_i^s, \quad v_y = 0$$
$$y \to \infty: \quad C_i = C_i^\circ, \quad v_y = -0.886(v\omega)^{1/2} \tag{42}$$

Following a similar mathematical procedure to that of Levich, the concentration distribution equation is

$$\frac{C_i - C_i^s}{C_i - C_i^\circ} = \frac{\int_0^\xi \exp[-\xi^3 + 0.885(D_i/v)^{1/3}\xi^4 - 0.394(D_i/v)^{2/3}\xi^5]\, d\xi}{\int_0^\infty \exp[-\xi^3 + 0.885(D_i/v)^{1/3}\xi^4 - 0.394(D_i/v)^{2/3}\xi^5]\, d\xi} \tag{43}$$

Graphically evaluating the integral of the denominator for D_i/v values between zero and 4×10^{-3}, Gregory and Riddiford found that, within 1 %, it can be represented by the empirical expression

$$\text{Integral}(\infty) = 0.8934 + 0.316(D_i/v)^{0.36} \tag{44}$$

Therefore, the rate of mass transport toward the disk is

$$j = -D_i(C_i^\circ - C_i^s)/\delta_d[0.8934 + 0.316(D_i/v)^{0.36}] \tag{45}$$

With the expression for δ_d given by the authors,[21,22]

$$\delta_d = 1.805(D_i/v)^{1/3}(v/\omega)^{1/2} \tag{46}$$

the maximum flux toward the disk is

$$j_{max} = 0.554 D_i^{2/3} v^{-1/6} \omega^{1/2} C_i^\circ / [0.8934 + 0.316(D_i/v)^{0.36}] \tag{47}$$

When $v/D_i \to \infty$, the numerical coefficient in equation (47) approaches that in the equation deduced by Levich.

Based on the approach already described, Kholpanov[23] has studied the rate of mass transport for the disk electrode, obtaining the following rate equation for the maximum flux:

$$j_{max} = D_i^{2/3} v^{-1/6} \omega^{1/2} C_i^\circ / 1.61[1 + 0.31(v/D_i)^{1/3}] \tag{48}$$

This equation is somewhat different from the one derived by Gregory and Riddiford both in the numerical coefficient and the exponent of the v/D_i term. Equation (48) without the $0.31(v/D_i)^{1/3}$ term equals equation (40).

(c) *Solution by Newman.*[8,24–27] The mathematical solution of the problem gives results analogous to those solutions already mentioned up to the convective-diffusion differential equation (31). As in the treatment of Gregory and Riddiford's, equation (31) is also solved for any flux. The resolution is expressed as follows:

$$C = C_i^s + (C_i^\circ - C_i^s) \frac{\int_0^y \exp\left[\int_0^y (v_y/D_i)\,dy\right] dy}{\int_0^\infty \exp\left[\int_0^y (v_y/D_i)\,dy\right] dy} \tag{49}$$

and then, the steady diffusional flow is given by

$$j = -D_i(C_i^\circ - C_i^s)/\int_0^\infty \exp\left[\int_0^y (v_y/D_i)\,dy\right] dy \tag{50}$$

After expressing the integrals in terms of $H(\xi)$ and ξ, and rearranging equation (50) by grouping all the terms independent of v/D_i in one term, we get

$$-\frac{j}{(C_i^\circ - C_i^s)(\omega v)^{1/2}} = \frac{1}{(v/D_i)\int_0^\infty \exp\left[(v/D_i)\int_0^\xi H(\xi)\,d\xi\right] d\xi} \tag{51}$$

This expression presents asymptotic values for $v/D_i \to 0$ and for $v/D_i \to \infty$. When $v/D_i \to 0$, the asymptote is 0.88447, while for $v/D_i \to \infty$, it is $0.62048(v/D_i)^{-2/3}$, as shown in Figure 3, according to the heat transfer results of Sparrow and Gregg,[20] as conveniently modified by Newman.[8]

After solving equation (51) by taking the three terms of the expression for $H(\xi)$ [equation (30)], the local rate of mass transport to

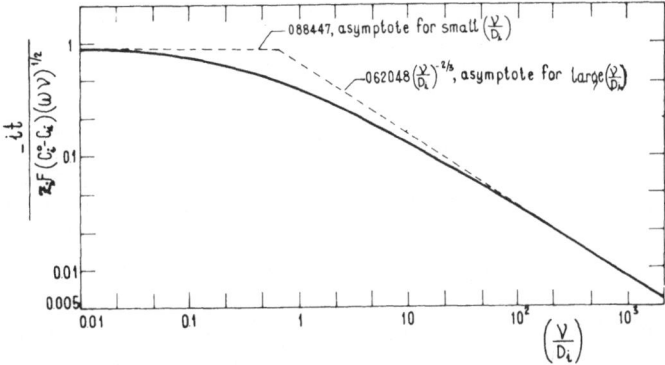

Figure 3. Asymptotic values of mass-transfer rate equation for the disk electrode for large and small v/D_i ratios, according to the results of Sparrow and Gregg.[20]

the disk surface is

$$j = -\frac{0.62048(v\omega)^{1/2}(v/D_i)^{-2/3}(C_i^\circ - C_i^s)}{1 + [0.2980/(v/D_i)^{1/3}] + [0.14514/(v/D_i)^{2/3}]} \tag{52}$$

This expression adequately represents the curve of Figure 3 for v/D_i larger than 100. The maximum error in this region is about 0.1 %, making this expression useful for electrochemical systems where v/D_i is of the order of 10^3.

For $y = 0$ and $C_i^s = 0$, the maximum flux is

$$j_{max} = -\frac{0.62048(v\omega)^{1/2}(v/D_i)^{-2/3}C_i^\circ}{1 + [0.2980/(v/D_i)^{1/3}] + [0.14514/(v/D_i)^{2/3}]} \tag{53}$$

Newman and Hsueh[26] extended further the study of mass transport at the rotating disk electrode by considering the effect of the variation of the transport properties within the diffusion layer. To solve the set of coupled nonlinear differential equations comprising the Navier–Stokes, the Fick, and the continuity equations, assumptions already considered by von Kárman[19] and Levich[6] such as steady laminar flow and no edge effect were taken into account. In addition, the variation of the properties and a finite velocity at the electrode surface were also considered. Starting from the velocity and concentration distribution functions, the mathematical problem is solved by iterative methods. Results confirm the earlier solutions for $v/D_i \to \infty$ obtained for two-dimensional boundary-layer flow.

(iii) Comparison of the Rate Equations

A useful way to show the differences among the three rate equations already considered is to calculate the diffusion coefficient of the reacting species, starting from experimental data, by means of equations (40), (47), and (53).

Let us suppose an electrochemical system containing a large excess of supporting electrolyte where any migration contribution is negligible and, consequently, the chemical potential within the diffusion layer is unaffected by concentration changes of the reacting species. Then, according to equation (40), the maximum flux, expressed as a limiting current density i_L, depends linearly on the square root of the angular velocity of the disk electrode. Thus, from equation (40),

$$D_i = [(\Delta i_L/\Delta\omega^{1/2})v^{1/6}/0.62z_iFC_i^{\circ}]^{3/2} \tag{54}$$

where z_i is the number of electrons per mole of diffusing species.

To obtain the diffusion coefficient from equation (47), one possibility is to solve the following equation:

$$0.5542z_iFv^{-1/6}D_i^{2/3}C_i^{\circ} - 0.316(i_L/\omega^{1/2})v^{-0.36}D_i^{0.36}$$

$$-0.8934(i_L/\omega^{1/2}) = 0 \tag{55}$$

which can be approximately transformed into a quadratic equation in terms of $D_i^{1/3}$ assuming that $D_i^{0.36} \approx D_i^{1/3}$. However, more rigorously, the equation deduced by Gregory and Riddiford must be solved either graphically or by trial and error.

If the same calculation is made with equation (53), it is convenient to write it in the following form:

$$(0.62048v^{-1/6}z_iFC_i^{\circ}\omega^{1/2} + 0.14514i_Lv^{-2/3})D_i^{2/3}$$

$$+ 0.2980i_Lv^{-1/3}D_i^{1/3} + i_L = 0 \tag{56}$$

Equation (56) is solved algebraically in the conventional way.

These equations were used to evaluate the diffusion coefficients of oxygen,[28] nitrite ion in a mixture of molten nitrates,[29] and hydrogen ion in solutions of hydrogen chloride in dimethylsulfoxide.[30] Results are assembled in Table 1. Equation (40) gives the

Table 1
Diffusion Coefficients Calculated with Eqs. (40), (47), and (53)

$C_i^\circ,$ M	D_i (40), cm^2/sec	D_i (47), cm^2/sec	D_i (53), cm^2/sec
Molecular oxygen in alkaline solutions, $T = 25°C^{28}$			
O_2 in KOH 2 M 1.12 × 10⁻⁵		1.19 × 10⁻⁵	1.18 × 10⁻⁵
Nitrite ion dissolved in nitrate melt, $T = 247°C^{29}$			
1.30 × 10⁻³	2.31 × 10⁻⁵	2.38 × 10⁻⁵	2.36 × 10⁻⁵
4.26 × 10⁻³	2.10 × 10⁻⁵	2.16 × 10⁻⁵	2.15 × 10⁻⁵
8.30 × 10⁻³	1.92 × 10⁻⁵	1.98 × 10⁻⁵	1.97 × 10⁻⁵
Solvated hydrogen ion in dimethylsulfoxide, $T = 30°C^{30}$			
1.33 × 10⁻³	2.28 × 10⁻⁶	2.33 × 10⁻⁶	2.32 × 10⁻⁶
2.81 × 10⁻³	2.37 × 10⁻⁶	2.43 × 10⁻⁶	2.42 × 10⁻⁶
5.47 × 10⁻³	2.52 × 10⁻⁶	2.58 × 10⁻⁶	2.52 × 10⁻⁶

lowest diffusion coefficients, while equation (47) gives somewhat higher values. However, the difference between these is 3%, a discrepancy which is within the experimental error.

2. Modified Diffusional Fluxes at a Rotating Disk Electrode

Kholpanov[23] has recently shown that interesting possibilities arise when the disk is considered at rest while the liquid rotates at some distance with an angular velocity ω. Under these conditions, the v_y-velocity distribution equation takes the form indicated by Bödewadt[31]

$$v_y = (\omega v)^{1/2}[-\tfrac{1}{3}(\omega/v)^{3/2}y^3 + 0.942(\omega/v)y^2] \tag{57}$$

With this velocity distribution function, the diffusional flow can be calculated starting from equation (31), taking into account the boundary conditions given by (32). The following maximum flow to the disk surface is found:

$$j_{max} = 0.54D_i^{3/4}\omega^{1/2}C_i^\circ/v^{1/4}[0.906 + 0.608(v/D_i)^{1/4}] \tag{58}$$

The case of the liquid flowing between two disks, a fixed one and a turning one, with a separation distance l between them has also been considered. The equation of the velocity distribution for this case is[32]

$$v_y = 2\omega l[-(\omega/20v)(y^3/l) + (\omega/30)(y^2/v)] \tag{59}$$

On the stationary disk near which a second disk rotates at some distance l, the maximum flux is

$$j_{max} = D_i C_i^{\circ} \omega^{1/2} / (40^{1/4} \times 0.906 D_i^{1/4} v^{1/4} + 0.222 l \omega^{1/2}) \qquad (60)$$

However, for high values of the Reynolds number ($Re > 10^4$), but keeping the flow laminar, and for a comparatively large axial separation between the disks, individual boundary layers are formed on both the rotating disk and the stationary one. The velocity distribution is then

$$v_y = 0.03(\omega^2/v)[-3\delta_h y^2 + 2y^2] \qquad (61)$$

and, from similar deductions, the maximum flux for this particular disk electrode is given by

$$j_{max} = 0.473 D_i^{1/3} \omega^{1/2} C_i^{\circ} / v^{1/6} \qquad (62)$$

Therefore, by examining the possible ways in which a liquid may flow around a disk electrode, such as in the combination of a stationary disk and a rotating disk electrode, the simultaneous determination of the physicochemical properties of an electrolytic solution, such as the kinematic viscosity v and the diffusion coefficient D_i, is possible.

3. The Rotating Ring Electrode

The mass transport at the rotating disk electrode has also been studied when a relaxation process takes place, as is the case when the center of the disk is, from the electrochemical standpoint, an inert region[22] and therefore, no concentration gradient exists there. The active area consists of a ring having a thickness smaller than the radius of the central inert disk, as shown in Figure 4. Although the hydrodynamic flow pattern for this system is the same as already described for the disk electrode, the local mass transport rate will now depend both on the radius of the inert central disk as well as on the thickness of the ring. That is, the condition of uniformly accessible surface is not fulfilled. The equation of maximum local flow for this case is[6]

$$j_{max} = D_i C_i^{\circ} / 1.61 (D_i/v)^{1/3} (v/\omega)^{1/2} [1 - (r_1^3/r^3)]^{1/3}, \quad r_1 < r < r_2 \qquad (63)$$

r being any radius of the ring, and r_1 the radius of the central inert zone. If the flow in the ring electrode is compared with that in one having the same dimensions in a disk electrode, and it is assumed

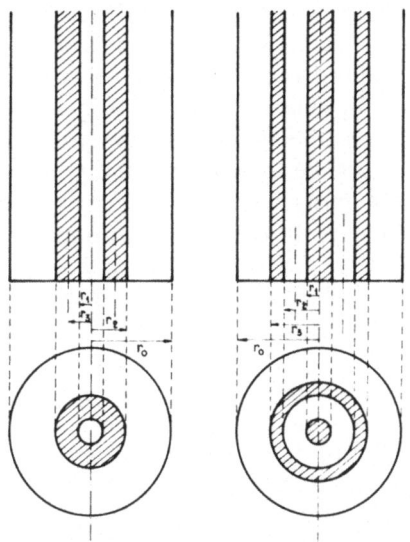

Figure 4. The ring and ring–disk electrodes.

that $(r_2 - r_1) \ll r_1$ is fulfilled, the result is

$$j_{max}/j_{max,d} = (3^{2/3}/2)[r_1/(r_2 - r_1)]^{1/3} \tag{64}$$

i.e., because of the radial flow, a larger flow is obtained for the ring electrode with a central inert region. If no radial mass transport on the electrode exists, or if it does not play an important role in the total amount of mass arriving at the electrode, as is the case, for instance, for a motionless fluid, the current density on the ring will remain practically constant for any ring thickness.

The study of relaxation processes in ring electrodes, e.g., by measuring the diffusional limiting current for oxygen reduction, shows that the current density noticeably increases when the ring width decreases.[6,22] The smaller the latter is, the larger is the active portion of the electrode surface advantageously placed to receive fresh solution. The experimental measurements obtained with this kind of electrode agree reasonably with the theory.[6]

4. The Rotating Ring–Disc Electrode

In order to confirm the reaction mechanisms of electrode processes, it is necessary to attain a direct confirmation of either the reaction products or the reaction intermediates. Frumkin and Nekrasov[33] described an elegant device for this direct confirmation, consisting of a rotating disk electrode applied as generating electrode and a concentric ring-shaped electrode as indicator. Through this arrangement, it is possible for the reaction products originated on the disk electrode surface to be carried away to the ring electrode and to be detected on its surface. To attain the highest efficiency, it is necessary to achieve the fastest possible transport of the reaction products formed on the disk electrode onto the indicating electrode. This can be achieved when the reaction products are generated close to the indicating electrode, and carried to it as fast as possible by the radial flow.

After completing the study of the rotating ring electrode, Levich and coworkers developed the theory of the rotating ring–disk electrode.[6,34,35] In this design, the disk surface is divided into the following regions, easily distinguishable on Figure 4. An active region (central disk), $0 < r < r_1$; an inactive region, $r_1 < r < r_2$; an active region (active ring), $r_2 < r < r_3$; and lastly, another inactive region, $r_3 < r < r_0$. The local mass transport rate at the ring is given by the expression

$$j_r = \frac{0.4j_d}{1 + (k_2\delta_{d,r}/D_2)} \frac{r_1^2 r_2}{r^3} \frac{[1 - \frac{3}{4}(r_1/r_2)^3]^{1/3}}{[1 - (r_2/r)^3]^{1/3}[1 - \frac{3}{4}(r_1/r)^3]} \tag{65}$$

where k_2 is the rate constant of the heterogeneous reaction occurring at the disk surface.

The problem of the rotating ring–disk electrode has also been considered by Albery,[36] who aimed at obtaining the mass transport rate equation in a more rigorous way, starting from the differential equation for mass transport in cylindrical coordinates. The equation, obtained through Laplace transformation, is in terms of the transformed variables,

$$\bar{u} = \sum_{n=0}^{\infty} \left(\frac{1}{z}\frac{\partial^2}{\partial z^2}\right)^n \frac{u_0}{s} - Ai(z)\left[\frac{1}{Ai'(z)}\sum_{n=0}^{\infty} \frac{\partial}{\partial z}\left(\frac{1}{z}\frac{\partial^2}{\partial z^2}\right)^n \frac{u_0}{s}\right]_{y=0}$$

$$= \sum_{n=0}^{\infty} \left(\frac{1}{sy}\frac{\partial^2}{\partial y^2}\right)^n \frac{u_0}{s} + \frac{Ai(z)}{[3^{1/3}\Gamma(\frac{1}{3})]^{-1}}\left[\sum_{n=0}^{\infty} \frac{\partial}{\partial y}\left(\frac{1}{sy}\frac{\partial^2}{\partial y^2}\right)^n \frac{u_0}{s^{4/3}}\right]_{y=0} \tag{66}$$

where $\bar{u}$ is the transformed variable related to concentration, s is the transformation factor, $Ai(z)$ are Airy functions, and u_0 is the concentration function when the radial coordinate is zero. The converse transformation must be made for each particular case.

Current/potential curves were calculated by Albery, Bruckenstein, and Napp[37] for reversible electrochemical systems at the ring electrode of a ring–disk system, applying a simultaneous current at the disk electrode. The corresponding equations were tested for the Ag/Ag(I) system in perchloric acid and were found to be in good agreement with the experiments. The ring–disk electrode was also used as a simple but sensitive analytical technique.[38] Arsenic(III) was titrated with bromine electrogenerated at the disk and the ring electrode was used to determine any excess which had not been consumed. Experimental results also agree with the equations derived from equation (66) for the rotating disk electrode. Equation (66) was applied to solve the kinetics of second-order electrochemical processes and also to determine the kinetics of electrochemical reactions with a preceding homogeneous chemical reaction either under a control by the heterogeneous or the homogeneous reaction.

5. The Translating Electrode in Stagnation-Point Flow

Another way to establish a diffusional boundary layer of constant thickness with motion of the working electrode is by means of the translating electrode, as described by Bopp and Mason.[39] A cylindrically shaped electrode is embedded in the leading edge of a larger nonconductive cylinder, the latter being mounted at the end of an L-shaped shaft as shown in Figure 5. The electrode turns with a constant angular velocity, establishing a laminar steady flow normal to the electrode, for a wide range of velocities. Further-

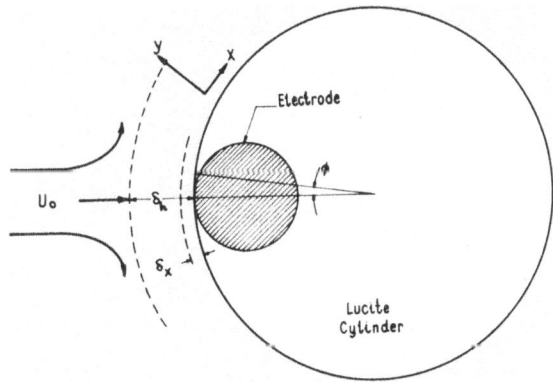

Figure 5. The translating electrode (from Ref. 39).

more, if the active part of the system is contained within an angle smaller than 20°, the thickness of the diffusional boundary layer may be considered constant all over the electrode surface.

To solve the hydrodynamic problem, the starting point, as usual, is to consider the Navier–Stokes and continuity equations, assuming an infinitely long cylinder ($z = \infty$) with the following boundary conditions:

$$y = 0: \quad v_x = 0, \quad v_y = 0$$
$$y \to \infty: \quad v_x = U_0 \tag{67}$$

Applying the same procedure already mentioned for other electrodes, the solution of the differential equations may be obtained by means of an adequately chosen dimensionless variable λ

$$\lambda = (4U_0/vd)^{1/2}y \tag{68}$$

The velocity distribution function corresponding to v_y is also expressed in terms of a function of λ, $f(\lambda)$

$$v_y = -(4vU_0/d)^{1/2}f(\lambda) \tag{69}$$

The $f(\lambda)$ can be expressed as a series expansion in terms of λ, and the relationship is plotted in Figure 6. For liquids, for the same

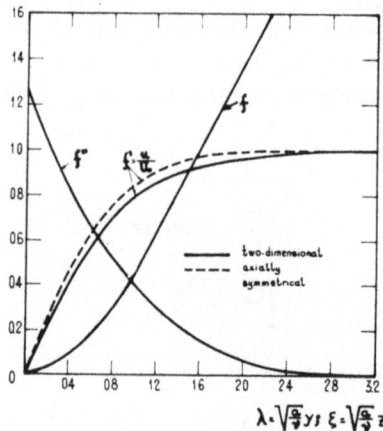

Figure 6. Velocity distribution of plane and rotationally symmetrical flow at a stagnation point.

reason as pointed out before, $f(\lambda)$ can be approximated by the following quadratic equation:

$$f(\lambda) = \alpha\lambda^2 \tag{70}$$

The proportionality constant α resulting from Figure 6 is roughly 0.575. The resolution of the new system of equations gives the following expression for the hydrodynamic boundary layer valid in the vicinity of the stagnation point:

$$\delta_h = 1.2(vd/\omega R_s)^{1/2} \tag{71}$$

Therefore, in the present case, the thickness of the hydrodynamic boundary layer is a function of the ratio between the cylinder diameter d and the radius of the rotating shaft R_s, which in the proposed design is smaller than one, and of the angular velocity ω of the translating electrode.

The convective diffusion equation for this system, considering that the mass transport process occurs in the direction normal to the tangent to the active surface, may be written as

$$v_y \, \partial C_i/\partial y = D_i \, \partial^2 C_i/\partial y^2 \tag{72}$$

with the following boundary conditions:

$$\begin{align} y = 0: \quad & C_i = C_i^s \\ y \to \infty: \quad & C_i = C_i^o \end{align} \tag{73}$$

The resolution equation (72) with boundary conditions (73) and the expression for v_y from (69) in terms of λ gives

$$\frac{C_i - C_i^s}{C_i^o - C_i^s} = 1.12\left(\frac{\alpha v}{3D_i}\right)^{1/3} \int_0^\lambda \exp\left[-\left(\frac{\alpha v}{3D_i}\right)\lambda^3\right] d\lambda \tag{74}$$

The maximum flux is obtained for $C_i^s = 0$, on the electrode surface. Then, from equation (74), expressing the result in terms of current density i

$$i = 1.292z_iFv^{-1/6}D_i^{2/3}(R_s/d)^{1/2}\omega^{1/2}(C_i^o - C_i^s) \tag{75}$$

According to Bopp and Mason,[39] and as can be seen by comparing equation (75) with equation (40), the translating electrode has advantages over the rotating disk electrode since the current, for the same potential applied to the electrode, can be modified either by increasing the rotation speed or by changing the ratio R_s/d.

The applications of the translating electrode are formally the same as those already stated for the rotating disk electrode. Thus, as an example, for a convective-diffusion-controlled electrochemical reaction, the expression for the diffusion-limited current ($C_i^s = 0$) is

$$i_L = 1.292 z_i F v^{-1/6} (R_s/d)^{1/2} D_i^{2/3} \omega^{1/2} C_i^\circ \tag{76}$$

As already indicated for the rotating disk electrode, equation (76) is valid when the migration contribution of the ith species is negligible.

III. CONVECTIVE DIFFUSION AT ELECTRODES WITH FLOWING SOLUTIONS

In the electrochemical systems studied previously, an increase of the mass transport rate is induced by means of the controlled motion of the working electrode. It was pointed out by different authors that both the rotating disk and the translating electrode in stagnation-point flow are useful in studying reactions with intermediate kinetics, because their surfaces are uniformly accessible from the diffusional standpoint.

Another way to achieve the same purpose is through a design of the electrochemical system where the transport of the reacting species toward the reaction surface is attained by means of a definite flow of the electrolytic solution at a static electrode.

The measurement of current/potential curves under different flow velocities permits the evaluation of kinetic parameters of electrochemical reactions with intermediate control, and the elucidation of reaction mechanisms. Jordan and Javick,[43,44] for the first time, used conical working electrodes to study the kinetics and mechanism of the electrochemical oxidation reaction of iodide in aqueous solutions on platinum.

The resolution of both the hydrodynamic and the mass transport differential equations was performed following the procedure previously used for moving electrode systems, since what matters from the hydrodynamic standpoint is the relative velocity between the electrode and the solution.

1. The Fixed Disk Electrode in Stagnation-Point Flow

The simplest electrode to be employed with flowing solutions is a disk on which a laminar flow incides normally to it, as shown in Figure 7a. This electrode was studied by Marchiano and Arvia.[45,46] An exact solution of the Navier–Stokes equation exists[7,40,41] for the hydrodynamics of the system provided both the distribution of radial velocity v_r and that of axial velocity v_y in terms of the following dimensionless variable ξ are known:

$$\xi = y(a/v)^{1/2} \tag{77}$$

where y is the coordinate normal to the disk surface, v is the kinematic viscosity of the fluid, and a is a constant, given by

$$U = ar \tag{78}$$

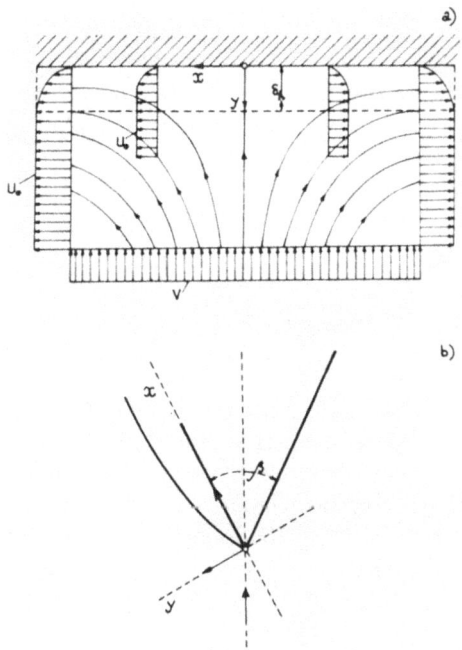

Figure 7. (a) Flow pattern in stagnation-point flow (from Ref. 7). (b) Flow past a wedge (from Ref. 45).

U is the potential flow and r the radial coordinate. The axial velocity distribution function, the interesting one in the mass transport problem, is

$$v_y = -2(av)^{1/2} f(\xi) \qquad (79)$$

$f(\xi)$ being the dimensionless velocity distribution function in terms of ξ, given by equation (79), which is plotted in Figure 6 as a function of ξ.

Since mass transport occurs only along the direction normal to the disk, the differential equation for the axially symmetrical system takes the following form:

$$v_y \, \partial C/\partial y = D_i \, \partial^2 C/\partial y^2 \qquad (80)$$

and the boundary conditions are

$$\begin{aligned} y = 0: & \qquad C_i = C_i^s \\ y \to \infty: & \qquad C_i \to C_i^\circ \end{aligned} \qquad (81)$$

Equation (80) is transformed into a new differential equation by writing it in terms of the dimensionless variable ξ, defined in equation (77). Thus

$$d^2 C_i/d\xi^2 + 2(v/D_i) f(\xi) \, dC_i/d\xi = 0 \qquad (82)$$

The boundary conditions in terms of variable ξ are

$$\begin{aligned} \xi = 0: & \qquad C_i = C_i^s \\ \xi \to \infty: & \qquad C_i \to C_i^\circ \end{aligned} \qquad (83)$$

Equation (82) is formally equal to the dimensionless differential equation for the translating electrode in stagnation-point flow[39] and therefore it is solved in a similar way. For liquids, the function $f(\xi)$ can be approximated by the following expression, as deduced from Figure 6, for $\xi \to 0$

$$f(\xi) = \alpha \xi^2 \qquad (84)$$

This function is satisfied within 1 % for ξ values lower than 0.5. The constant α is obtained from the same plot and equals 0.515.

From equations (82)–(84),

$$\frac{C_i - C_i^s}{C_i^o - C_i^s} = \frac{\int_0^\xi \exp\left[-2(v/D_i)\int_0^\xi \alpha\xi^2\, d\xi\right] d\xi}{\int_0^\infty \exp\left[-2(v/D_i)\int_0^\xi \alpha\xi^2\, d\xi\right] d\xi} \tag{85}$$

By integrating the denominator, the concentration distribution is found to be

$$(C_i - C_i^s)/(C_i^o - C_i^s) = 1.12(2\alpha v/D_i)^{1/3}\int_0^\xi \exp[-\tfrac{2}{3}(v/D_i)\alpha\xi^3]\, d\xi \tag{86}$$

and the maximum flux is

$$j_{max} = 1.12(2\alpha/3)^{1/3}D_i C_i^o (U/vr)^{1/2}(v/D_i)^{1/3} \tag{87}$$

Therefore, grouping the numerical coefficients in equation (87),

$$j_{max} = 0.780 D_i C_i^o (U/vr)^{1/2}(v/D_i)^{1/3} \tag{88}$$

This equation establishes that the maximum flux is directly proportional to the concentration of the diffusing species, to the square root of the fluid velocity, and to the 2/3-power of the diffusion coefficient, and is inversely proportional to the square root of the disk radius and to the 1/6-power of the kinematic viscosity. Equation (88) keeps the same form as those already described for the moving electrodes.

As for the case of the rotating disk electrode, the thickness of the hydrodynamic boundary layer is determined by considering the value of ξ corresponding to 99% of the limiting value of the function $\phi(\xi)$. The result is

$$\delta_h = \xi_0(vr/U)^{1/2} \tag{89}$$

This thickness is independent of the radius, since $r/U = 1/a$, where a is a constant given by the potential flow equation. Therefore, the surface of the disk electrode under stagnation-point flow is uniformly accessible from the diffusion standpoint, since the thickness of the hydrodynamic boundary layer is independent of the disk radius. Thus,

$$\delta_d \approx (D_i/v)^{1/3}\delta_h \tag{90}$$

2. Conical Electrodes in an Axial Laminar Flow

The mass transport rate equation initially used for conical electrodes was deduced by analogy with the heat transfer equations.[43,47] A theoretical rate equation for these electrodes was derived by Marchiano and Arvia[45] starting from the solutions of the hydrodynamic problem.

To obtain an approximate solution to the mass transport differential equation applied to conical electrodes, the two-dimensional case comprising flow past a wedge, schematically shown in Figure 7(b), was first considered. Falkner and Skan[48] obtained a dimensionless Navier–Stokes equation for the laminar regime, which was afterward numerically solved by Hartree.[49,50] This approximation leads to the following velocity distribution equations in the hydrodynamic boundary layer:

$$v_x = U(x)f'(\xi^*) = u_1 x^m f'(\xi^*) \tag{91}$$

and

$$v_y = -[\tfrac{1}{2}(m + 1)vu_1]^{1/2}x^{(m-1)/2}\{f(\xi^*) + [(m - 1)/(m + 1)]\xi^*f'(\xi^*)\} \tag{92}$$

where v_x and v_y are the velocity components, respectively, along directions x and y; $U(x)$ is the potential flow; and $f(\xi^*)$ is the streaming function and $f'(\xi^*)$ its first derivative, both given in terms of the dimensionless variable ξ^*, defined by

$$\xi^* = y[\tfrac{1}{2}(m + 1)(u_1/v)]^{1/2}x^{(m-1)/2} \tag{93}$$

and m is a number related to the angle β by the following expression:

$$m = \beta/(2 - \beta) \tag{94}$$

The mass transport differential equation for the two-dimensional flow, under stationary conditions and with diffusion occurring along the y axis, is

$$v_y \,\partial C_i/\partial y + v_x \,\partial C_i/\partial x = D_i \,\partial^2 C_i/\partial y^2 \tag{95}$$

Transformation of equation (95) into a total differential equation is performed by expressing it in terms of ξ^* and substituting v_x and v_y by expressions (91) and (92), respectively. Applying Mangler transformation equations,[7] the differential equation for axially symmetrical flow derived from (95) is

$$(d^2 C_i/d\xi^{*2}) + (av/D_i)f(\xi^*)\,dC_i/d\xi^* = 0 \qquad (96)$$

where a is a factor depending on m. Solving equation (96) for the condition of maximum flux ($C_i^s = 0$), the local mass transport rate corresponding to the distance x is

$$j_{max} = 1.12(a\alpha'/3)^{1/3}[U'(x')/vx']^{1/2}(v/D_i)^{1/3}D_i C_i^o \qquad (97)$$

where α' is derived from the functions plotted in Figure 8 and $U'(x')$ and x' are the transformed potential flow and coordinate, respectively. Then, the average maximum flux $\bar{j}_{max}$ over all the surface is

$$\bar{j}_{max} = 0.80 D_i C_i^o (\bar{U}'/vx')^{1/2}(v/D_i)^{1/3} \qquad (98)$$

which is independent, as a first approximation, of the cone angle. Equation (98) resembles the one employed earlier in hydrodynamic voltammetry with conical platinum microelectrodes[43] based on an

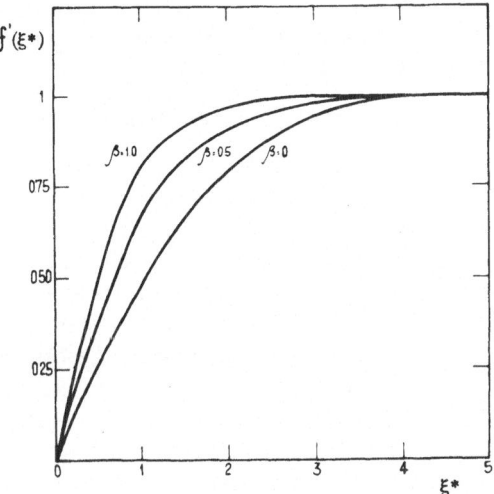

Figure 8. Velocity distribution at a wedge (from numerical tables of Ref. 49).

extension to mass transport of the equations deduced for the corresponding heat transfer problem. Taking as the characteristic length the generatrix of the conical electrode, X, the expression deduced for this case is

$$j_{max} = (4/3\sqrt{3})D_i C_i^o (U/vX)^{1/2}(v/D_i)^{1/3} \tag{99}$$

The numerical coefficient of this equation is 0.769, a value included within those calculated through the solution of the relevant differential equation. Hence, the equation obtained for a fixed disk in stagnation-point flow and the equation for conical electrodes in an axial laminar flow involve practically the same numerical coefficient after a proper definition of flow velocity.

3. **Application of Fixed Disk Electrodes in Stagnation-Point Flow and of Conical Electrodes to Convective Diffusion Controlled Processes**

For a component i participating in an electrode reaction, the mass transport rate in the steady state can be expressed in terms of current density, since

$$i = z_i F D_i (\partial C_i/\partial y)_{y=0} \tag{100}$$

Thus, the maximum mass transport rate on the fixed disk in stagnation-point flow is given by

$$i_L = 0.780 z_i F D_i (U/vr)^{1/2}(v/D_i)^{1/3} C_i^o \tag{101}$$

and likewise, for the conical electrodes,

$$i_L = 0.80 z_i F D_i (\overline{U}'/vx')^{1/2}(v/D_i)^{1/3} C_i^o \tag{102}$$

These equations are based upon the models where the counter-electrode is assumed to be symmetrically placed at an infinite distance. Thus, a steady axial flow pattern is obtained.

To test the theoretical rate equations, the cell designs must approach the ideal models. Diagrams of two cells are given in Figure 9. Cell (a) was used with different flow velocities, cone angles, electrode lengths, and electrodes involving different lengths of the inert zone. The entrance length, the concentration of the diffusing species, and the ratio v/D_i were also changed.[51] Three reactions were studied: the reduction of ferricyanide ion and the oxidation of

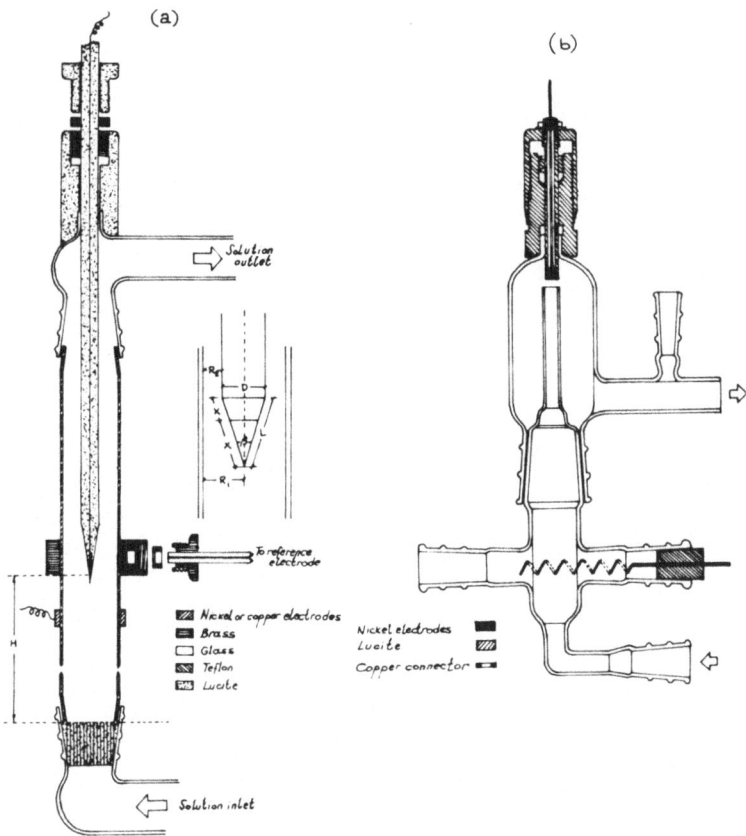

Figure 9. Electrolysis cells (a) with a conical electrode, and (b) with a fixed disk electrode.[51]

ferrocyanide ion, both on activated nickel electrodes, and the electrodeposition of copper ions on copper electrodes. These reactions have often been used to study mass transport processes in electrochemistry. The reactions were studied in the presence of an excess of supporting electrolyte, covering a wide range of potential. Up to a rather high flow velocity, these reactions are useful for studying mass transport processes.

To apply equation (102), the velocity must be properly defined. If the velocity profile is fully developed, the maximum fluid velocity

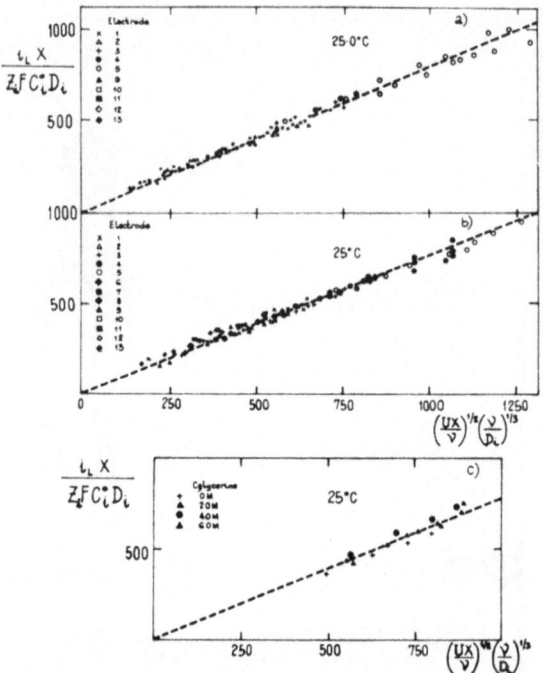

Figure 10. Results obtained with conical and fixed disk electrodes. Dotted line corresponds to the theoretical equation. Data taken from Refs. 46 and 51. Electrodes 1–13: nickel; electrode 14: copper. (a) Reduction of ferricyanide ion. (b) Oxidation of ferrocyanide ion. (c) Electrodeposition of copper.

Electrode	Angle, deg	X, cm	X_1, cm	S, cm^2
1	180	0.63	—	1.24
2	135	0.68	—	1.35
3	90	0.89	—	1.76
4	75	1.03	—	2.04
5	30	2.46	—	4.87
6	30	0.50	2.00	0.27
7	30	0.91	1.54	0.73
8	30	1.57	0.80	2.10
9	80	0.75	—	1.12
10	80	0.56	—	0.64
11	60	0.55	—	0.45
12	60	0.43	—	0.30
13	60	0.35	—	0.15
14	80	0.97	—	1.92

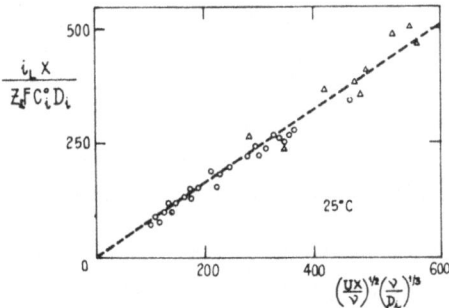

Figure 11. Results obtained with fixed disk electrodes. The dotted line corresponds to the theoretical equation. The data were obtained with electrodes at the end of the duct. Oxidation of ferrocyanide ion and reduction of ferricyanide ion on activated nickel electrodes. (From Ref. 46.)

must be considered. Its value, for a Poiseuille profile, is twice that of the average velocity, which can be obtained by measuring the volume of fluid flowing through the electrolysis cell.

Figure 10 shows data plotted according to equation (102) obtained with different electrodes, after systematically changing the variables. The three reactions studied prove the validity of the mass transport rate equations for conical electrodes. The largest deviation from the theoretical curve is less than 10 % in value.

The fixed disk electrode in stagnation-point flow is shown in Figure 9(b). The validity of equation (101) has been tested by using the electrode reactions previously mentioned. Figure 11 shows the experimental results as compared to the theoretical curve. Using static disk electrodes axially placed at the end of a duct, the average fluid velocity must now be employed in the rate equation, since as a first approximation, the velocity profile in the region where the liquid impinges on the electrode surface tends toward a constant value which approaches the velocity measured by a flowmeter placed in the hydrodynamic circuit. Similar reasoning is also valid for static conical electrodes similarly placed at the end of the conduit.

4. Tubular and Canal-Shaped Electrodes

The hydrodynamics of the laminar flow in ducts comprises two cases: (i) flow in the entrance length, and (ii) Poiseuille flow. Friction exerted over the wall by the fluid entering a duct gives rise to a boundary layer which increases starting from the entrance length section. When the boundary-layer thickness equals the duct radius or the canal half-width, a fully developed velocity profile or Poiseuille profile is attained. The system under study in this section consists of a working electrode embedded in the duct wall at the region of Poiseuille flow. The counterelectrode is assumed to be placed, as before, at infinity.

The maximum mass transport rate toward the wall of a cylindrical duct was first studied by Lévêque,[52] who did not define the zone where the deduced expression could be applied. Several studies of tubular and canal-shaped electrodes were later made, mainly aimed at establishing empirical or semiempirical relationships of eventual application in electrochemical technology.[53-62]

Levich[6] distinguished the three zones originating within a duct when a convective-diffusion process occurs on its wall. In the first zone, both the hydrodynamic and diffusion boundary layers begin to be formed and the diffusional flow to the duct surface is produced in the same way as in a flat plate,[6] since the diffusion boundary-layer thickness is smaller than the duct radius and, consequently, the surface curvature can be neglected. In the second zone, the Poiseuille profile is fully developed, but, since $v \gg D_i$, the thickness of the diffusion boundary-layer can still be considered to be smaller than the radius. Levich deduced the transport equation for this section. In the last zone, both profiles are fully developed, although this region may not be easily attainable.

In the zone of the Poiseuille profile, the velocity components can be written in the following form:

$$v_x \approx 2(U_0/R)y, \qquad v_y = 0 \qquad (103)$$

where the variable y is defined by

$$y = R - r \qquad (104)$$

R is the duct radius and r the radial coordinate. Following the reasoning adopted for the previous problems, the differential equation for convective diffusion is solved with the following boundary conditions:

$$y = 0: \qquad C_i = 0$$
$$y \to \infty: \qquad C_i \to C_i^\circ \qquad (105)$$

and, therefore, the local maximum flux is

$$j_{max} = D_i(\partial C_i/\partial y)_{y=0}$$

$$= (D_i C_i^\circ/x^{1/3})(U_0/D_i R)^{1/3} \int_0^\infty \exp\{-(2/9)\zeta^3\}\, d\zeta \tag{106}$$

where

$$\zeta = (U_0/D_i R)^{1/3} y/x^{1/3} \tag{107}$$

Then,

$$j_{max} = 0.67(D_i C_i^\circ/x^{1/3})(U_0/D_i R)^{1/3} \tag{108}$$

Although this expression does not explicitly contain viscosity, it is not independent of it, as its deduction is based on the condition $v \gg D_i$. Equation (108) was tested by Blaedel and coworkers.[63] From the previous expression, the thickness of the diffusional boundary layer is

$$\delta_d = (1/0.67)(D_i/v)^{1/3}(v/U_0 R)^{1/3}(R^2 x)^{1/3} \tag{109}$$

Current/potential curves for tubular electrodes were first derived by Blaedel and Klatt.[64] Gerischer et al.[65] also studied transport processes on flat electrodes embedded in a canal wall, attempting to apply this device to the kinetic study of electrochemical reactions. Recently, Matsuda[66] has also referred to canal-shaped electrodes, deducing a general mass transport rate equation for this type of electrode.

5. Electrodes Consisting of Solids of Revolution

Matsuda[67] has developed a generalized treatment of the stationary current/potential curves for electrodes formed by solids of revolution for various reaction mechanisms, the simplest corresponding to a redox electrochemical reaction with no preceding chemical step.

In order to find the concentration distribution in the vicinity of the working electrode, the convective-diffusion differential equation (72) expressed in curvilinear coordinates is solved. The x coordinate, as shown in Figure 12, corresponds to OP' measured along the meridian curve M, starting from the stagnation point of the solid of revolution. The y coordinate corresponds to the PP' measured along

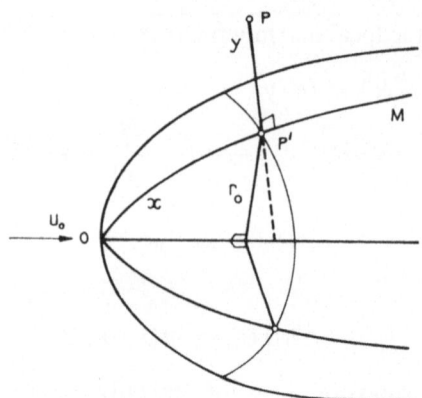

Figure 12. Generalized scheme of a body of
revolution under axially impinging flow (from
Ref. 67).

the normal to the surface of the solid of revolution. v_x and v_y are the
flow velocity components at point P.

A general redox reaction is expressed by the following equation:

$$\sum_i p_i S_i + z_i e = 0 \tag{110}$$

z_i being the number of electrons entering the total reaction and p_i the
stoichiometric factor of S_i. This factor is positive for the reducing
species and negative for the oxidizing species. If the reaction rate is
expressed in terms of the current density, the boundary conditions
are

$$
\begin{array}{lll}
x > 0, & y = 0 & i = (z_i/p_i)FD_i(\partial C_i/\partial y) \\
x = 0, & y > 0 \ \Big\} & \\
x > 0, & y \to \infty \Big\} & C_i = C_i^\circ
\end{array} \tag{111}
$$

To solve equation (72) in curvilinear coordinates, it is expressed in
terms of the function ψ, which is given by

$$\psi = [r_0 \tau(x)/2\mu]y^2 \tag{112}$$

where $\tau(x)$ is the shear stress and μ is the viscosity coefficient. The
velocity component v_x is

$$v_x = [2\tau(x)/r_0\mu]^{1/2}\psi^{1/2} \tag{113}$$

After introducing this equation into the convective-diffusion differential equation and considering the corresponding boundary conditions, the former converts, by means of a Laplace transformation, into a modified Bessel differential equation of 1/3rd order. By solving the latter and returning to the original coordinates, the following expression is derived for the concentration of S_i at the electrode surface:

$$C_i^s = C_i^o - [(\tfrac{1}{3})^{1/3}/\Gamma(\tfrac{2}{3})]D_i^{-2/3}\mu^{1/3}$$

$$\times \int_0^x \left\{ (p_i i/z_i F) r_0(x_1) \bigg/ \left[\int_{x_1}^x r_0(x_2)^{3/2} \tau(x_2)^{1/2} \, dx_2 \right]^{2/3} \right\} dx_1 \tag{114}$$

In the vicinity of the stagnation point, the following relationship are fulfilled[68]:

$$r_0(x) = x \tag{115}$$

$$\tau(x) = 1.312(\mu\rho u_1^3/x)^{1/2} \tag{116}$$

where ρ is the specific gravity of the fluid and u_1 is the velocity of the potential flow, which, near the stagnation point, is

$$u_1 = U_0(x/L) \tag{117}$$

U_0 is the axial flow velocity in a region infinitely far away from the solid of revolution, and L is the characteristic length, depending on the shape of the solid of revolution. Expressions for L for particular solids of revolution are given in Table 2.

Table 2
Expressions for L for Some Solids of Revolution[a]

Sphere	$L = \tfrac{2}{3}R_s$
Disk	$L = (\pi/2)R_d$
Ellipsoid of revolution (oblate)	$L = \dfrac{\varepsilon}{\varepsilon^2 - 1}\left\{ \dfrac{\varepsilon}{(\varepsilon^2 - 1)^{1/2}}\sin^{-1}\dfrac{(\varepsilon^2 - 1)^{1/2}}{\varepsilon} - 1 \right\}a; \quad (\varepsilon > 1)$
Ellipsoid of revolution (prolate)	$L = \dfrac{\varepsilon}{1 - \varepsilon^2}\left\{ 1 - \dfrac{\varepsilon^2}{(1 - \varepsilon^2)^{1/2}}\log\dfrac{1 + (1 - \varepsilon^2)^{1/2}}{\varepsilon} \right\}a; \quad (\varepsilon < 1)$

[a] From Matsuda.[67] R_s is the radius of the sphere; R_d the radius of the disk; a the equatorial radius of the ellipsoid of revolution; b the polar radius of the ellipsoid of revolution; and $\varepsilon = a/b$.

Therefore, considering equations (115)–(117) with (114), after calculating the numerical factors, we get

$$C_i^s = C_i^o - [3^{1/3}/(1.312)^{1/3}\Gamma(\tfrac{2}{3})](U_0/L)^{-1/2}v^{1/6}D_i^{-2/3}$$

$$\times \int_0^x [(p_i i/z_i F)x_1/(x^3 - x_1^3)^{2/3}]\,dx_1 \qquad (118)$$

In electrodes formed by solids of revolution, this equation is of interest when the diffusion boundary layer approaches a uniform thickness. To satisfy this condition, certain geometric relationships of the working electrode should be fulfilled. The latter are obtained from the expressions deduced in hydrodynamics for computing the process of boundary-layer formation.[7] So, for spherical electrodes,

$$r_0 = R_s \sin(x/R_s) \approx x\{1 - \tfrac{1}{6}(x/R_s)^2 + \cdots\} \qquad (119)$$

and

$$u_1 = \tfrac{3}{2}U_0 \sin(x/R_s) \approx U_0\{x/\tfrac{2}{3}R_s\}\{1 - \tfrac{1}{6}(x/R_s)^2 + \cdots\} \qquad (120)$$

where R_s is the sphere radius. Under the condition that $x < \tfrac{2}{3}R_s$, values of r_0 and u_1 fall within a limiting error of about 3% when $L = \tfrac{2}{3}R_s$. Therefore, the electrode length must be shorter than $\tfrac{2}{3}R_s$. Analogously, for a disk electrode, the corresponding equations are

$$r_0 = x \qquad (121)$$

and

$$u_1 = (2/\pi)U_0 x/(R_d^2 - x^2)^{1/2}$$

$$\approx U_0(x/\tfrac{1}{2}\pi R_d)\{1 + \tfrac{1}{2}(x/R_d)^2 + \cdots\} \qquad (122)$$

Table 3
Conditions for the Application of the Equations
Derived for Stagnation-Point-Flow Electrodes[a]

Sphere	$1 < \tfrac{2}{3}R_s$	
Disk	$1 < \tfrac{1}{4}R_d$	
Ellipsoids of revolution	$1 < \tfrac{2}{5}\varepsilon a$,	$1 < \tfrac{2}{5}\varepsilon \lvert 3\varepsilon^2 - 4\rvert^{-1/2}a$

[a] From Matsuda.[67]

where R_d is the disk radius. The condition $1 < R_d/4$ results from this analysis. Table 3 contains the conditions for some solid-of-revolution electrodes with stagnation-point flow.

IV. ELECTRODES UNDER FREE CONVECTION

The different theoretical and experimental treatments of transport phenomena in electrode reactions hitherto studied correspond to electrochemical systems kept at a constant temperature, which, from the hydrodynamic standpoint, corresponds to conditions of forced convection, that is, subject to the action of an external force on the electrolytic solution or on the electrode, resulting in an increase of the mass transport rate. The periodic motion of the electrode, as in the case of the dropping mercury electrode,[69,70] the volume of which is a periodic function of time, or of vibrating electrodes,[71,72] produces similar effects.

Another way of increasing the mass transport rate is by provoking a density gradient through the establishment of either a concentration or a temperature gradient. This leads to the creation of a convective motion due to the gravitational field, known as free convection.

In electrochemistry, the problem of isothermal free convection has been theoretically and experimentally studied for the case when the density difference is due to a concentration gradient produced by the electrode reaction.[73–79] The various proposed solutions, as well as the different experimental correlations developed for a flat, vertical plate electrode, lead formally to the same equation, with a numerical coefficient the value of which depends on the approximations made while deriving the equations. The general equation for the maximum flux when no temperature gradient exists between the electrodes takes the form

$$j_{max} = \text{const } D_i(v/D_i)^{1/4}(g\gamma/v^2)^{1/4}C_i^\circ x^{-1/4} \qquad (123)$$

where x is the electrode height, g is the gravitational acceleration, and γ is the densification coefficient, to be defined later.

Bearing in mind the cause of natural convection, if a temperature gradient is superimposed upon the concentration gradient, flow velocity and, consequently, the rate of mass transport should change with the temperature gradient.

A system under nonisothermal free convection is easily obtained with a fixed working electrode dipped in an unstirred solution, under a thermal gradient between the working electrode and a point within the bulk of the solution.

The effect of thermal convection in static electrochemical systems was qualitatively displayed by Ducret and Cornet[80] using two different electrodes heated by the Joule effect. One of these was an electrode with liquid circulation and the other an electrode with electrical heating. These electrodes were tested with two different electrochemical systems, and they showed that the transport rate increases with the amount of heat supplied during the process.

The results indicate that the thermal convection effect is reproducible and the current/potential curves of the tested systems can be compared to those recorded by other methods, thus suggesting possible applications to electrochemical analysis.

1. Theory of the Thermal Convective Electrode

An approach to the theory of the thermal convective electrode was developed by Marchiano and Arvia,[81] choosing as a model an ideal electrode consisting of a flat vertical plate. This model had been already studied for a simple case of isothermal free convection and of pure thermal convection.[73–79] The equations derived from the theory of the thermal convective electrode contain, as limiting cases, the equations for the electrode model under isothermal free convection.

The differential equations of heat transfer for the corresponding model were solved several decades ago[82,83] and, for the case of mass transport, the local flow equations, *mutatis mutandis*, are of the same form as equation (123).

The effects of density gradients caused by simultaneous gradients of temperature and concentration were considered with the purpose of obtaining information on the influence of heat transfer related to binary diffusion and heat transfer within the boundary layer in laminar free convection over a vertical plate.[84–87]

In order to obtain the mass transport rate equation under a combined effect of both temperature and concentration gradients, it is necessary to solve simultaneously the equations related to momentum, energy, and mass transport.

For the particular case of a vertical flat plate subject to buoyancy forces only, the momentum equation is

$$v_x \, \partial v_x/\partial x + v_y \, \partial v_x/\partial y = v \, \partial^2 v_x/\partial y^2 + g[(\rho_0 - \rho)/\rho_0] \qquad (124)$$

where x is the coordinate defined along the plate, and y is the coordinate normal to it; v_x and v_y are the velocity components in the x and y directions, respectively; v is the kinematic viscosity; g is the gravitational acceleration; ρ_0 is a reference density; and ρ is a density which depends on both concentration and temperature. The mass and energy transport equations, respectively, are

$$v_x \, \partial C_i/\partial x + v_y \, \partial C_i/\partial y = D_i \, \partial^2 C_i/\partial y^2 \qquad (125)$$

$$v_x \, \partial T/\partial x + v_y \, \partial T/\partial y = (k_T/\rho C_p) \, \partial^2 T/\partial y^2 \qquad (126)$$

where T is the temperature, k_T is the thermal conductivity coefficient, and C_p is the specific heat at constant pressure. The boundary conditions for solving the present problem are

$$
\begin{aligned}
y = 0: & \quad v_x = v_y = 0, \quad C_i = 0, \quad T = T_e \\
y \to \infty: & \quad v_x = v_y = 0, \quad C_i = C_i^\circ, \quad T = T_0
\end{aligned}
\qquad (127)
$$

Considering that density is a weak function of concentration and temperature, we have

$$\rho = \rho_0 + (\partial\rho/\partial C_i)(C_i - C_i^\circ) + (\partial\rho/\partial T)(T - T_0) \qquad (128)$$

The densification and the expansion coefficients γ and σ, respectively, are introduced as is usually done when the cases of isothermal mass-transport and heat-transfer free convection are considered independently. These coefficients are defined by

$$\gamma = (C_i^\circ/\rho_0) \, \partial\rho/\partial C_i \qquad (129)$$

and

$$\sigma = (\Delta T/\rho_0) \, \partial\rho/\partial T \qquad (130)$$

where $\Delta T = T_0 - T_e$, T_0 being a reference temperature.

To solve equations (124)–(126) simultaneously, with the boundary conditions (127), the set of partial differential equations must be transformed into a set of total differential equations.

The dimensionless independent variable μ and the concentration and temperature profiles φ and τ are defined respectively as

$$\mu = (g/4v^2)^{1/4} y/x^{1/4} \tag{131}$$

$$\varphi = (C_i^\circ - C_i)/C_i^\circ \tag{132}$$

$$\tau = (T_0 - T)/(T_0 - T_e) \tag{133}$$

and the velocity components are expressed in terms of μ through the streaming function ψ as follows:

$$v_x = \partial\psi/\partial y = 4v(g/4v^2)^{1/2} x^{1/2} f'(\mu) \tag{134}$$

$$v_y = -\partial\psi/\partial x = v(g/4v^2)^{1/4} [\mu f'(\mu) - 3f(\mu)]/x^{1/4} \tag{135}$$

The streaming function is defined by

$$\psi = 4v(g/4v^2)^{1/4} x^{3/4} f(\mu) \tag{136}$$

$f(\mu)$ is a function of the variable μ to be determined, which must satisfy the following equation:

$$f'''(\mu) + 3f(\mu)f''(\mu) - 2f'^2(\mu) + \gamma\varphi + \sigma\tau = 0 \tag{137}$$

Similarly, the total differential equations for mass and energy transport, in terms of φ and τ, respectively, are

$$\varphi'' + 3(v/D_i)f(\mu)\varphi' = 0 \tag{138}$$

$$\tau'' + (3v\rho C_p/k_T)f(\mu)\tau' = 0 \tag{139}$$

where the primes correspond to the order of the derivatives of the dimensionless functions $f(\mu)$, φ, and τ. The boundary conditions related to equations (137)–(139) are

$$\begin{aligned}\mu = 0: &\quad f(\mu) = f'(\mu) = 0, \quad \varphi = 1, \quad \tau = 1 \\ \mu \to \infty: &\quad f'(\mu) = 0, \qquad\qquad \varphi = 0, \quad \tau = 0\end{aligned} \tag{140}$$

In order to obtain the distribution functions related to non-isothermal, free convection, it is first convenient to consider the following limiting cases: (i) isothermal free convection under a concentration gradient; and (ii) pure thermal convection.

The first case (case i) corresponds to $\Delta T = 0$, and, consequently, the whole set of total differential equations is reduced to equations (137) and (138), which have already been analyzed by different authors.[6,14] To establish the concentration distribution, it is assumed that function $f(\zeta)$ can be expressed as a series expansion in terms of the variable ζ in the following way:

$$f(\zeta) = \tfrac{1}{2}A_\zeta\zeta^2 - \tfrac{1}{6}\zeta^3 + [\tfrac{1}{2}A_\zeta(v/D_i)]^{1/3}(\zeta^4/24 \times 0.89) \qquad (141)$$

where

$$\zeta = (g\gamma/4v^2)^{1/4}y/x^{1/4} \qquad (142)$$

Taking into account the boundary conditions, the constant A_ζ is

$$A_\zeta = 0.48(D_i/v)^{1/4} \qquad (143)$$

The dimensionless concentration distribution φ is therefore given by the following expression

$$\varphi = 1 - [(0.48/2)^{1/3}(v/D_i)^{1/4}/0.89]\zeta \qquad (144)$$

In a similar manner, for the second case (case ii), when there is no concentration gradient, the set of equations is that corresponding to the heat transfer problem. Following basically the assumptions already made for the previous case, which means that to a first approximation the convergence of the τ-integrals, for ratios $v\rho C_p/k_T$ of about 10 or higher, can be comparable to those values given in terms of φ, the result is

$$f(\lambda) = \tfrac{1}{2}A_\lambda\lambda^2 - \tfrac{1}{6}\lambda^3 + (A_\lambda v\rho C_p/2k_T)^{1/3}(\lambda^4/24 \times 0.89) \qquad (145)$$

where

$$\lambda = (g\sigma/4v^2)^{1/4}y/x^{1/4} \qquad (146)$$

and

$$A_\lambda = 0.48(k_T/v\rho C_p)^{1/4} \qquad (147)$$

Similarly, the temperature distribution is

$$\tau = 1 - [(0.48/2)^{1/3}(v\rho C_p/k_T)^{1/4}/0.89]\lambda \qquad (148)$$

The limiting cases having been discussed, the general equation containing the combined contribution of both temperature and concentration gradients can be inferred (case iii). Taking into account that the motion of the fluid is produced in the present case by superposition of the two effects, it is convenient to postulate a new streaming function represented by the algebraic sum of those obtained for both single cases. This function is given by

$$\psi = \psi_\lambda + \psi_\varphi \tag{149}$$

Then, according to (136), (141), and (145), the streaming function is

$$
\psi = 4v\left(\frac{g}{4v^2}\right)^{1/4} x^{3/4}\left\{\left[A_\zeta \gamma^{3/4} \pm A_\lambda \sigma^{3/4}\right]\frac{\mu^2}{2} - \frac{\gamma \pm \sigma}{6}\mu^3\right.
$$
$$
\left. + \left[\left(A_\zeta\frac{v}{D_i}\right)^{1/3}\gamma^{5/4} \pm \left(A_\lambda\frac{v\rho C_p}{k_T}\right)^{1/3}\sigma^{5/4}\right]\frac{\mu^4}{2^{1/3}\times 24 \times 0.89}\right\} \tag{150}
$$

Comparing (150) and (136), the term between brackets now corresponds to $f(\mu)$. Therefore, the distribution equation for the velocity component along direction x is immediately obtained from (134) as

$$
v_x = 4v\left(\frac{g}{4v^2}\right)^{1/2} x^{1/2}\left\{0.48\left[\frac{\gamma^{3/4}D_i^{1/4}}{v^{1/4}} \pm \frac{\sigma^{3/4}k_T^{1/4}}{(v\rho C_p)^{1/4}}\right]\mu - \frac{\gamma \pm \sigma}{2}\mu^2\right.
$$
$$
\left. + \frac{(0.24)^{1/3}}{6 \times 0.89}\left[\left(\frac{v}{D_i}\right)^{1/4}\gamma^{5/4} \pm \left(\frac{v\rho C_p}{k_T}\right)^{1/4}\sigma^{5/4}\right]\mu^3\right\} \tag{151}
$$

It is interesting now to evaluate for a unitary length ($X = 1$) and under conveniently chosen conditions of temperature and concentration the velocity profile v_x along the normal direction y for all three previous cases. For this purpose, the variables are given values close to those found in real systems. These velocity profiles are shown in Figure 13. Figure 13(a) corresponds to two different temperatures and two concentrations for isothermal free convection. The curves show the already known maximum, decreasing and shifting toward higher y values when the concentration of the diffusing species decreases. The profile is fully developed in the same y range as the concentration profile. Figure 13(b) corresponds to the velocity profile in the boundary layer for thermal free convection. Maximum velocities are higher than those corresponding to the previous case, and the profiles extend to higher values of y. Figure 13(c) shows velocity profiles for two different temperatures and two concentrations. For each concentration, two particular cases are considered. In the first, buoyancy forces arising from the temperature and the concentration gradients are added; in the second, these forces are subtracted. For the first case, when the factor γ equals σ, the maximum becomes an inflection point because

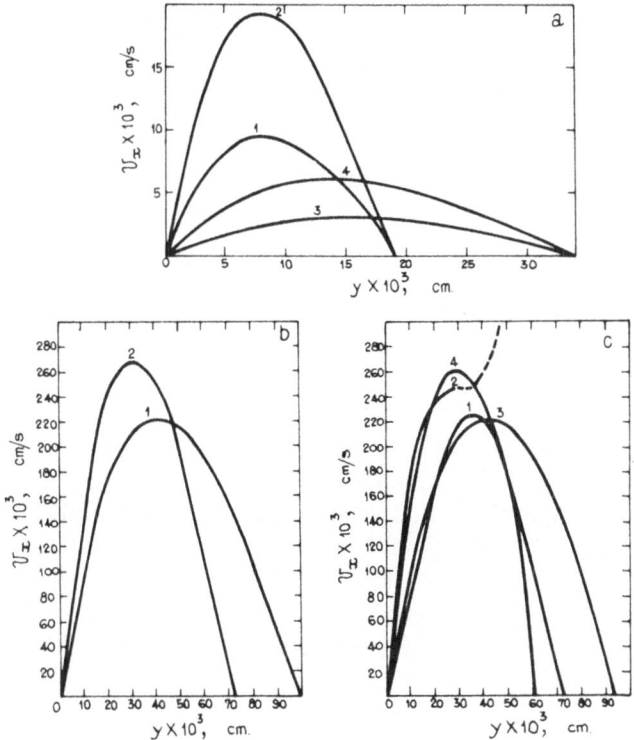

Figure 13. Velocity profiles deduced for the thermal convective electrode, according to Refs. 81 and 88, $X = 1$ cm. a: (1) 5°C, $C = 0.05\ M$; (2) 35°C, $C = 0.05\ M$; (3) 5°C, $C = 0.005\ M$; (4) 35°C, $C = 0.005\ M$. b: (1) $\Delta T = -15$°C, $T_0 = 12.5$°C; (2) $\Delta T = +15$°C, $T_0 = 27.5$°C. c: (1) $\Delta T = -15$°C, $T_0 = 12.5$°C, $C = 0.05\ M$; (2) $\Delta T = +15$°C, $T_0 = 27.5$°C, $C = 0.05\ M$; (3) $\Delta T = -15$°C, $T_0 = 12.5$°C, $C = 0.005\ M$; (4) $\Delta T = +15$°C, $T_0 = 27.5$°C, $C = 0.005\ M$.

of the influence of the function related to $f(\mu)$ for higher values of y, which are not taken into account in the simple case. In this particular situation, the boundary condition $f'(\mu) = 0$ when $\mu = \mu_0$ is not fulfilled, but the concentration distribution is not much affected, because, to evaluate it, only the first term in the series expansion of $f(\mu)$ is considered. Figure 13(c) reveals a more marked influence of the thermal gradient on the magnitude of the velocity component for the combined case.

Solving equations (138) and (139), bearing in mind the boundary conditions (140) and assuming for $f(\mu)$ the type of approximation stated for isothermal free convection and for heat transfer to be valid in thermal convection, the concentration distribution is

$$\varphi = 1 - \frac{(0.24)^{1/3}}{0.89}\left\{\left[\frac{\gamma^{3/4}D_i^{1/4}}{\nu^{1/4}} \pm \frac{\sigma^{3/4}k_T^{1/4}}{(\nu\rho C_p)^{1/4}}\right]^{1/3}\left(\frac{\nu}{D_i}\right)^{1/3}\right\}\mu \quad (152)$$

Returning to the initial variables,

$$C_i = 0.70C_i^\circ\left[\frac{\gamma^{3/4}D_i^{1/4}}{\nu^{1/4}} \pm \frac{\sigma^{3/4}k_T^{1/4}}{(\nu\rho C_p)^{1/4}}\right]^{1/3}\left(\frac{\nu}{D_i}\right)^{1/3}\left(\frac{g}{4\nu^2}\right)^{1/4}\frac{y}{x^{1/4}} \quad (153)$$

The concentration profile is represented in terms of φ and y in Figure 14 both for isothermal free convection and thermal free convection. It is quite clear that the gradient of the concentration function is higher for the case of the combined contribution. Therefore, the thickness of the diffusion boundary layer for this case, $(y)_{\varphi=0} = \delta_d$, at constant electrode height, decreases. Then, from equation (153), the maximum local laminar flux under the simultaneous effect of

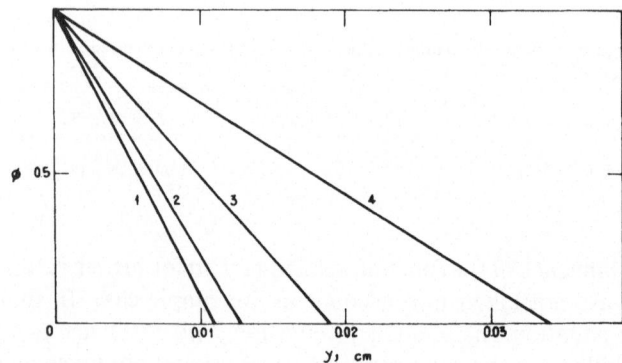

Figure 14. Concentration profiles deduced for a thermal convective electrode (from Refs. 81 and 88). (1) Case (iii), $C = 0.05\ M, \Delta T = +15°C$, $T_0 = 27.5°C$. (2) Case (iii), $C = 0.005\ M,\ \Delta T = +15°C,\ T_0 = 27.5°C$. (3) Case (i), $C = 0.05\ M, \Delta T = 0°C, T_0 = 5°C$. (4) Case (i), $C = 0.005\ M$, $\Delta T = 0°C,\ T_0 = 5°C$.

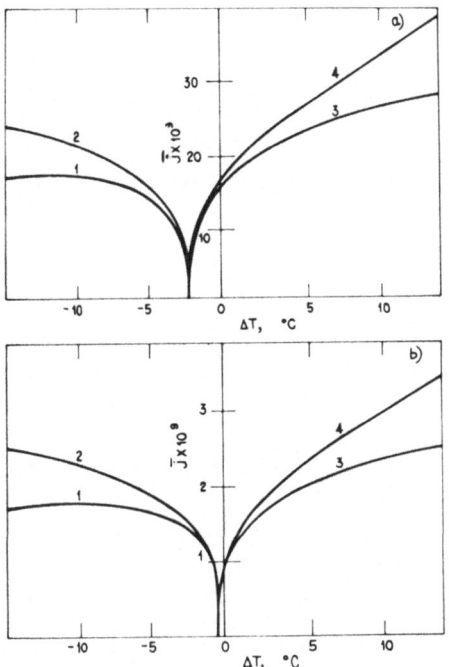

Figure 15. Dependence of the maximum flow
rate on the thermal gradient. (a) $C_i^\circ = 0.05\,M$,
(b) $C_i^\circ = 0.005\,M$.

concentration and temperature gradients at a vertical plate is

$$j_{max} = D_i\left(\frac{\partial C_i}{\partial y}\right)_{y=0} = 0.70 D_i C_i^\circ \left[\frac{\gamma^{3/4}D_i^{1/4}}{\nu^{1/4}} \pm \frac{\sigma^{3/4}k_T^{1/4}}{(\nu\rho C_p)^{1/4}}\right]^{1/3}$$

$$\times \left(\frac{\nu}{D_i}\right)^{1/3}\left(\frac{g}{4\nu^2}\right)^{1/4} x^{-1/4} \qquad (154)$$

and the average maximum flux is

$$\bar{j}_{max} = 0.90 D_i C_i^\circ \left[\frac{\gamma^{3/4}D_i^{1/4}}{\nu^{1/4}} \pm \frac{\sigma^{3/4}k_T^{1/4}}{(\nu\rho C_p)^{1/4}}\right]^{1/3}\left(\frac{\nu}{D_i}\right)^{1/3}\left(\frac{g}{4\nu^2}\right)^{1/4} X^{-1/4}$$

$$(155)$$

where X is the electrode height.

Figure 15 shows the plot of equation (155) for different combinations of variations in both the plate temperature and the average temperature, for positive and negative ΔT. The curves exhibit several interesting features. Thus, when $\Delta T = 0$, the maximum flux corresponds to a process under isothermal laminar free convection, whose rate can be calculated with the equation found by Levich.[6] As ΔT decreases, j_{max} continuously decreases, too, approaching zero for $\Delta T = (\Delta T)_{j=0}$. According to equation (155), such a situation occurs when

$$\gamma^{3/4} D_i^{1/4}/v^{1/4} = \sigma^{3/4} k_T^{1/4}/(v \rho C_p)^{1/4} \tag{156}$$

The fact that j_{max} becomes zero at $(\Delta T)_{j=0}$ means that the x velocity component due to the thermal gradient is exactly balanced by the component due to the concentration gradient. Consequently, the process approaches pure diffusion. The latter, as was already studied,[87] corresponds to a nonstationary process whose rate approaches zero when the time tends to infinity. At $\Delta T < (\Delta T)_{j=0}$, j_{max} begins again to grow, because, under these circumstances, there is once more a finite velocity component.

The value of $(\Delta T)_{j=0}$ depends on the concentration of the diffusing species, and when this decreases, the former approaches zero. This is immediately clear from Figure 15, since the curves correspond to electrolytic solutions whose concentrations vary by a factor of ten.

Another feature shown in the graphs is the asymmetry relative to the j axis in $(\Delta T)_{j=0}$. The temperature effect on viscosity and on the diffusion coefficient contributes to the occurrence of a slight maximum on one of the branches of the curve, at negative ΔT, when both the plate and reference temperature decreases.

2. Verification of the Theory

For an electrochemical system comprising a working electrode consisting of a vertical, flat plate, where the reaction occurs under nonisothermal, free-convection–diffusion control, the average maximum flux in terms of the limiting current density is

$$i_L = 0.90 z_i F D_i C_i^\circ \left[\frac{\gamma^{3/4} D_i^{1/4}}{v^{1/4}} \pm \frac{\sigma^{3/4} k_T^{1/4}}{(v \rho C_p)^{1/4}} \right]^{1/3} \left(\frac{v}{D_i} \right)^{1/3} \left(\frac{g}{4 v^2} \right)^{1/4} X^{-1/4} \tag{157}$$

The relation between i_L and ΔT, as obtained from equation (157), agrees with previous observations,[81] although this conclusion has only a qualitative character, so far as the hydrodynamic standpoint is concerned, because of the arbitrary shape of the electrodes used.

Equation (157) was checked in detail[86,88] employing an electrolysis cell which was designed on the basis of the equation implied in the theory, where the maximum anode-to-cathode distance was taken to be at least twice the distance of the thermal boundary layer.

When no temperature gradient exists between the electrodes, the diffusion boundary layer in free convection begins at the lower edge of the working electrode. On the other hand, when there is a temperature gradient and temperature is kept constant over all the surface contacting the electrolytic solution, two possibilities arise. In the first, the origin of the active surface coincides with the origin of the vertical wall; in this case, both the diffusion and thermal boundary layers start at the same point and consequently X is taken directly as the electrode height. In the second case, this coincidence does not occur, i.e., the boundary-layer origins are located at different points on the surface; thus, a relaxation process exists and it is not included in the deduction of equation (157).

The experimental results are shown in Figure 16, where they are compared with the predictions of equation (157). For this

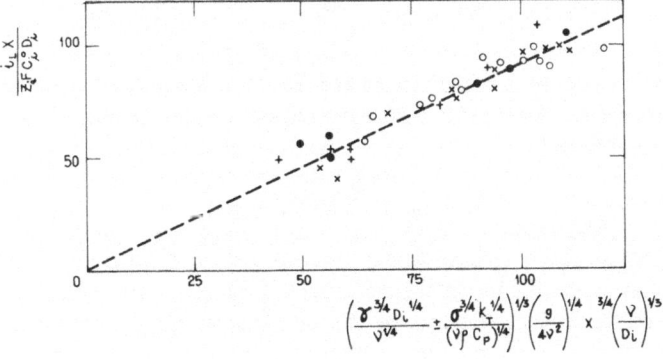

Figure 16. Comparison of experimental results with the equation deduced for the thermal convective electrode. Electrodeposition of copper on copper electrodes. (From Refs. 81, 88, and 90.)

purpose, the equation is written as follows:

$$\frac{i_L X}{z_i F C_i^0 D_i} = 0.9 \left[\frac{\gamma^{3/4} D_i^{1/4}}{\nu^{1/4}} \pm \frac{\sigma^{3/4} k_T^{1/4}}{(\nu \rho C_p)^{1/4}} \right]^{1/3} \left(\frac{g}{4\nu^2} \right)^{1/4} X^{3/4} \left(\frac{\nu}{D_i} \right)^{1/3} \quad (158)$$

Data obtained with positive and negative ΔT values and for in-creasing and decreasing average temperature $T_m = (T_0 + T_e)/2$ have been included. Near the point defined for $(\Delta T)_{j=0}$, the predictable instability can be observed; therefore, these points have not been plotted. The values of the physicochemical properties of the solutions were taken from the literature.[88-90] in calculating the theoretical curve. The theoretical deductions are acceptably matched by the experimental results.

3. Comparison between the Mass Transport Rate Due to Nonisothermal Free Convection and That Due to Forced Convection in Laminar Flow

From the standpoint of transport processes, the thermal gradi-ent gives rise to an increase in the entire rate process which is comparable to the effect of an externally driven stirring. If a laminar flow prevails in the system, the equivalence between the effect of the thermal gradient and the stirring efficiency can be estimated. For this purpose, a system of simple geometry such as the rotating disk electrode can be considered. Thus, by comparing equation (157) with equation (152) and solving for ΔT, we find

$$\Delta T = (\nu \rho C_p / k_T)^{1/3} (1/\sigma^*) [5.2 + 10^{-3} \omega^{3/2} X^{3/4} - (\gamma^{3/4} D_i^{1/4} / \nu^{1/4})]^{4/3} \quad (159)$$

If the second term in the square brackets in this equation is neglected, a quadratic relationship between temperature gradient and rotation speed is found

$$\Delta T \approx K' \omega^2 \quad (160)$$

where K' is a proportionality constant including distance X, expansion coefficient σ^* equal to $1/\rho(\partial \rho/\partial T)$, the ratio $\nu \rho C_p / k_T$, and the numerical coefficients of the rate equations. Therefore, under certain circumstances, the existence of the thermal gradient in the mass-transport-controlled process gives rise to a noticeable change in its rate, similar to what occurs when the system is stirred.

V. APPLICATIONS TO ELECTROCHEMICAL KINETICS

The application developed in Sections I–IV of hydrodynamic concepts to mass transport at electrodes of different geometries constitutes the basis of the so-called hydrodynamic voltammetry. From the kinetic standpoint, the most important relationship is the one between current density and electrode potential, which definitely depends on the type of mechanism prevailing in the electrode process.

Previous rate equations for limiting current density at different electrodes have been derived exclusively for mass-transport-controlled electrode processes. These equations are independent of the electrode potential, as the surface concentration of the reacting species on the electrode surface is always supposed zero. This assumption is fulfilled when any activation overpotential is negligible, that is, when the velocity of the interfacial reaction at equilibrium, represented by the exchange current density, is much larger than the diffusion limiting current density, $i_0 \gg i_L$, and the potential applied to the electrode is high enough to maintain an equilibrium concentration approaching zero.

In electrochemical kinetics, there are also electrode processes under mixed or intermediate control and under pure activation control. In the former case, the concentration of the reacting species at the electrode surface has a finite value which depends both on the transport rate of the reacting species to the surface and on the rate of the electrochemical reaction, while in the latter case ($i_0 \ll i_L$), this concentration is practically the concentration of the reacting species in the bulk of the solution.

To study electrode processes under mixed control, hydrodynamic voltammetry is adequate because it permits the separate interpretation of the mass transport and activation contributions. Obviously, if the electrode reaction is under a net activation control, the hydrodynamics of the electrochemical system can, in principle, be ignored.

1. General Expressions for Current–Potential Curves

For a reversible electrochemical reaction occurring in the presence of an inert supporting electrolyte, the rate equation in terms of current density, for any electrode potential, can be written as follows:

$$i = i_L[1 - (C_i^s/C_i^\circ)] \tag{161}$$

To obtain the current–potential curve, C_i^s should be substituted by the expression derived from Nernst's equation for a simple concentration galvanic cell involving C_i^s and C_i°, corresponding to a concentration overpotential $E - E_{rev}$,

$$C_i^s = C_i^\circ \exp[\mp z_i F(E - E_{rev})/RT] \tag{162}$$

where E is the electrode potential and E_{rev} is the reversible electrode potential, both measured against the same reference electrode. The positive sign is valid for the cathodic process ($E < E_{rev}$) while the negative sign is for the anodic process ($E > E_{rev}$). Then, when no ohmic overpotential is considered, the steady current–potential relation is expressed by

$$i = i_L\{1 - \exp[\mp z_i F(E - E_{rev})/RT]\} \tag{163}$$

For a particular electrode, the limiting current density i_L is substituted by the corresponding expression derived earlier for the maximum average flux on the basis of convective-diffusion concepts.

Equations for electrochemical systems consisting of binary electrolytes have been dealt with by Levich[6] and Newman.[8] The rate equations and the expression for the current–potential curve involve in this case the effective diffusion coefficient, defined at the beginning of this chapter, and a ratio of the number of charges of the two ionic species involved.

For an electrochemical process under intermediate control, the rate of the interfacial reaction has to be considered in the derivation of the equation for the stationary current–potential curve. As an example, let us take a simple first-order redox reaction involving the electron transfer between the soluble species R and O, with the assumption that no preceding homogeneous chemical reaction occurs. The electrode process is represented by

$$O + z_i e = R \tag{164}$$

The reaction rate equation in terms of current density is

$$i = z_i F(k_f C_O^s - k_b C_R^s) \tag{165}$$

where k_f and k_b are the formal rate constants for the forward and backward electrode reactions, respectively; C_O^s and C_R^s represent the concentrations of O and R, respectively, at the electrode surface ($y = 0$). The rate constants depend on electrode potential as follows:

$$k_f = k_0 \exp[-\alpha F(E - E_{rev})/RT] \tag{166}$$

and

$$k_b = k_0 \exp[(1 - \alpha)F(E - E_{rev})/RT)] \tag{167}$$

k_0 is the formal rate constant at equilibrium ($E = E_{rev}$) and α is the transfer coefficient assisting the reaction in the cathodic direction.

Substituting the expressions for k_f and k_b in the rate equation (165), a general expression for the rate of the electrode process involving a potential–current-density relationship is obtained

$$i = z_i F k_0 \left\{ C_O^s \exp\left[-\frac{\alpha F(E - E_{rev})}{RT} \right] \right.$$
$$\left. - C_R^s \exp\left[\frac{(1 - \alpha)F(E - E_{rev})}{RT} \right] \right\} \tag{168}$$

When applying this equation to an electrochemical system of particular geometry, the corresponding concentration distribution functions obtained before at $y = 0$ should be employed for C_O^s and C_R^s. Equation (168) becomes the current density/potential relationship of an activation-controlled electrochemical process represented by reaction (164) when $C_O^s = C_O^\circ$ and $C_R^s = C_R^\circ$.

2. Applications for Intermediate Kinetics

The application of the rotating disk electrode to electrode processes under intermediate kinetics for calculating rate constants from potential–current-density curves was developed by Frumkin and Tedoradse,[93] Koutecky and Levich,[94] Randles,[95] Jordan,[96] and Jahn and Vielstich.[97]

Frumkin and Tedoradse[93] studied the kinetics of the chlorine electrode in aqueous solutions at the rotating disk electrode under intermediate control. For a redox reaction such as that one represented by equation (164), the mass transport through the diffusion layer for the two reacting species is given by

$$i = z_i F D_O(C_O^\circ - C_O^s)/\delta_{d,0} \tag{169}$$

and

$$i = z_i F D_R (C_R^o - C_R^s)/\delta_{d,R} \qquad (170)$$

$\delta_{d,O}$ and $\delta_{d,R}$ being the diffusion boundary-layer thicknesses for the species O and R, respectively.

In the case of the rotating disk electrode, equation (41) can be written in the form

$$\delta_d = A_i/\sqrt{\omega} \qquad (171)$$

where $A_i = 1.61 D_i^{1/3} \nu^{1/6}$. From equations (169) and (170), the concentrations at the electrode surface are

$$C_O^s = C_O^o - (i A_O/z_i F D_O \sqrt{\omega}) \qquad (172)$$

and

$$C_R^s = C_R^o - (i A_R/z_i F D_R \sqrt{\omega}) \qquad (173)$$

Introducing these equations in the rate equation (165), after rearranging terms we find

$$\frac{1}{i} = \frac{1}{z_i F(k_f C_O^o - k_b C_R^o)}\left[1 + \left(\frac{k_f A_O}{D_O} + \frac{k_b A_R}{D_R}\right)\frac{1}{\sqrt{\omega}}\right] \qquad (174)$$

This equation can be applied to the evaluation of the exchange current density i_0 and the transfer coefficient α. Plotting $1/i$ against $1/\sqrt{\omega}$ for constant values of the overpotential, two equations for k_f and k_b are deduced from the slope and the ordinate at $1/\sqrt{\omega} = 0$. From the dependence of the rate constants on electrode overpotential, one can calculate both i_0 and α, in the usual way, according to equations (166) and (167). Values of the standard exchange current densities i_0/C_i^o up to nearly $10\ \mathrm{A\ liter\ cm^{-2}\ mole^{-1}}$ might be measured. Evaluation of higher standard exchange current densities is no longer possible if the ratio of the slope to the ordinate intercept becomes too high. By lowering the bulk concentration C_i^o, both the slope and the intercept for $1/\sqrt{\omega} = 0$ increase. Hence, the concentration effect on the accuracy of the evaluated parameters is small. An overpotential change produces a similar effect. At overpotentials lower than $RT/z_i F$, an overpotential increase yields a decrease of both the slope and the intercept, while for overpotentials much larger than $RT/z_i F$, the slope reaches a constant value, as should be expected for a purely diffusion-controlled polarization.

With the method developed by Frumkin and Tedoradse,[93] it is also possible to evaluate the order p_i of an irreversible electrode reaction. When the concentration at the disk surface is not zero, the current density equation is, according to Levich's theory,[6]

$$i = 0.62 z_i F D_i^{2/3} v^{-1/6} (C_i^o - C_i^s) \omega^{1/2} = B_i (C_i^o - C_i^s) \omega^{1/2} \quad (175)$$

where $B_i = 0.62 z_i F D_i^{2/3} v^{-1/6}$ and C_i^s is the surface concentration assuming that no influence of adsorption exists. Under steady-state conditions, the mass transport rate toward the interface is equal to the rate of the interfacial reaction

$$D_i (\partial C_i / \partial y)_{y=0} = k_f (C_i^s)^{p_i} = i/z_i F \quad (176)$$

When increasing the number of revolutions at constant potential, C_i^s approaches the value of C_i^o; then, the current density reaches a kinetic limiting value $(i_k)_L$, which is independent of the rotation speed and is given by

$$(i_k)_L = z_i F k_f (C_i^o)^{p_i} \quad (177)$$

Let ω_0 be the value for ω at which the limiting diffusion current becomes equal to the kinetic limiting current

$$(i_k)_L = B_i (C_i^o) \omega_0^{1/2} \quad (178)$$

The current density i_k of the process under mixed control is

$$i_k = B_i (C_i^o - C_i^s) \omega_0^{1/2} = z_i F k_f (C_i^s)^{p_i} \quad (179)$$

If equation (179) is divided by equation (178), taking into account equation (177), the result is

$$i_k = (i_k)_L [1 - (C_i^s/C_i^o)] = (i_k)_L \{1 - [i_k/(i_k)_L]^{1/p_i}\} \quad (180)$$

and the expression for the reaction order, then, is

$$p_i = \frac{\log(i_k)_L - \log(i_k)}{\log(i_k)_L - \log[(i_k)_L - i_k]} \quad (181)$$

In order to apply equation (181), the values of $(i_k)_L$ and i_k are required.

For a first-order process ($p_i = 1$), from equations (175) and (176) the following expression results:

$$1/i = [1/(i_k)_L] + (1/B_i C_i^o \omega^{1/2}) \quad (182)$$

To obtain $(i_k)_L$, experimental data are plotted by taking $1/i$ versus $1/\omega^{1/2}$ and extrapolating the function at $1/\omega^{1/2} = 0$. For this case,

$(i_k)_L = 2i_k$. The B_i value can be obtained from the slope at the origin of the plot of i versus $\omega^{1/2}$, or else by starting from the diffusion coefficient and the physicochemical properties of the electrolytic solutions employed.

For $p_i = 1/2$, equations (175) and (176) give

$$i^{1/2} = (i_k)_L^{1/2} - [i(i_k)_L^{1/2}/B_i C_i^\circ \omega^{1/2}] \tag{183}$$

and, for $p_i = 2$, we obtain

$$i^2 = (i_k)_L^2 - [i(i_k)_L^2/B_i C_i^\circ \omega^{1/2}] \tag{184}$$

As an extension of the basic theory of the rotating disk electrode, Koutecky and Levich[94,98] have attempted applications to the kinetic study of electrode reactions preceded by homogeneous chemical reactions and to catalytic reactions. The concept of the reaction boundary layer, as it is referred to in polarography and chronopotentiometry,[16] has also been extended to the rotating disk electrode technique.[99] These types of reactions are important in electrode processes involving organic compounds.[100–103]

For a list of studies in which the rotating disk electrode has been applied, the reader is referred to Riddiford's publication.[22] More recently, this electrode has also been applied to solve the kinetics of electrode processes in nonaqueous solvents, such as the iodine/iodide/triiodide,[104–105] bromide/tribromide,[106] and hydrogen electrodes in dimethylsulfoxide,[107,108] and iodine/iodide/triiodide and bromide/tribromide electrodes in acetonitrile.[109,110] The oxidation of nitrite ion in molten sodium nitrate–potassium nitrate eutectic[29] and the oxidation of sulfite ion in molten thiocyanate[111,112] have also been investigated with this technique. An application of the rotating disk electrode to study the catalytic wave produced by the reduction of iodine in the presence of iodate in acid perchlorate media has also been published.[113]

Similar treatments were developed by Bopp and Mason[39] for the translating electrode in stagnation-point flow, with an electrochemical reaction with intermediate kinetics the reaction order of which for a single-step mechanism is p_i. In this case, from equations (76) and (176), we get

$$k_f \left[C_i^\circ - \frac{i}{1 \cdot 3(z_i F D_i/d)\left(\dfrac{v}{D_i}\right)^{1/3}\left(\dfrac{R_s d}{v}\right)^{1/2}\omega^{1/2}} \right]^{p_i} = \frac{i}{z_i F} \tag{185}$$

This equation allows us to evaluate k_f and p_i, i being known at different ω values, following the methods already given. As deduced from equation (185), the plot of i versus $\omega^{1/2}$ will show a tendency to deviate to a larger or smaller degree from the straight line expected for convective-diffusion kinetic control, depending on the magnitude of k_f. For large values of k_f, the kinetics of the reaction will be controlled by the transport process. The other extreme case occurs when the rate of the heterogeneous reaction is very slow; here, the concentration of species i on the electrode surface will approach its concentration in the bulk of the solution. Then, for a first-order reaction, there is a kinetic limiting current density given by

$$(i_k)_L = z_i F k_f C_i^\circ \qquad (186)$$

Therefore, the rate of the electrode process is independent of the rotation speed of the working electrode.

Bopp, Stonehart, and Mason[114] obtained potential–current-density relationships for the ionization of hydrogen, in stagnation-point flow, using translating electrodes made of single-crystal and of polycrystalline bright platinum and platinized polycrystalline platinum, at rotational speeds up to 256 rpm, and studied the kinetics and mechanism of that reaction.

The diffusion boundary layer is also constant along the surface of a fixed disk electrode under stagnation-point flow, as already described in Section III. Consequently, this electrode is also suitable for study of mixed-control electrode processes. It was applied to determine the kinetic parameters of the ferro/ferricyanide redox system in $1M$ potassium hydroxide solutions using platinum and graphite electrodes.[46] Results obtained with this technique agree with those reported earlier.[115–117]

Conical electrodes have been also employed in hydrodynamic voltammetry to study various electrode processes such as the iodide/triiodide and ferro/ferricyanide redox couples in aqueous solutions.[44,118]

3. Stationary Current–Potential Relations Derived from the Generalized Treatment of Hydrodynamic Voltammetry

Matsuda[66,67] also developed a generalized theory of hydrodynamic voltammetry, deriving the equations for stationary current–potential curves based on an electrode reaction represented by (164).

By substituting in equation (165) C_O^s and C_R^s by the corresponding expressions obtained from equation (118), the relation for the current–potential characteristic is obtained after solving the resulting second-class Volterra integral equation. The final expression for the total current i is[66,67]

$$i = \frac{z_i F(k_f C_O^\circ - k_b C_R^\circ)}{1 + [\Gamma(\tfrac{1}{3})/(1.312)^{1/3} 3^{2/3}](U_0/L)^{-1/2} v^{1/6}[(k_f/D_O^{2/3}) + (k_b/D_R^{2/3})]} \tag{187}$$

If the electrode potential is far from the equilibrium value, either in the cathodic or anodic direction, the expressions for the cathodic and anodic limiting current densities are, respectively,

$$(i_L)_c = 0.850 z_i F C_O^\circ D_O^{2/3} v^{-1/6} U_0^{1/2} L^{-1/2} \tag{188}$$

and

$$(i_L)_a = -0.850 z_i F C_R^\circ D_R^{2/3} v^{-1/6} U_0^{1/2} L^{-1/2} \tag{189}$$

From equations (187)–(189) and (166) and (167), the total current I is found to be

$$I = \left[\frac{(I_L)_c}{1 + e^f} + \frac{(I_L)_a}{1 + e^f} \right]$$
$$\times \frac{(k_0/D^{2/3})(U_0/L)^{-1/2} v^{1/6}(e^{-\alpha f/z_i} + e^{(1-\alpha)f/z_i})}{0.850 + (k_0/D^{2/3})(U_0/L)^{-1/2} v^{1/6}(e^{-\alpha f/z_i} + e^{(1-\alpha)f/z_i})} \tag{190}$$

where

$$f = (z_i F/RT)(E - E_{1/2}^{rev}) \tag{191}$$

$$D = D_O^{1-\alpha} D_R^\alpha \tag{192}$$

and

$$E_{1/2}^{rev} = E^\circ - (RT/z_i F) \ln(D_O/D_R)^{2/3} \tag{193}$$

For a fast electrochemical reaction, the stationary current–potential relation obtained from (190) has the simple form

$$E = E_{1/2}^\circ + (RT/z_i F) \ln\{[(I_L)_c - I]/[I - (I_L)_a]\} \tag{194}$$

For an irreversible reaction, the cathodic relation is represented by the following expression:

$$E = (E_{1/2}^{irr})_c - (RT/\alpha F) \ln[I/(I_L)_c - I] \tag{195}$$

where

$$(E^{irr}_{1/2})_c = (RT/\alpha F)\{\ln[k_0 \exp(\alpha F E_{rev}/RT)] - \tfrac{2}{3} \ln D_O$$
$$+ \tfrac{1}{6} \ln v - \tfrac{1}{2} \ln(U_0/L) + 0.163\} \tag{196}$$

and, similarly, the anodic relation is given by

$$E = (E^{irr}_{1/2})_a - [RT/(1 - \alpha)F] \ln\{[I - (I_L)_a]/(-I)\} \tag{197}$$

with

$$(E^{irr}_{1/2})_a = [RT/(1 - \alpha)F]\{-\ln\{k_0 \exp[-(1 - \alpha)FE_{rev}/RT]\}$$
$$+ \tfrac{2}{3} \ln D_R - \tfrac{1}{6} \ln v + \tfrac{1}{2} \ln(U_0/L) - 0.163\} \tag{198}$$

If a higher-order electrochemical reaction involving soluble species is considered, the rate equation can be conveniently expressed by

$$i = z_i F \left\{ k_f \prod_{p_i > 0} (C^s_O)^{p_i} - k_b \prod_{p_i < 0} (C^s_R)^{-p_i} \right\} \tag{199}$$

Following the mathematical procedure already described and assuming that the current density is independent of the x coordinate, the current density is given by

$$\frac{i}{z_i F} = k_f \left\{ \prod_{p_i > 0} (C^o_i)^{p_i} \right\} \left\{ \prod_{p_i > 0} \left[1 - \frac{i}{(i_L)_i} \right]^{p_i} \right\}$$
$$- k_b \left\{ \prod_{p_i < 0} (C^o_i)^{-p_i} \right\} \left\{ \prod_{p_i < 0} \left[1 - \frac{i}{(i_L)_i} \right]^{-p_i} \right\} \tag{200}$$

where the convective-diffusion limited current is

$$(i_L)_i = 0.850(z_i F/p_i)C^o_i D^{2/3}_i v^{-1/6} U^{1/2} L^{-1/2} \tag{201}$$

For a reversible electrochemical reaction, the current/potential relation is

$$E = E_{rev} - (RT/z_i F)(\tfrac{2}{3} \sum p_i \ln D_i) - (RT/z_i F)(\sum p_i)\{\ln z_i F A$$
$$- \tfrac{1}{6} \ln v + \tfrac{1}{2} \ln(U_0/L) - 0.163\}$$
$$+ (RT/z_i F) \sum \{p_i \ln[p_i(I_{L,i} - I)]\} \tag{202}$$

where A is the electrode area.

As mentioned earlier, for an irreversible electrochemical reaction, the anodic curve is defined by

$$E = [RT/(1 - \alpha)F]\bigg(-\ln\{z_iFAk_0\exp[(1 - \alpha)FE_{rev}/RT]\}$$

$$-\tfrac{2}{3}\sum_{p_i<0}(p_i\ln D_i) - \bigg(\sum_{p_i<0}p_i\bigg)[\ln z_iFA - \tfrac{1}{6}\ln v + \tfrac{1}{2}\ln(U_0/L)$$

$$-0.163]\bigg) + [RT/(1 - \alpha)F]\ln\bigg[(-I)/\prod_{p_i<0}\{p_i(I_{L,i} - I)\}^{-p_i}\bigg]$$

$$(203)$$

and the cathodic curve is given by

$$E = (RT/\alpha F)\{\ln[z_iFAk_0\exp(\alpha FE_{rev}/RT)] - \tfrac{2}{3}\sum_{p_i>0}(p_i\ln D_i)$$

$$-\bigg(\sum_{p_i>0}p_i\bigg)[\ln z_iFA - \tfrac{1}{6}\ln v + \tfrac{1}{2}\ln(U_0/L) - 0.163]\}$$

$$-(RT/\alpha F)\ln\bigg[I/\prod_{p_i>0}\{p_i(I_{L,i} - I)\}^{p_i}\bigg]$$

$$(204)$$

When a preceding chemical reaction which supplies the electrochemically active substance exists, the differential equations and boundary conditions which would lead to the concentration distribution are modified according to the order of the foregoing reaction. These cases have been studied in detail by Matsuda[67] using for the mathematical development a substitution procedure proposed by Koutecky and Brdicka[119,120] for preceding homogeneous chemical reactions of first and higher orders. Matsuda derived explicit relationships for stationary current/potential curve and the half-wave potentials for reversible and irreversible electrochemical processes.

4. Collection Efficiency at Neighboring Electrodes

Experimental setups comprising two neighboring electrodes used simultaneously, one of them the generating electrode and the second the detecting electrode, are important for studying reaction intermediates produced in an electrochemical reaction. It is of course interesting to get expressions for the collection efficiency of neighboring electrodes, such as, for example, the rotating ring–disk electrode.

The collection efficiency of the system of neighboring electrodes N is defined as the ratio of material produced electrochemically on the first electrode to the amount that reaches the second electrode

$$N = \frac{\text{current in the detecting electrode}}{\text{current in the generating electrode}} \tag{205}$$

Ivanov and Levich[34] have deduced an approximate equation for calculating the collection efficiency for the rotating ring–disk electrode, assuming that the reaction occurring at the generating electrode is

$$A + z_A e = B^* \tag{206}$$

where the intermediate B^* has a half-life long enough for it to be carried by the liquid flow to the detecting electrode surface, reacting according to

$$B^* = C + z_B e \tag{207}$$

The time required for B^* to reach the detecting electrode depends on the rotation speed of the ring–disk electrode. To solve the mass transport differential equation for B^*, Ivanov and Levich supposed that either the intermediate is electrically neutral or the solution contains a large excess of an inert electrolyte. Furthermore, it is assumed that the rate of formation of B^* is equal to the rate of diffusion of A toward the disk surface, so that its rate of formation is given by Levich equation for the rotating disk electrode, (40).

If r_1, r_2, and r_3 define the regions of the rotating ring disk electrode (Figure 4), the total flux of B^*, j_B, at the ring electrode is [equation (65)]

$$j_B = \frac{0.8[1 - \tfrac{3}{4}(r_1/r_2)^3]j_A}{1 + (k_B\delta_B/D_B)} \int_1^{r_3/r_2} \frac{y^2\,dy}{(y^3 - 1)^{1/3}[y^3 - \tfrac{3}{4}(r_1/r_2)^3]} \tag{208}$$

where j_A is the maximum convective-diffusion current of A at the rotating disk electrode; k_B is the heterogeneous rate constant for the disappearance of B^* at the disk surface; δ_B is the diffusion boundary layer of B^* given by Levich's equation for the rotating disk electrode; D_B is the diffusion coefficient of B^*; and the variable

y is the ratio r/r_2. Equation (208) is valid when $(r_2 - r_1) \ll r_1$ and $(r_3 - r_2) \ll r_2$.

Equation (208) can be written in terms of current density as follows:

$$i_B = N \frac{z_B}{z_A} \frac{(i_L)_A}{1 + (k_B \delta_B/D_B)}$$

(209)

where N, the collection efficiency, is a factor involving the r-dependent terms

$$N = 0.8 \left[1 - \frac{3}{4} \left(\frac{r_1}{r_2} \right)^3 \right] \int_1^{r_3/r_2} \frac{y^2 \, dy}{(y^3 - 1)^{1/3} [y^3 - \frac{3}{4}(r_1/r_2)^3]}$$

(210)

The reactant A at the disk generates B^*, which in part diffuses away from the disk and in part undergoes a heterogeneous reaction characterized by the rate constant k_B. Then,

$$k_B C_B^s + D_B (C_B^s/\delta_B) = i_A/z_A F$$

(211)

From equations (209) and (210), the expression for N is

$$N = i_B/[z_B F D_B (C_B/\delta_B)]$$

(212)

Values of N calculated with equation (210) are about 8% higher than the experimental values reported by Frumkin et al.[35] after investigating the oxygen reduction on an amalgamated gold disk-ring electrode in 0.1 M sodium hydroxide solutions.

Albery and Bruckenstein[121-124] have also derived a collection efficiency equation for the rotating ring–disk electrode by applying the more rigorous mathematical treatment developed by Albery.[36] The theoretical expression for N is

$$N = 1 - F \frac{(r_2/r_1)^3 - 1}{(r_3/r_1)^3 - (r_2/r_1)^3} + \left[\left(\frac{r_3}{r_1} \right)^3 - \left(\frac{r_2}{r_1} \right)^3 \right]^{2/3}$$

$$\times \left\{ 1 - F \left[\left(\frac{r_2}{r_1} \right)^3 - 1 \right] \right\}$$

$$- \left(\frac{r_3}{r_1} \right)^2 \left\{ 1 - F \left[\frac{(r_2/r_1)^3 - 1}{(r_3/r_1)^3 - (r_2/r_1)^3} \left(\frac{r_3}{r_1} \right)^3 \right] \right\}$$

(213)

where the function $F(\theta)$ is given by

$$F(\theta) = \frac{3^{1/2}}{2\pi} \int_0^\theta \frac{d\lambda}{\lambda^{2/3}(1 + \lambda)}$$

$$= \frac{3^{1/2}}{4\pi} \ln \frac{(1 + \theta^{1/3})^3}{1 + \theta} + \frac{3}{2\pi} \arctan\left(\frac{2\theta^{1/3} - 1}{3^{1/2}}\right) + \frac{1}{4} \qquad (214)$$

Values of N calculated with equation (213) were tabulated by Albery and Bruckenstein[121] for different r_3/r_2 and r_2/r_1 ratios. A comparison of the theoretical values with the experimental ones gives a better agreement than the earlier approximate treatment. Another interesting conclusion is that, for large gaps $(r_2 \gg r_1)$ and/or for large ring widths $(r_3/r_1) \gg (r_2/r_1)$, the percentage difference between values of N predicted by equations (210) and (213) is small and approaches the experimental error.

The collection efficiency was also considered for static electrodes exposed to flowing solutions, as, for instance, in the work of Gerischer *et al.*[65] on the canal-shaped electrode, and a general equation for the collection efficiency in any geometric system was obtained by Matsuda[125] considering two neighboring electrodes embedded in the surface of a solid of revolution with axial flow, as shown in Figure 17.

The resolution of the mathematical problem to obtain the surface concentration is analogous to that already dealt with in

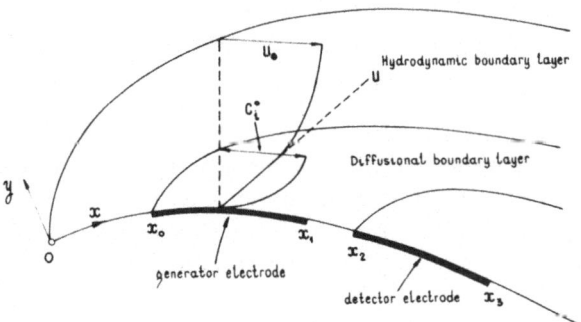

Figure 17. Model of two neighboring electrodes used to calculate the collection efficiency (from Ref. 125).

Section III. The surface concentration for a two-dimensional flow is expressed in a general form as follows:

$$C_i^s = C_i^\circ - \frac{(\frac{1}{3})^{1/3}}{\Gamma(\frac{2}{3})} D_i^{-2/3} \mu^{1/3} \int_{x_0}^x \frac{f_i(x')\,dx'}{\left\{ \int_{x'}^x [\tau(x'')]^{1/2}\,dx'' \right\}^{2/3}} \tag{215}$$

where the flux is given in terms of the function $f_i(x')$.

The current I_1 flowing over the generating electrode corresponds to an electrode reaction represented by

$$S_1 \pm z_1 e = S_2 \tag{216}$$

where S_1 and S_2 are the species participating in the electrode reaction and z_1 is the number of electrons entering the total reaction. The sign convention is the same as already mentioned. The condition of limiting current density related to S_1 is given by $C_1^s = 0$. When the generating electrode extends from x_0 to x_1, as shown in Figure 17, the total current at the generating electrode is

$$I_1 = \pm [3^{4/3}/2\Gamma(\tfrac{1}{3})] z_1 F C_1^\circ D_1^{2/3} \mu^{-1/3} \left[\int_{x_0}^{x_1} \tau^{1/2}\,dx \right]^{2/3} \tag{217}$$

where $\Gamma(1/3)$ is the gamma function of 1/3. For a three-dimensional flow, we get for I_1

$$I_1 = \pm [3^{4/3}\pi/\Gamma(\tfrac{1}{3})] z_1 F C_1^\circ D_1^{2/3} \mu^{-1/3} \left(\int_{x_0}^{x_1} [r_0^3\tau]^{1/2}\,dx \right)^{2/3} \tag{218}$$

The reaction product S_2 formed on the surface of the generating electrode is afterwards reduced or oxidized on the surface of the indicating electrode yielding S_3, according to the reaction

$$S_2 \pm z_2 e = S_3 \tag{219}$$

z_2 being the number of electrons participating in the reaction. The condition of limiting current at the indicating electrode is $C_2^s = 0$. To find the concentration distribution of S_2 in the vicinity of the indicating electrode surface, equation (215) is now solved taking into account the following boundary conditions for S_2:

$$
\begin{aligned}
x_0 < x < x_1: \quad & D_2(\partial C_2/\partial y)_{y=0} = -D_1(\partial C_1/\partial y)_{y=0} \\
x_1 < x < x_2: \quad & D_2(\partial C_2/\partial y)_{y=0} = 0 \\
x_2 < x: \quad & C_2^s = 0
\end{aligned}
\tag{220}
$$

After solving equation (215) and further assuming that the concentration of S_2 in the bulk of the solution is zero, the concentration gradient at the indicating electrode is derived. Then, as the current I_2 at the indicating electrode is proportional to the concentration gradient of S_2, it is expressed for a two-dimensional flow as

$$
I_2 = \pm \frac{3^{4/3}}{2\Gamma(\frac{1}{3})} z_2 FC_1^{\circ} D_1^{2/3} \mu^{-1/3} \left\{ \left(\int_{x_2}^{x_3} \tau^{1/2}\, dx \right)^{2/3} G\left(\frac{\int_{x_0}^{x_1} \tau^{1/2}\, dx}{\int_{x_1}^{x_2} \tau^{1/2}\, dx} \right) \right.
$$

$$
+ \left(\int_{x_0}^{x_1} \tau^{1/2}\, dx \right)^{2/3} G\left(\frac{\int_{x_2}^{x_3} \tau^{1/2}\, dx}{\int_{x_1}^{x_2} \tau^{1/2}\, dx} \right) - \left(\int_{x_0}^{x_3} \tau^{1/2}\, dx \right)^{2/3}
$$

$$
\times G\left(\frac{\int_{x_0}^{x_1} \tau^{1/2}\, dx}{\int_{x_0}^{x_3} \tau^{1/2}\, dx} \frac{\int_{x_2}^{x_3} \tau^{1/2}\, dx}{\int_{x_1}^{x_2} \tau^{1/2}\, dx} \right) \right\}
\tag{221}
$$

Similarly, for a three-dimensional flow, the following result is obtained:

$$
I_2 = \pm \frac{3^{4/3}\pi}{\Gamma(\frac{1}{3})} z_2 FC_1^{\circ} D_1^{2/3} \mu^{-1/3} \left\{ \left[\int_{x_2}^{x_3} (r_0^3\tau)^{1/2}\, dx \right]^{2/3} G\left(\frac{\int_{x_0}^{x_1} (r_0^3\tau)^{1/2}\, dx}{\int_{x_1}^{x_2} (r_0^3\tau)^{1/2}\, dx} \right) \right.
$$

$$
+ \left[\int_{x_0}^{x_1} (r_0^3\tau)^{1/2}\, dx \right]^{2/3} G\left(\frac{\int_{x_2}^{x_3} (r_0^3\tau)^{1/2}\, dx}{\int_{x_1}^{x_2} (r_0^3\tau)^{1/2}\, dx} \right)
$$

$$
- \left[\int_{x_0}^{x_3} (r_0^3\tau)^{1/2}\, dx \right]^{2/3} G\left(\frac{\int_{x_0}^{x_1} (r_0^3\tau)^{1/2}\, dx}{\int_{x_0}^{x_3} (r_0^3\tau)^{1/2}\, dx} \frac{\int_{x_2}^{x_3} (r_0^3\tau)^{1/2}\, dx}{\int_{x_1}^{x_2} (r_0^3\tau)^{1/2}\, dx} \right) \right\}
\tag{222}
$$

where the function $G(\theta) = 1 - F(\theta)$, and $F(\theta)$ is defined by equation (214).

Relating equations (215) and (221), the general expression for the collection efficiency in a two-dimensional flow results

$$
N = \left| \frac{I_2}{I_1} \right| = \frac{z_2}{z_1} \left\{ \left(\frac{\int_{x_2}^{x_3} \tau^{1/2}\,dx}{\int_{x_0}^{x_1} \tau^{1/2}\,dx} \right)^{2/3} G\!\left(\frac{\int_{x_0}^{x_1} \tau^{1/2}\,dx}{\int_{x_1}^{x_2} \tau^{1/2}\,dx} \right) + G\!\left(\frac{\int_{x_2}^{x_3} \tau^{1/2}\,dx}{\int_{x_1}^{x_2} \tau^{1/2}\,dx} \right) \right.
$$
$$
\left. - \left(\frac{\int_{x_0}^{x_3} \tau^{1/2}\,dx}{\int_{x_0}^{x_1} \tau^{1/2}\,dx} \right)^{2/3} G\!\left(\frac{\int_{x_0}^{x_1} \tau^{1/2}\,dx}{\int_{x_0}^{x_3} \tau^{1/2}\,dx} \frac{\int_{x_2}^{x_3} \tau^{1/2}\,dx}{\int_{x_1}^{x_2} \tau^{1/2}\,dx} \right) \right\} \tag{223}
$$

Analogously, relating equations (216) and (222), the collection efficiency in three-dimensional flow is obtained as follows:

$$
N = \left| \frac{I_2}{I_1} \right| = \frac{z_2}{z_1} \left\{ \left(\frac{\int_{x_2}^{x_3} (r_0^3 \tau)^{1/2}\,dx}{\int_{x_0}^{x_1} (r_0^3 \tau)^{1/2}\,dx} \right)^{2/3} G\!\left(\frac{\int_{x_0}^{x_1} (r_0^3 \tau)^{1/2}\,dx}{\int_{x_1}^{x_2} (r_0^3 \tau)^{1/2}\,dx} \right) \right.
$$
$$
+ G\!\left(\frac{\int_{x_2}^{x_3} (r_0^3 \tau)^{1/2}\,dx}{\int_{x_1}^{x_2} (r_0^3 \tau)^{1/2}\,dx} \right) - \left(\frac{\int_{x_0}^{x_3} (r_0^3 \tau)^{1/2}\,dx}{\int_{x_0}^{x_1} (r_0^3 \tau)^{1/2}\,dx} \right)^{2/3}
$$
$$
\left. \times\, G\!\left(\frac{\int_{x_0}^{x_1} (r_0^3 \tau)^{1/2}\,dx}{\int_{x_0}^{x_3} (r_0^3 \tau)^{1/2}\,dx} \frac{\int_{x_2}^{x_3} (r_0^3 \tau)^{1/2}\,dx}{\int_{x_1}^{x_2} (r_0^3 \tau)^{1/2}\,dx} \right) \right\} \tag{224}
$$

For flows where the shear stress τ in the wall is known, Matsuda[125] derived from equation (224) expressions for the collection efficiency of electrodes with a particular geometry. Thus, for rotating ring–disk electrodes,

$$
N = \frac{z_2}{z_1} \left\{ \left(\frac{x_3^3 - x_2^3}{x_1^3 - x_0^3} \right)^{2/3} G\!\left(\frac{x_1^3 - x_0^3}{x_2^3 - x_1^3} \right) + G\!\left(\frac{x_3^3 - x_2^3}{x_2^3 - x_1^3} \right) \right.
$$
$$
\left. - \left(\frac{x_3^3 - x_0^3}{x_1^3 - x_0^3} \right)^{2/3} G\!\left(\frac{x_1^3 - x_0^3}{x_3^3 - x_0^3} \frac{x_3^3 - x_2^3}{x_2^3 - x_1^3} \right) \right\} \tag{224}
$$

This relationship is the same as that deduced by Albery and Bruckenstein[121] [equation (213)] when it is expressed with the same notation. For canal-shaped and tubular electrodes with Poiseuille velocity profiles, the following common equation for the collection efficiency is obtained:

$$
N = \frac{z_2}{z_1} \left\{ \left(\frac{x_3 - x_2}{x_1 - x_0} \right)^{2/3} G\left(\frac{x_1 - x_0}{x_2 - x_1} \right) + G\left(\frac{x_3 - x_2}{x_2 - x_1} \right) \right.
$$
$$
\left. - \left(\frac{x_3 - x_0}{x_1 - x_0} \right)^{2/3} G\left(\frac{x_1 - x_0}{x_3 - x_0} \frac{x_3 - x_2}{x_2 - x_1} \right) \right\} \tag{226}
$$

For a flow past a wedge, the following expression results:

$$
N = \frac{z_2}{z_1} \left\{ \left(\frac{x_3^{3(m+1)/4} - x_2^{3(m+1)/4}}{x_1^{3(m+1)/4} - x_0^{3(m+1)/4}} \right)^{2/3} G\left(\frac{x_1^{3(m+1)/4} - x_0^{3(m+1)/4}}{x_2^{3(m+1)/4} - x_1^{3(m+1)/4}} \right) \right.
$$
$$
+ G\left(\frac{x_3^{3(m+1)/4} - x_2^{3(m+1)/4}}{x_2^{3(m+1)/4} - x_1^{3(m+1)/4}} \right) - \left(\frac{x_3^{3(m+1)/4} - x_0^{3(m+1)/4}}{x_1^{3(m+1)/4} - x_0^{3(m+1)/4}} \right)^{2/3}
$$
$$
\left. \times G\left(\frac{x_1^{3(m+1)/4} - x_0^{3(m+1)/4}}{x_3^{3(m+1)/4} - x_0^{3(m+1)/4}} \frac{x_3^{3(m+1)/4} - x_2^{3(m+1)/4}}{x_2^{3(m+1)/4} - x_1^{3(m+1)/4}} \right) \right\} \tag{227}
$$

Linear flow along a plate and flow with a two-dimensional stagnation point correspond, respectively, to $m = 0$ ($\beta = 0$) and $m = 1$ ($\beta = 1$).

For circular conical electrodes, expression (224) yields, after substituting $r_0(x)$ and $\tau(x)$ and taking into account the Leuteritz and Mangler transformation,[126] the equation

$$
N = \frac{z_2}{z_1} \left\{ \left(\frac{x_3^{3(m'+3)/4} - x_2^{3(m'+3)/4}}{x_1^{3(m'+3)/4} - x_0^{3(m'+3)/4}} \right)^{2/3} G\left(\frac{x_1^{3(m'+3)/4} - x_0^{3(m'+3)/4}}{x_2^{3(m'+3)/4} - x_1^{3(m'+3)/4}} \right) \right.
$$
$$
+ G\left(\frac{x_3^{3(m'+3)/4} - x_2^{3(m'+3)/4}}{x_2^{3(m'+3)/4} - x_1^{3(m'+3)/4}} \right) - \left(\frac{x_3^{3(m'+3)/4} - x_0^{3(m'+3)/4}}{x_1^{3(m'+3)/4} - x_0^{3(m'+3)/4}} \right)^{2/3}
$$
$$
\left. \times G\left(\frac{x_1^{3(m'+3)/4} - x_0^{3(m'+3)/4}}{x_3^{3(m'+3)/4} - x_0^{3(m'+3)/4}} \frac{x_3^{3(m'+3)/4} - x_2^{3(m'+3)/4}}{x_2^{3(m'+3)/4} - x_1^{3(m'+3)/4}} \right) \right\} \tag{228}
$$

Three-dimensional flow with a stagnation point corresponds to $m' = 1$.

5. Steady Current–Potential Characteristics for Simple Electrode Processes under Laminar Isothermal Free Convection

The current density equations given in Section IV for both the isothermal and nonisothermal free-convection-controlled processes were obtained disregarding the influence of the electrode potential, since they were derived for maximum flux at the electrode. To establish the current–potential relationships, it is necessary to solve the convective-diffusion differential equation for any flux.

To begin with, we consider an isothermal free-convection process at a vertical, flat plate electrode in a gravitational field. It is assumed[91] that the electrochemical reaction rate is very large compared to the mass transport rate, the reaction taking place in the presence of a large excess of an inert supporting electrolyte. The concentration of the reacting species on the electrode surface is potential dependent.[92] The set of partial differential equations now comprises equations (124) and (125), and the boundary conditions, assuming that the electrode potential is lower than that corresponding to maximum flux, are

$$
\begin{aligned}
y = 0: &\qquad v_x = v_y = 0, \quad C_i = C_i^s \\
y \to \infty: &\qquad v_x = v_y = 0, \quad C_i = C_i^o
\end{aligned}
\tag{229}
$$

For a reversible electrode process, the concentration of the reacting species on the surface is related to the electrode potential by the Nernst equation

$$
C_i^s = C_i^o \exp[z_i F(E - E_{rev})/RT] \tag{230}
$$

According to this dependence, the densification coefficient γ also becomes potential-dependent as the concentration gradient goes from zero, when no potential is applied, to a maximum value equal to the bulk concentration of the ith species, when the electrode potential approaches infinity. The densification coefficient is defined as follows:

$$
\gamma = [(C_i^o - C_i^s)/\rho_0]\, \partial\rho/\partial C_i \tag{231}
$$

and, according to equation (230),

$$
\gamma = \gamma_\infty \{1 - \exp[z_i F(E - E_{rev})/RT]\} \tag{232}
$$

γ_∞ is the densification coefficient defined by equation (129). It is also convenient to write the dimensionless concentration function φ as

$$\varphi = (C_i^\circ - C_i)/(C_i^\circ - C_i^s)$$

$$= (C_i^\circ - C_i)/C_i^\circ \{1 - \exp[z_i F(E - E_{rev})/RT)\} \qquad (233)$$

To transform the partial differential equations into ordinary differential equations, the former are expressed in terms of the dimensionless variable ζ given by

$$\zeta = (g\gamma/4v^2)^{1/4} y/x^{1/4}$$

$$= \zeta_\infty \{1 - \exp[z_i F(E - E_{rev})/RT]\}^{1/4} \qquad (234)$$

In this way, the dimensionless differential equations and boundary conditions are the same as those given for the limiting case, as are the dimensionless velocity and concentration distribution functions. However, after returning to the original variables, these functions are electrode-potential-dependent.

Thus, after applying Fick's law, the following local current density expression for the reversible process at any electrode potential results

$$i = 0.70 z_i F C_i^\circ D_i (g\gamma_\infty/4v^2)^{1/4} (v/D_i)^{1/4} x^{-1/4}$$

$$\times \{1 - \exp[z_i F(E - E_{rev})/RT]\}^{5/4} \qquad (235)$$

Rearranging this equation, we get

$$i = i_L \{1 - \exp[z_i F(E - E_{rev})/RT]\}^{5/4} \qquad (236)$$

where i_L is given by

$$i_L = 0.70 z_i F D_i C_i^\circ (g\gamma_\infty/4v^2)^{1/4} (v/D_i)^{1/4} x^{-1/4} \qquad (237)$$

A set of current–potential curves calculated with equation (235) is shown in Figure 18.

For an electrode reaction involving the deposition of an insoluble substance under intermediate kinetics, the free-convection differential equation previously given for the case of a reversible reaction involving a single reacting species is also valid, as are the boundary conditions. However, after the steady state is reached, the rate of the heterogeneous reaction equals the flow rate at the

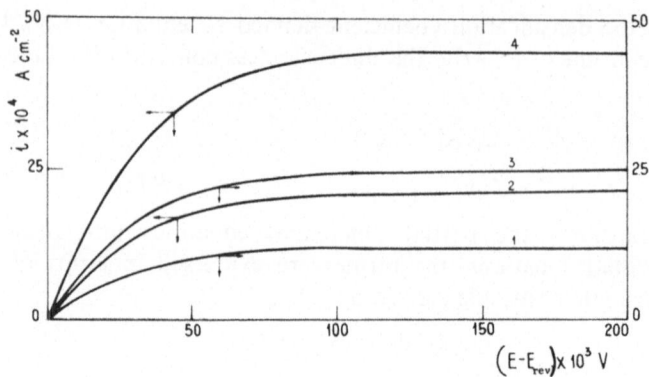

Figure 18. Current/potential curves calculated with equation (235).
(1) $C_i^o = 5 \times 10^{-3} M$; $T_e = 5°C$. (2) $C_i^o = 5 \times 10^{-2} M$; $T_e = 5°C$.
(3) $C_i^o = 5 \times 10^{-3} M$; $T_e = 35°C$. (4) $C_i^o = 5 \times 10^{-2} M$; $T_e = 35°C$.

surface, and the following additional boundary condition should also be considered:

$$D_i(\partial C_i/\partial y)_{y=0} = k_f C_i^s \exp[-\alpha F(E - E_{rev})/RT]$$
$$- k_b \exp[(1 - \alpha)F(E - E_{rev})/RT] \qquad (238)$$

where α is the transfer coefficient already defined. The concentration can also be expressed in terms of current density

$$C_i^s = (i_L/A\gamma_\infty^{1/4}) - (i/A\gamma^{1/4}) \qquad (239)$$

where

$$A = 0.70 z_i F D_i C_i^o (g/4v^2)^{1/4}(v/D_i)^{1/4} x^{-1/4} \qquad (240)$$

Introducing the surface concentration given by equation (239) into equation (238) and considering the exchange current density i_0 $(E = E_{rev})$,

$$z_i F k_f C_i^o = z_i F k_b = i_0 \qquad (241)$$

the following expression results:

$$i = \frac{i_k\{1 - \exp[z_i F(E - E_{rev})/RT]\}^{5/4}}{\{1 - \exp[z_i F(E - E_{rev})/RT]\}^{1/4} + [i_k/(i_L)_c]} \qquad (242)$$

The expression for i_k is

$$i_k = i_0 \exp[-\alpha F(E - E_{rev})/RT] \qquad (243)$$

Equation (242) gives the shape of the complete steady current–potential curve of an electrode reaction involving the deposition of an insoluble substance under intermediate kinetics. If a process involving two soluble species is considered, the current–potential curve is given by

$$i = \{\exp[-\alpha F(E - E_{rev})/RT] - \exp[(1 - \alpha)F(E - E_{rev})/RT]\}$$

$$\times \left\{ \frac{1}{i_0} + \frac{\exp[-\alpha F(E - E_{rev})/RT]}{(i_L)_c\{1 - \exp[z_i F(E - E_{rev})/RT]\}^{1/4}} \right.$$

$$\left. - \frac{\exp[(1 - \alpha)F(E - E_{rev})/RT]}{(i_L)_a\{1 - \exp[-z_i F(E - E_{rev})/RT]\}^{1/4}} \right\}^{-1}$$

$$(244)$$

Obviously, equation (242) is a particular case contained in equation (244). The same result as equation (242) is obtained from equation (244) when the anodic limiting current is large, approaching infinity. In the same way, if the cathodic limiting current is large enough, the equation of the anodic current–potential curve is derived. Cathodic current–potential curves calculated with equation (242) assuming different exchange current density values are depicted in Figure 19. They change shape in the usual way, from a nearly reversible polarization wave when i_0 is about 1 A cm^{-2} or even larger, to a net irreversible polarization wave when i_0 is about 10^{-5} A cm^{-2} or lower. The same conclusions are obviously also reached for the anodic steady current–potential curves.

6. Steady Current–Potential Characteristics for Simple Electrode Processes under Nonisothermal Free Convection

For an electrochemical reaction under nonisothermal, free convection, the concentration of the reacting species is assumed again to be dependent on the electrode potential. The simultaneous set of partial differential equations related to the mass transport problem under a combined effect of temperature and concentration gradients, considering again a vertical, flat plate electrode, is the same as that already indicated [equations (124) to (126)]. However,

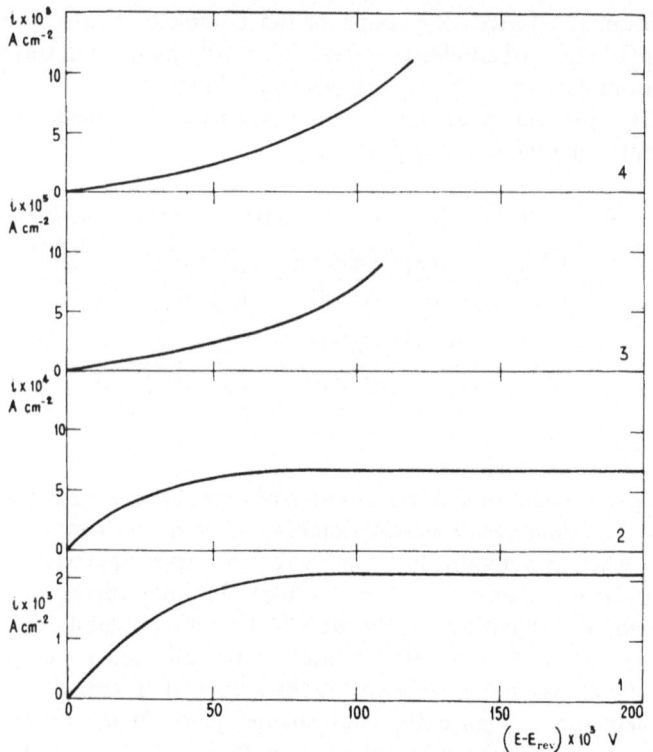

Figure 19. Current/potential curves calculated with equation (242) assuming different i_o values. $C_i^o = 5 \times 10^{-2}\,M$; $T_m = 12.5°C$. (1) $i_0 = 1\,\text{A/cm}^2$. (2) $i_0 = 10^{-3}\,\text{A/cm}^2$. (3) $i_0 = 10^{-5}\,\text{A/cm}^2$. (4) $i_0 = 10^{-8}\,\text{A/cm}^2$.

the boundary conditions are now

$$y = 0: \quad v_x = v_y = 0, \quad C_i = C_i^s, \quad T = T_e$$
$$y \to \infty: \quad v_x = v_y = 0, \quad C_i = C_i^o, \quad T = T_0 \tag{245}$$

With the dimensionless variable μ defined in equation (131) and the densification coefficient given by equation (232), the set of partial differential equations can be solved as already described to obtain the equations for the velocity and concentration profiles.[91]

Applying Fick's law, the rate equation for the cathodic electrode position of an insoluble species is, in terms of the current density as a function of the electrode potential,

$$i = 0.70 z_i F D_i C_i^{\circ} \left\{ \frac{\gamma_{\infty}^{3/4} \{1 - \exp[z_i F(E - E_{rev})/RT]\}^{3/4} D_i^{1/4}}{\nu^{1/4}} \right.$$
$$\left. \pm \frac{\sigma^{3/4} k_T^{1/4}}{(\nu \rho C_p)^{1/4}} \right\}^{1/3} \left(\frac{\nu}{D_i} \right)^{1/3} \left(\frac{g}{4 \nu^2} \right)^{1/4} x^{-1/4} \left[1 - \exp \frac{z_i F(E - E_{rev})}{RT} \right]$$

$$(246)$$

When $E - E_{rev}$ approaches infinity, equation (246) yields the maximum current density. Cathodic current–potential curves calculated with this equation for different values of electrode temperature T_e are plotted in Figure 20.

Figure 21 corresponds to the variation of maximum flux with ΔT at two different electrode potentials. Curves (1, 1') are calculated for negative ΔT values and a simultaneous decrease of the plate temperature and the reference temperature of the system. Curves (2, 2') are also given for negative ΔT values, at constant plate temperature and increasing reference temperature of the system.

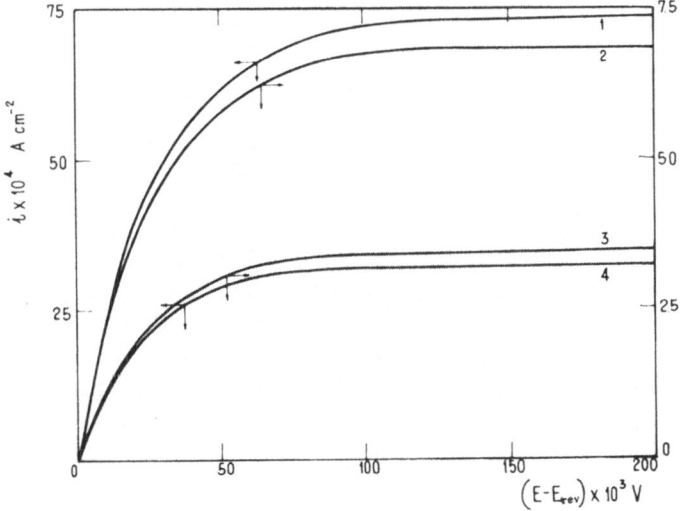

Figure 20. Current/potential curves calculated with equation (246) for different concentrations and electrode temperatures. (1) $C_i^{\circ} = 5 \times 10^{-2} M$; $T_e = 35°C$; $\Delta T = 15°C$. (2) $C_i^{\circ} = 5 \times 10^{-3} M$; $T_e = 35°C$; $\Delta T = 15°C$. (3) $C_i^{\circ} = 5 \times 10^{-3} M$; $T_e = 5°C$; $\Delta T = -15°C$. (4) $C_i^{\circ} = 5 \times 10^{-2} M$; $T_e = 5°C$; $\Delta T = -15°C$.

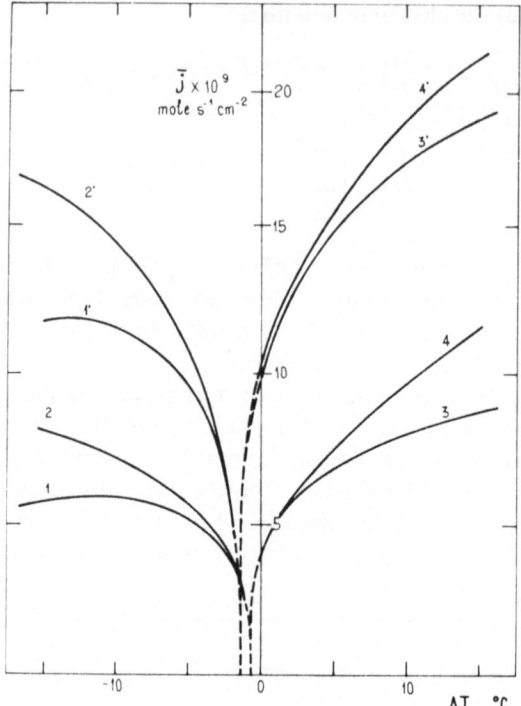

Figure 21. Variation of the flow rate with ΔT at $C_i^\circ =$ 5×10^{-2} M. Curves 1, 2, 3, and 4 correspond to $E - E_{\text{rev}} =$ 0.01 V, and curves 1′, 2′, 3′, and 4′ correspond to $E - E_{\text{rev}} =$ 0.03 V.

Curves $(3, 3')$ correspond to positive ΔT values, constant plate temperature, and decreasing reference temperature. Finally, curves $(4, 4')$ also correspond to positive ΔT when both the plate and reference temperature increase. Figure 21 also shows the same features already reported for the case of maximum flux, but, in addition, shows that the point where the thermal and concentration gradients cancel each other shifts toward more negative ΔT values as the electrode potential increases.

For the case of the deposition of an insoluble substance, the boundary conditions represented by (238) must also be taken into account, together with (245), to solve the set of differential equations. Following the procedure already mentioned for intermediate kinetics under isothermal free convection, the local current/potential expression of an electrode reaction involving the deposition of an insoluble substance is

$$i = \frac{i_k\{1 - \exp[z_i F(E - E_{rev})/RT]\}}{1 + [i_k/(i_L)_c][(S_\infty \pm P)/(S \pm P)]^{1/3}} \quad (247)$$

where

$$S_\infty = \gamma_\infty^{3/4} D_i^{1/4}/v^{1/4}, \quad S = \gamma^{3/4} D_i^{1/4}/v^{1/4}, \quad P = \sigma^{3/4} k_T^{1/4}/(v\rho C_p)^{1/4} \quad (248)$$

and i_k has the meaning already indicated.

For an electrode reaction involving two soluble species, the following equation results:

$$i = \{\exp[-\alpha F(E - E_{rev})/RT] - \exp[(1 - \alpha)F(E - E_{rev})/RT]\}$$
$$\times \left\{\frac{1}{i_0} + \frac{\exp[-\alpha F(E - E_{rev})/RT]}{(i_L)_c}\left(\frac{S_\infty \pm P}{S \pm P}\right)^{1/3}\right.$$
$$\left. - \frac{\exp[(1 - \alpha)F(E - E_{rev})/RT]}{(i_L)_a}\left(\frac{S_\infty \pm P}{S \pm P}\right)^{1/3}\right\}^{-1} \quad (249)$$

Equation (247) is deduced from (249) when, as before, the anodic limiting current approaches infinity. When this is the case for the cathodic limiting current, the local anodic steady current–potential curve is obtained.

Current–potential curves calculated with equation (247) are shown in Figure 22 for $\Delta T = +15°C$ and in Figure 23 for $\Delta T = -15°C$. In both cases, curves were calculated for three i_0 values, 10^{-5}, 10^{-3}, and $1 \, A/cm^2$. A change in shape with i_0 is observed when going from a near reversible curve for $i_0 = 1 \, A \, cm^{-2}$ to a near irreversible wave for $i_0 = 10^{-5} \, A \, cm^{-2}$. In this sense, for the same i_0, the thermal convective electrode shows a more marked distortion of current–potential curves than that reported above for an electrode process with mixed kinetics occurring on a vertical plate electrode under isothermal laminar free convection.

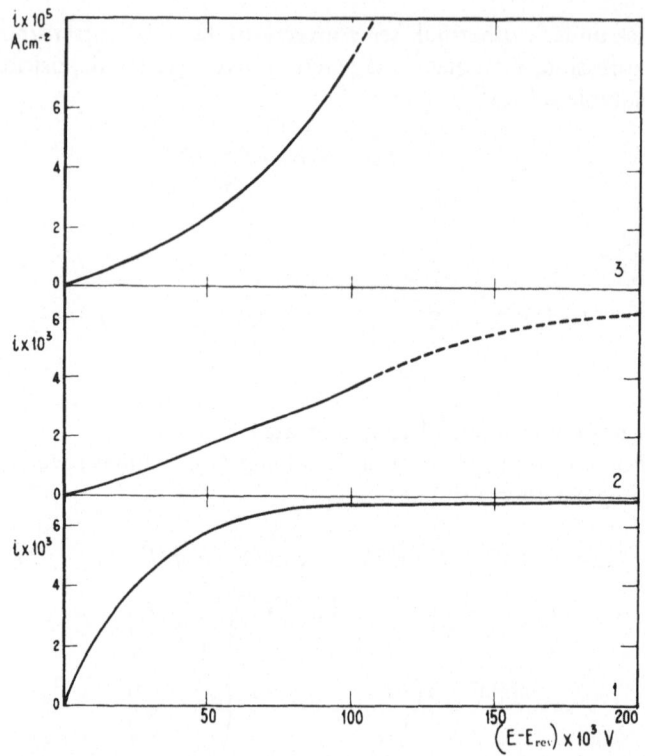

Figure 22. Cathodic current/potential curves calculated with equation (247), assuming different i_0 values. $C_i^\circ = 5 \times 10^{-2} M$; $T_e = 35°C$; $\Delta T = 15°C$. (1) $i_0 = 1$ A/cm^2. (2) $i_0 = 10^{-3}$ A/cm^2. (3) $i_0 = 10^{-5}$ A/cm^2.

7. The Influence of Electrode Height on the Rate Equation

In the study of isothermal, free convection as well as non-isothermal free convection, the limiting rate equation corresponding to an infinite electrode potential yields a fourth-root dependence of the flow rate on the electrode height. The influence of the latter on the rate equation at potentials lower than those corresponding to the maximum flux was studied in a system involving the electrodeposition of copper on copper electrodes.[127-130] These results show that current density is virtually constant if the applied current is less than 1/3 of the limiting current for exclusive cathodic deposition of copper. Later, these results were theoretically interpreted and it was shown that the effect occurs under most conditions,[92] implying that the concentration difference between the bulk of the solution and the electrode surface is proportional to the fifth root of the distance from the leading edge of the electrode. This conclusion is based upon results derived from previous work on the corresponding heat transfer problem.[129] It was concluded that the rate of mass transport becomes independent of the electrode height when the kinetics of the electrode reaction correspond to an activation process. The arguments employed to arrive at these conclusions are not always clear.

The theoretical equations reported above either for isothermal or nonisothermal free convection, assuming uniform concentration and temperature prevailing at the electrode surface, indicate that

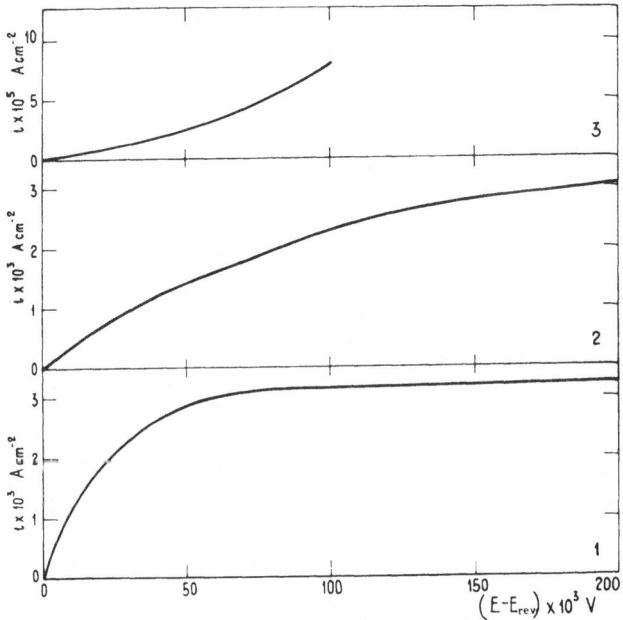

Figure 23. Cathodic current/potential curves calculated with equation (247) assuming different i_0 values. $C_i^0 = 5 \times 10^{-2}\,M$; $T_e = 5°C$; $\Delta T = -15°C$. (1) $i_0 = 1\,A/cm^2$. (2) $i_0 = 10^{-3}\,A/cm^2$. (3) $i_0 = 10^{-5}\,A/cm^2$.

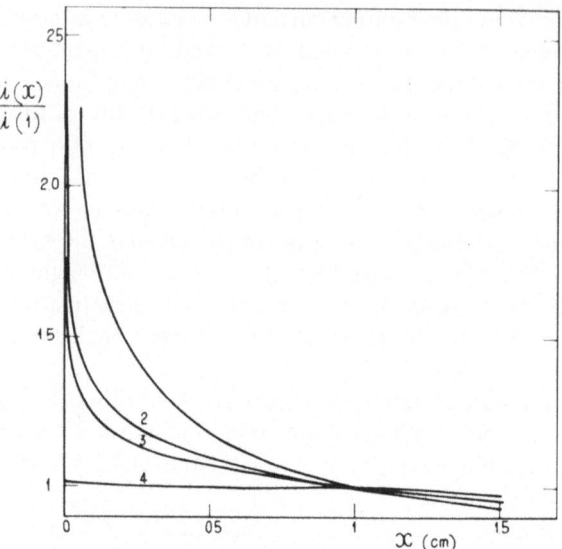

Figure 24. Current density ratio as a function of electrode height for different overpotentials and exchange current densities (isothermal free convection). $C_i^\circ = 5 \times 10^{-2}\,M$; $T_m = -12.5°C$. (1) $i_0 = \infty$ and at any overpotential. (2) $i_0 = 10^{-3}\,A/cm^2$; $E - E_{rev} = 0.05\,V$. (3) $i_0 = 10^{-3}\,A/cm^2$; $E - E_{rev} = 0.01\,V$. (4) $i_0 = 10^{-5}\,A/cm^2$; $E - E_{rev} = 0.03\,V$. (5) $i_0 = 10^{-5}\,A/cm^2$; $E - E_{rev} = 0.01\,V$.

the relationship between the current density and the electrode height depends principally on the rate of the electrochemical reaction at the interface. Thus, for a reversible reaction ($i_0 \to \infty$), the fourth-root dependence on the current density is still valid at electrode potentials lower than that corresponding to the maximum flux. As the electrochemical reaction becomes slower, that dependence is less significant, and in the limiting case ($i_0 \to 0$), the current density is independent of the electrode height.

From equations (235) and (242), it is possible to visualize the effect of electrode height on current density as presented in Figure 24. This plot corresponds to the ratio of the current density at different x values to the current density at $x = 1$, the latter being taken as a reference. For the reversible case, at any electrode potential, the function is in good agreement with the $-1/4$ power of x. When the value of i_0 decreases, the local current density becomes less dependent on x. Thus, when i_0 is equal to 10^{-5} A/cm^2, the ratio $i(x)/i(1)$ is practically independent of x. It is also concluded that, at $x = 0$, the local current density changes from infinity, for a reversible reaction, toward finite values which decrease with the exchange current density. Consequently, the rate equations permit a reasonable interpretation of the effect of the electrode height on the rate of mass transport at a vertical plate electrode under free convection.

On applying these conclusions to results reported in previous work on the electrodeposition of copper under laminar isothermal free convection at a vertical plate electrode, the weaker dependence found must be attributed to the i_0 value related to that electrode reaction, which, as reported by Bockris and Mattsson,[134] is about 10^{-2} A/cm^2 at 25°C.

VI. CONCLUSION

This chapter is intended as an introduction to the theory of hydrodynamic voltammetry and concisely covers the systematic deduction of the mass transport rate equations for solid electrodes.

The theory of the rotating disk electrode was the first to be considered, this device being an important tool often used in electrochemical kinetics. The theory of other solid electrodes developed later offers interesting possibilities for kinetic applications, as deduced from the corresponding rate equations.

The knowledge of flow patterns at solid electrodes is essential for the study of various particular problems of electrode kinetics, including cell and electrode design. First, these electrodes are suitable once the diffusion coefficients are known, and the diffusion coefficients of reacting species are easily evaluated with good precision by applying the theoretical mass transport rate equations.

Second, the kinetic study of electrode processes under intermediate control is also achieved with these solid electrodes. The evaluation of the exchange current density, the transfer coefficient, and the reaction order corresponding to the activation process is feasible because a separation and independent analysis of convective-diffusion and activation process contributions can be made. Rate constants up to the order of 0.1 cm/sec can be determined under laminar flow conditions.

As far as the above mentioned applications are concerned, the electrodes under free convection offer interesting possibilities in electrochemical kinetics which still have not been fully exploited. The application of hydrodynamic voltammetry using neighboring electrodes is also becoming of increasing importance, as it permits not only the detection of free radicals or intermediates arising from electrode reactions, but also the kinetic study of their reactions. Besides, it is of relevance for the straightforward and unambiguous establishment of electrode reaction mechanisms.

REFERENCES

[1] V. G. Levich, *Acta Physicochim. U.R.S.S.* **17** (1942) 257.
[2] V. G. Levich, *Acta Physicochim. U.R.S.S.* **19** (1944) 117.
[3] V. G. Levich, *Acta Physicochim. U.R.S.S.* **19** (1944) 133.
[4] V. G. Levich, *Zh. Fiz. Khim.* **18** (1944) 335; **22** (1948) 575.
[5] V. G. Levich, *Disc. Faraday Soc.* **1** (1947) 37.
[6] V. G. Levich, *Physicochemical Hydrodynamics*, Prentice-Hall, Englewood Cliffs, N.J., 1962.
[7] H. Schlichting, *Boundary Layer Theory*, McGraw-Hill Book Co., New York, 1968.
[8] J. Newman, in *Advances in Electrochem. and Electrochem. Engineering*, Vol. 5, p. 87. Interscience. New York, 1967.
[9] R. B. Bird, W. E. Steward, and E. N. Lightfoot, *Transport Phenomena*, John Wiley and Sons, New York, 1962.
[10] N. Curle, *The Laminar Boundary Layer Equations*, Oxford University Press, Oxford, 1962.
[11] S. Goldstein, *Modern Developments in Fluid Dynamics*, Clarendon Press, Oxford, 1938.
[12] M. ten Bosch, *Die Warmeübertragung*, III ed., Springer, Berlin, 1936.
[13] D. A. Frank-Kamenetzkii, *Diffusion and Heat Exchange in Chemical Kinetics*, Princeton University Press, Princeton, N.J., 1955.
[14] E. R. G. Eckert, *Introduction to the Transfer of Heat and Mass*, McGraw-Hill Book Co., New York, 1950.
[15] W. Jost, *Diffusion*, Academic Press, New York, 1952.
[16] P. Delahay, *New Instrumental Methods in Electrochemistry*, 3rd ed., Interscience, New York, 1962.
[17] J. Proszt, V. Cieleszky, and K. Gyórbiro, *Polarographie*, Akadémiai Kiadó, Budapest, 1967.
[18] W. G. Cochran, *Proc. Cambridge Phil. Soc.* **30** (1934) 365.
[19] T. von Kárman, *Z. Angew. Math. Mech.* **1** (1921) 233.
[20] E. M. Sparrow and J. L. Gregg, *J. Heat Transfer* **81C** (1959) 249.
[21] D. P. Gregory and A. C. Riddiford, *J. Chem. Soc.* (1956) 3765.
[22] A. C. Riddiford, in *Advances in Electrochem. and Electrochem. Engineering*, Vol. 4, p. 47, Interscience, New York, 1966.

[23] L. P. Kholpanov, *Russ. J. Phys. Chem.* **41** (1967) 1085.
[24] J. Newman, *J. Electrochem. Soc.* **113** (1966) 1235.
[25] J. Newman, *J. Electrochem. Soc.* **114** (1967) 239.
[26] J. Newman and L. Hsueh, *Electrochim. Acta* **12** (1967) 417.
[27] L. Hsueh and J. Newman, *Electrochim. Acta* **12** (1967) 429.
[28] R. E. Davis, G. L. Horvath, and C. W. Tobias, *Electrochim. Acta* **12** (1967) 287.
[29] M. E. Martins, A. J. Calandra, and A. J. Arvia, *Electrochim. Acta* **15** (1970) 111.
[30] J. A. Olabe and A. J. Arvia, *Electrochim. Acta* **14** (1969) 785.
[31] U. Bödewadt, *Z. Angew. Math. Mech.* **20** (1940) 17.
[32] D. Gröhne, *Nachr. Akad. Wiss. Göttingen, Math-physik. Kl.* **12** (1965).
[33] A. N. Frumkin and L. N. Nekrasov, *Dokl. Akad. Nauk SSSR* **126** (1959) 115.
[34] Yu. B. Ivanov and V. G. Levich, *Dokl. Akad. Nauk SSSR* **126** (1959) 1029.
[35] A. N. Frumkin, L. N. Nekrasov, V. G. Levich, and Yu. B. Ivanov, *J. Electroanal. Chem.* **1** (1959/60) 84.
[36] W. J. Albery, *Trans. Faraday Soc.* **62** (1966) 1915.
[37] W. J. Albery, S. Bruckenstein, and D. T. Napp, *Trans. Faraday Soc.* **62** (1966) 1932.
[38] W. J. Albery, S. Bruckenstein, and D. C. Johnson, *Trans. Faraday Soc.* **62** (1966) 1938.
[39] G. R. Bopp and D. M. Mason, *Electrochem. Tech.* **2** (1964) 129.
[40] L. Howarth, *Aeronaut. Res. Council (Great Britain), Rep. and Mem.* (1935) 1632.
[41] H. Goertler, *Deut. Versuchsanstaltf. Luftahrt Rep. Nr.* 34 (1957).
[42] N. Froessling, *Lunds Univ. Arsskr. N. F. Avd.* **35** (1940) 2.
[43] J. Jordan, R. A. Javick, and W. E. Ranz, *J. Am. Chem. Soc.* **80** (1958) 3846.
[44] J. Jordan and R. A. Javick, *Electrochim. Acta* **6** (1962) 23.
[45] S. L. Marchiano and A. J. Arvia, *Electrochim. Acta* **12** (1967) 801.
[46] J. C. Bazán, S. L. Marchiano, and A. J. Arvia, *Electrochim. Acta* **12** (1967) 821.
[47] H. A. Johnson and M. W. Rubesin, *Trans. Am. Soc. Mech. Eng.* **71** (1949) 447.
[48] V. M. Falkner and S. W. Skan, *Phil. Mag.* **12** (1931) 865.
[49] D. R. Hartree, *Proc. Cambridge Phil. Soc.* **33** (1937) 223.
[50] K. Hiemenz, *Dinglers Polytechn. J.* **92** (1911) 321.
[51] J. S. W. Carrozza, S. L. Marchiano, J. J. Podestá, and A. J. Arvia, *Electrochim. Acta* **12** (1967) 809.
[52] J. Lévêque, *Ann. Mines* **12** (1928) 201, 305, 381.
[53] C. S. Lin, E. B. Denton, H. S. Gaskill, and G. L. Putnam, *Ind. Eng. Chem.* **43** (1951) 2136.
[54] W. L. Friend and A. B. Metzner, *Am. Inst. Chem. Eng. J.* **4** (1958) 393.
[55] G. Wranglén, *Acta Chim. Scand.* **12** (1958) 1143.
[56] G. Wranglén, *Acta Chim. Scand.* **13** (1959) 830.
[57] G. Wranglén and O. Nilsson, *Electrochim. Acta* **7** (1962) 121.
[58] J. C. Bazán and A. J. Arvia, *Electrochim. Acta* **9** (1964) 17.
[59] J. C. Bazán and A. J. Arvia, *Electrochim. Acta* **9** (1964) 667.
[60] T. K. Ross and A. A. Wragg, *Electrochim. Acta* **10** (1965) 1093.
[61] R. E. Sioda, *Electrochim. Acta* **13** (1968) 375.
[62] R. E. Sioda, *Electrochim. Acta* **13** (1968) 1559.
[63] W. J. Blaedel, C. L. Olson, and L. R. Sharma, *Anal. Chem.* **35** (1963) 2100.
[64] W. J. Blaedel and L. N. Klatt, *Anal. Chem.* **38** (1966) 879.
[65] H. Gerischer, I. Mattes, and R. Braun, *J. Electroanal. Chem.* **10** (1965) 553.
[66] H. Matsuda, *J. Electroanal. Chem.* **15** (1967) 325.
[67] H. Matsuda, *J. Electroanal. Chem.* **15** (1967) 109.
[68] L. F. Crabtree, D. Küchemann, and L. Sowerby, in *Laminar Boundary Layers*, Ed., L. Rosenhead, Oxford Univ. Press, Oxford, 1963.
[69] D. Ilkovic, *Coll. Czech. Chem. Comm.* **6** (1934) 498.

[70]D. Ilkovic, *J. Chim. Phys.* **35** (1938) 129.
[71]E. D. Harris and A. J. Lindsey, *Nature* **162** (1948) 413.
[72]A. J. Lindsey, *J. Phys. Chem.* **56** (1952) 439.
[73]J. N. Agar, *Disc. Faraday Soc.* **1** (1947) 26.
[74]C. R. Wilke, *Chem. Eng. Prog.* **45** (1949) 219.
[75]C. R. Wilke, *Chem. Eng. Prog.* **46** (1949) 95.
[76]S. Ostrach, *Nat. Adv. Comm. Aeronaut. Tech. Rep. Nr.* 2635 (1952).
[77]N. Ibl and R. H. Müller, *J. Electrochem. Soc.* **105** (1958) 346.
[78]N. Ibl and U. Braun, *Chimia* **21** (1967) 395.
[79]L. Heerman and C. Feneau, *Ind. Chim. Belge* **32** (1967) 530.
[80]L. Ducret and C. Cornet, *J. Electroanal. Chem.* **11** (1966) 317.
[81]S. L. Marchiano and A. J. Arvia, *Electrochim. Acta* **13** (1968) 1657.
[82]E. Pohlhausen, *Z. Angew. Math. Mech.* **1** (1921) 115.
[83]G. H. Keulegan, *J. Res. Nat. Bur. Stand.* **47** (1951) 156.
[84]W. G. Mathers, A. J. Madden, Jr., and E. L. Piret, *Ind. Eng. Chem.* **49** (1957) 961.
[85]N. G. Hill, E. del Casal, and D. W. Zeh, *Int. J. Heat and Mass Transfer* **8** (1965) 1135.
[86]W. R. Wilcox, *Chem. Eng. Sci.* **13** (1961) 113.
[87]J. A. De Leeuw den Bouter, B. de Munnik, and P. M. Heertjes, *Chem. Eng. Sci.* **23** (1968) 1185.
[88]S. L. Marchiano and A. J. Arvia, *Electrochim. Acta* **14** (1969) 741.
[89]H. A. Laitinen and I. M. Kolthoff, *J. Am. Chem. Soc.* **61** (1939) 3344.
[90]S. L. Marchiano, J. S. W. Carrozza, and A. J. Arvia, *Anal. Asoc. Quím. Arg.* **56** (1968) 123.
[91]S. L. Marchiano and A. J. Arvia, *Electrochim. Acta* **15** (1970) 325.
[92]C. Wagner, *J. Electrochem. Soc.* **104** (1957) 129.
[93]A. Frumkin and G. Tedoradse, *Z. Elektrochem.* **62** (1958) 251.
[94]J. Koutecky and B. G. Levich, *Zh. Fiz. Khim. USSR* **32** (1958) 1565.
[95]J. E. B. Randles, *Can. J. Chem.* **37** (1959) 238.
[96]J. Jordan, *Anal. Chem.* **27** (1955) 1708.
[97]D. Jahn and W. Vielstich, *J. Electrochem. Soc.* **109** (1962) 849.
[98]J. Koutecky and B. G. Levich, *Dokl. Akad. Nauk SSSR* **117** (1957) 441.
[99]Z. Galus and R. N. Adams, *J. Electroanal. Chem.* **4** (1962) 248.
[100]T. Mizoguchi and R. N. Adams, *J. Am. Chem. Soc.* **84** (1952) 2058.
[101]D. Hawley and R. N. Adams, *J. Electroanal. Chem.* **8** (1964) 163.
[102]L. S. Marcoux, R. N. Adams, and S. W. Feldberg, *J. Phys. Chem.* **73** (1969) 2611.
[103]P. A. Malachevsky, L. S. Marcoux, and R. N. Adams, *J. Phys. Chem.* **70** (1966) 4068.
[104]M. C. Giordano, J. C. Bazán, and A. J. Arvia, *Electrochim. Acta* **11** (1966) 1553.
[105]A. J. Arvia, M. C. Giordano, and J. J. Podestá, *Electrochim. Acta* **14** (1969) 389.
[106]C. Martinez, J. A. Wargon, and A. J. Arvia, to be published.
[107]D. Posadas, J. J. Podesta, and A. J. Arvia, *Electrochim. Acta* **15** (1969) 1225.
[108]J. A. Olabe and A. J. Arvia, *Electrochim. Acta*, in press.
[109]V. A. Macagno, M. C. Giordano, and A. J. Arvia, *Electrochim. Acta* **14** (1969) 335.
[110]T. Iwasita and M. C. Giordano, *Electrochim. Acta* **14** (1969) 1045.
[111]M. E. Martins, G. Paus, A. J. Calandra, and A. J. Arvia, *Anal. Asoc. Quím. Arg.* **57** (1969) 91.
[112]A. J. Arvia, A. J. Calandra, and M. E. Martins, to be published.
[113]P. Beran and S. Bruckenstein, *J. Phys. Chem.* **72** (1968) 3630.
[114]G. R. Bopp, P. Stonehart, and D. M. Mason, *Electrochem. Tech.* **4** (1966) 416.
[115]A. Rejner and J. Balej, *Coll. Czech. Chem. Comm.* **26** (1961) 237.

[116] J. E. B. Randles and K. W. Sommerton, *Trans. Faraday Soc.* **48** (1952) 937.

[117] N. Tanaka and R. Tamamushi, *Electrochim. Acta* **9** (1964) 963.

[118] J. Jordan and R. A Javick, *J. Am. Chem. Soc.* **80** (1958) 1264.

[119] J. Koutecky and R. Brdicka, *Coll. Czech. Chem. Comm.* **12** (1947) 337.

[120] J. Koutecky, *Nature* **174** (1954) 233.

[121] W. J. Albery and S. Bruckenstein, *Trans. Faraday Soc.* **62** (1966) 1920.

[122] W. J. Albery and S. Bruckenstein, *Trans. Faraday Soc.* **62** (1966) 1946.

[123] W. J. Albery and S. Bruckenstein, *Trans. Faraday Soc.* **62** (1966) 2584.

[124] W. J. Albery and S. Bruckenstein, *Trans. Faraday Soc.* **62** (1966) 2596.

[125] H. Matsuda, *J. Electroanal. Chem.* **16** (1968) 153.

[126] R. Leuteritz and W. Mangler, *Untersuch. Mitt. Deut. Luftahrf.* (1945) Nr. 3226.

[127] C. W. Tobias, M. Eisenberg, and C. R. Wilke, *J. Electrochem. Soc.* **99** (1952) 359C.

[128] K. Asada, F. Hino, S. Yoshizawa, and S. Okada, *J. Electrochem. Soc.* **167** (1960) 242.

[129] E. M. Sparrow and J. L. Gregg, *Trans. Am. Soc. Tech. Eng.* **78** (1956) 435.

[130] N. Ibl, W. Ruegg, and G. Trumpler, *Helv. Chim. Acta* **36** (1953) 1624.

[131] E. Mattson and J. O'M. Bockris, *Trans. Faraday Soc.* **55** (1959) 1586.

4

The Mechanism of Charge Transfer from Metal Electrodes to Ions in Solution

Dennis B. Matthews

Union Carbide Australia, Ltd., Chemicals Division
Rhodes, N.S.W., Australia

John O'M. Bockris

Electrochemistry Laboratory
John Harrison Laboratory of Chemistry
University of Pennsylvania, Philadelphia, Pennsylvania

I. INTRODUCTION

The essential distinguishing element of electrode, compared with chemical, kinetics is the characteristic dependence of electrode reaction rates on the electrode potential. The basic theory of this dependence is, however, one of the less developed areas of electrode kinetics. Two facts are outstanding: (i) at 25°C, the symmetry factor β is frequently close to 1/2; (ii) the symmetry factor is independent of electrode potential over large ranges of potential, though it varies with it at the highest current densities. Second-order dependencies of β on temperature, solution concentration, and solution composition are issues concerning which there are as yet insufficient data to justify further theoretical development.

For a given step of a reaction sequence, the dependence of current density i on electrode potential is characterized by the symmetry factor β. The forward- and reverse-current densities are given by

$$\vec{i} = i_0 \exp(-r\eta F/RT) \tag{1}$$

and

$$\ddot{i} = i_0 \exp[(1 - \beta)r\eta F/ET] \tag{2}$$

respectively, where i_0 is the exchange-current density, r is the number of electrons transferred in the given step (r is either unity or zero), and η is the overpotential. The net rate is given by

$$i = \vec{i} - \ddot{i} \tag{3}$$

Equations (1) and (2) involve the commonly accepted, yet arbitrary, division of the electrochemical potential η into electrical and chemical parts,[1]

$$\bar{\mu}_j^\alpha = \mu_j^\alpha + z_j F\phi^\alpha \tag{4}$$

where $\bar{\mu}_j^\alpha$ is the electrochemical potential of the species j in the phase α, z_j is the valence of the species j, μ_j^α is the chemical potential of the species j in the phase α, and ϕ^α is the inner potential of the phase α.

The inner potential ϕ is further divided[1] into the outer potential ψ and the dipole potential χ,

$$\phi^\alpha = \psi^\alpha + \chi^\alpha \tag{5}$$

The real potential ρ_j^α of a species j in the phase α is defined by

$$\rho_j^\alpha = \mu_j^\alpha + z_j F\chi^\alpha \tag{6}$$

The electronic work function Φ_F of a metal and the solvation energy L of an ion are related to the real potential by

$$\Phi_F = \rho_e^{\text{metal}} \tag{7}$$

$$L = \rho_i^{\text{solution}} \tag{8}$$

The real potential is a measurable quantity. The surface potential χ is not experimentally measurable for a single phase.[2] The contribution to the dipole-potential difference of a phase α in contact with another phase β is denoted by $^\beta g^\alpha$.

The central position of the symmetry factor in electrode kinetics, as expressed by

$$d(\Delta \bar{G}_0^{\ddagger})/d(F \Delta\phi) = \beta \tag{9}$$

where $\Delta \bar{G}_0^{\ddagger}$ is the standard electrochemical free energy of activation, cannot be underestimated, and theories of charge transfer at electrodes must inevitably involve theories of the symmetry factor. Equation (9) is a particular case of the more general relation[8,108]

$$d(\Delta \bar{G}_0^{\ddagger})/d(\Delta \bar{G}_0) = \beta \qquad (10)$$

where $\Delta \bar{G}_0$ is the standard electrochemical free energy of reaction. The dependence of the rate of the overall reaction on electrode potential is expressed by the transfer coefficient α. The relation of the transfer coefficient to the symmetry factor is given by (cf. Appendix I),

$$\alpha = s/v + \beta r \qquad (11)$$

where s is the number of electrons transferred per act of the overall reaction, in steps preceding the rate-determining step (r.d.s.), v is the stoichiometric number (cf. Appendix I), and r is the number of electrons transferred in one act of the r.d.s. No account has been taken in equation (11) of double-layer effects or of effects arising from the adsorption of reactants or products at the electrode.

II. GURNEY'S QUANTUM MECHANICAL THEORY OF CHARGE TRANSFER

Gurney's pioneer paper[3] in charge-transfer theory was an attempt to rationalize the experimental results obtained by Bowden[4] in his investigations of hydrogen overpotential.

1. Neutralization of Gaseous Ions

Gurney uses as his starting point the neutralization of gaseous ions, which occurs by electron tunneling.[5] The reaction considered is of the type

$$\text{ion} + e^-(\text{M}) \rightarrow \text{atom} + \text{M} \qquad (12)$$

where $e^-(\text{M})$ represents an electron in the metal M. A potential energy–distance diagram for this electron-transfer reaction is obtained by comparing the energy of the electron in the metal to the energy of the electron in the atom, taking cognizance of the environment (and associated interaction energies) in which M, the ion, and the atom are situated. The difference in these two electron energies, i.e., the energy change of reaction (12), gives the energy separation between the initial state and the final state. It was assumed by Gurney that there is no interaction of either the ion or the atom with the metal.† Hence, the standard enthalpy change of the electron in going from the metal to the ion is given by

$$\Delta H_0(e) = \Phi - J \tag{13}$$

where Φ is the work function of the metal and J is the ionization potential of the atom.‡ The change in potential energy U_e of the electron as it travels from the metal to the ion is given by the full line in Figure 1. Near the metal, the potential energy of the electron is determined (partly) by the force between the electron and its image in the metal, and near the ion, the potential energy of the electron is determined by the Coulombic force exerted on the electron by the ion. Because the image potential and the Coulombic potential in the curves of Figure 1 merge at the energy level corresponding to infinite separation of metal and electron, we conclude that Gurney considered large metal–ion separations. (In Figure 1, MM represents the energy levels of the metal that, at $0°K$, are fully occupied by electrons. The uppermost level MN is the Fermi level and AB represents the energy level of the valence electron in the atom.)

According to classical theory, in order for the electron to pass from the metal to the ion, it must surmount the potential–energy barrier NCA. As a result of the wavelike nature of the electron, penetration into, and transition through (tunneling) the potential–energy barrier is possible and was found long ago to be highly probable in gases.[5] If the electron transfer occurs without release or absorption of radiant energy, then such transfer can occur only between levels of equal electron energy. Hence, electron transfer

---

†This simplification by Gurney proved to be unfortunate, perhaps one might say disastrous. Thus, it caused the predictions of Gurney's view to become markedly discrepant with the order of magnitude of the heat of activation observed for some electrochemical reactions. This led electrochemists to repudiate the quantum mechanical theory of Gurney's (one of the earliest applications of quantum mechanics to chemistry). The quantum mechanics of electrode processes had to wait until the 1960's to be rediscovered.
‡Presumably, the first ionization potential is meant to correspond to the removal of the outer or valence electron from the atom.

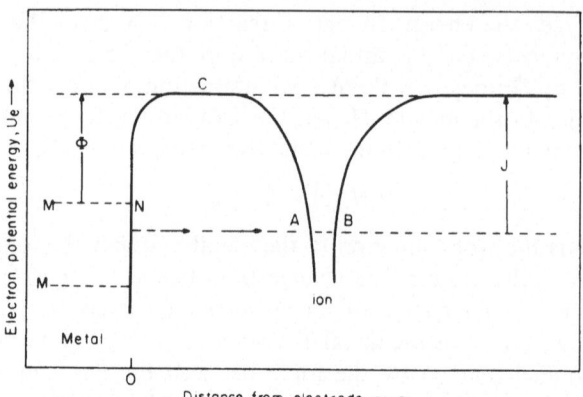

Figure 1. Model for electron transfer from a metal to an ion in vacuum according to Gurney.[3] Φ is the electronic work function of the metal; J is the ionization potential of the atom; AB is the energy level of the valence electron in the atom; MM are the filled electron-energy levels of the metal; MN is the Fermi level of the metal; NCA is the potential-energy barrier for electron transfer.

through the potential–energy barrier may occur only if (see Figure 1) the ionization potential J is greater than, or equal to, the work function Φ of the metal. This is the basic condition for electron tunneling at the metal–gas interface. A possible transfer of an electron by tunneling is illustrated in Figure 1 by the arrows going from an electron energy level in the metal to an empty electron level of equal energy in the atom. An impossible transfer by electron tunneling at $0°K$ would be one where the level AB is above the Fermi level MN.

2. Neutralization of Ions in Solution, $0°K$ Approximation

Gurney's next step is to extend consideration of the basic condition for electron tunneling from neutralization of ions in the gas phase to the neutralization of ions in electrolytes. The electron-transfer reaction now under consideration is of the type

$$\text{ion–solvent} + e^-(M) \rightarrow \text{atom} + \text{solvent} + M \qquad (14)$$

The energy of the electron in the metal is again given by $-\Phi$ as in the vacuum case, but the energy of the electron in the atom involves another term, the solvation energy of the ion. Thus, the energy of the electron in the atom is given by the energy change of the process

$$\text{ion–solvent} + e^-(\text{vac}) \to \text{atom} + \text{solvent} \qquad (15)$$

The total energy change of reaction (15) is $-J - L$, where L is the solvation energy of the ion. As in the vacuum case, the ion or atom interactions with the metal are ignored and the atom–solvent interactions are also ignored at first. An important refinement is introduced later to allow for the atom–solvent interaction. The standard enthalpy change of the electron in going from the metal to the ion is given by

$$\Delta H_0(e) = \Phi - J - L \qquad (16)$$

Utilizing the same assumptions regarding conservation of energy as in the vacuum case (radiationless transition), we obtain that *electron tunneling through the potential–energy barrier (Figure 2) can occur only if the electron energy of the atom equals the electron energy of the metal.* The basic electron-tunneling condition is thus

$$\Phi \leq J + L \qquad (17)$$

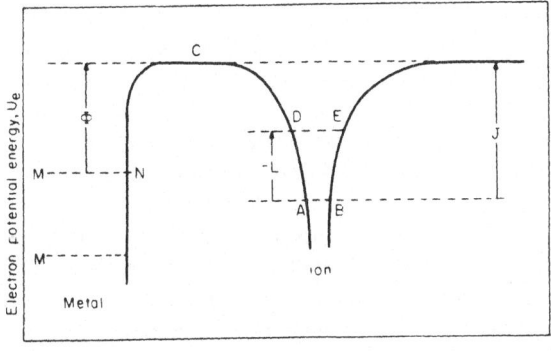

Figure 2. Model for electron transfer from a metal to an ion in solution according to Gurney.[3] Metal–atom and atom–solvent interactions have been assumed negligible. Symbols as in Figure 1, except for L, which is the ion–solvation energy, and DE, which is the energy level of the valence electron in the atom.

Since L is a negative quantity and J is positive, then

$$J + L \leq J \tag{18}$$

If the value of L is such that $J + L < \Phi$, then the tunneling condition (17) is not satisfied and electron tunnel transfer cannot occur. The effect of solvation is thus to make tunnel transfer of the electron more difficult. Figure 2 illustrates a case where electron tunneling from the metal to the ion in solution is not possible, because the empty electron energy level DE in the ion is *above* the electron energy level MN in the metal.

However, electron tunneling may be brought about by increasing the cathodic potential of the metal with respect to the solution, which changes the energy of the electron in the metal to $-\Phi - F\psi^m$ and the energy of the electron in the atom to $-J - L - F\psi^s$, where ψ is the outer potential of the given phase. It is necessary at this point to digress somewhat in order to justify the use of ψ in these expressions.

In his original paper, Gurney[3] speaks of the electrode potential as acting to increase or decrease the electrode work function. Nowadays, this view cannot be accepted, for several reasons. In the first place, we recognize that the term electrode potential, as used by Gurney, refers to the potential of the test electrode with respect to some reference electrode, the potential of an electrode, by itself, being an unmeasurable quantity. Furthermore, the potential of a single electrode is a potential difference between the electrode and the electrolyte. Lastly, we are placed in the position of asking what sort of potentials constitute this potential difference. In the present situation, we initially considered the case of a metal and a solution in contact, in which case, we utilized the terms Φ and L, which are real potentials. If we neglect differences between χ and g, then the situation corresponds to zero charge on the electrode, so that $\Delta\psi = 0$. For a given metal and solution, we can change the single-electrode potential from that of the above to some new value by altering the charge on the electrode, which, if we assume χ (or g) to be independent of charge, results in a change $\Delta\psi$ in single-electrode potential. This, then, is the justification for using $\Delta\psi$ and not, say, $\Delta\phi$, in the above expressions.†

†An exact solution requires the use of terms $\phi - \chi$ instead of ψ, where $\phi = \psi + g$, leading to a term $\Delta\phi - \Delta\chi$ instead of $\Delta\psi$ in equation (20).

At a charged electrode, then, the electron tunneling condition is

$$\Phi + F\psi^m < J + L + F\psi^s \qquad (19)$$

or

$$\Phi + F\,\Delta\psi < J + L \qquad (20)$$

where

$$\Delta\psi = \psi^m - \psi^s \qquad (21)$$

If $\Delta\psi$ is negative, then the effect of the electrode potential is to make electron tunneling easier. Electron tunneling at $0°K$ becomes possible when $\Delta\psi$ is sufficiently negative that

$$F\,\Delta\psi < J + L - \Phi \qquad (22)$$

It was thus concluded by Gurney[3] that "the overpotential is not in this case due to an obstructive film of gas on the electrode, but, on the contrary is an essential condition without which decomposition of the electrolyte cannot proceed."†

It should be mentioned at this stage that the Born–Oppenheimer and adiabatic approximations‡ are used implicitly by Gurney in calculating electron energies. The assumption is that, *during* the electron transition, the various atoms, ions, and molecules do not move. The electron transition may be preceded or followed by movement of other particles. This separation of motions is justified in view of the small mass of the electron compared to the mass of the H atom, water molecules, etc.

3. Neutralization of Ions in Solution at Temperatures above 0°K

At temperatures above $0°K$, associated with the ion–solvent system are rotation–vibration levels distributed according to Boltzmann's law. It is assumed that these vibration–rotation levels are sufficiently blurred by interaction with adjacent solvent molecules to be treated as forming a continuous spectrum of levels. For each

---

†Up to the time of Gurney's theory (1931), the necessity of making the electrode potential depart from its thermodynamic equilibrium value before a net current could pass was not clearly understood. The phenomenon of overpotential was often explained as due to some special difficulty, e.g., the presence of a gaseous film, or of a layer of dipoles.

‡For a discussion of these approximations, see Appendix II.

rotation–vibration level, the electron energy of the atom will be different. Hence, the electron energy of the atom is not single-valued but possesses a distribution of values. The condition for electron tunneling is thus

$$\Phi + F\,\Delta\psi < J + L_n \tag{23}$$

where L_n is the hydration energy of an ion in the nth rotation–vibration level. The number of ions with hydration energy L is given by

$$N(L) = N_0 \exp[-(L - L_0)/RT] \tag{24}$$

where N_0 is the number of ions in the ground state, $L = L_0$.

At temperatures above $0°K$, the number of filled electron states in the metal at a given energy level E is given by

$$N_b(E) = AE^{1/2}\{1 + \exp[E - E_F)/RT]\}^{-1} \tag{25}$$

where E_F is the value of E at the Fermi level of the metal and A is defined by[33]

$$A = [4\pi V(2m_e)^{3/2}/h^2]\,dE \tag{26}$$

where V is the volume and m_e is the electron mass.

Let E_n and E_0 be total electron energies (with respect to $E = 0$ when the electron is at infinite distance from the metal *in vacuo*), corresponding to $L = L_n$ and $L = L_0$, respectively. Since $L_n < L_0$ (L being negative), then $E_n < E_0$ (cf. Figure 3). Equation (24) may be rewritten

$$N(E) = N_0 \exp[+(E - E_0)/RT] \tag{27}$$

At temperatures above $0°K$, electrons in the metal with energy E greater than E_F may tunnel to ions with energy E less than E_0, the electron transfer being limited, as before, to radiationless transitions. The number of electrons tunneling at a given energy level E is given by

$$\int_{E_F - F\Delta\psi}^{E_0} N_b(E)N(E)W_e(E)\,dE \tag{28}$$

where $N_b(E)$ and $N(E)$ are the number of electron states with energy E in the metal and in the atoms, respectively, and $W_e(E)$ is the probability of electron tunneling.

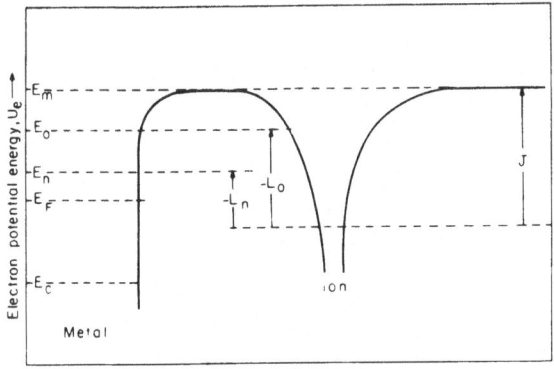

Figure 3. Relation of solvation energy L to the electron energy E; E_F is the Fermi level; E_c is the energy level at the bottom of the conduction band; L_0 and L_n are the solvation energies for the ground and nth vibration-rotation levels, respectively; E_0 and E_n are the electron energies for the ion with solvation energies L_0 and L_n, respectively; E_m is the energy of the top of the electron-transfer barrier. (After Gurney.[3])

Expression (28) ignores the contributions from electrons that surmount the barrier and from electrons that tunnel below the Fermi level E_F (or $E_F - F\,\Delta\psi$ in the presence of a metal–solution potential difference $\Delta\psi$). The first approximation is reasonable in view of the fact that the barrier height (cf. Figure 3) is of the order of magnitude of the work function ($\simeq 5\,\mathrm{eV}$ or 115 kcal). At energies E below E_0, the number of electron energy levels in the solution falls off exponentially, while the number of electrons in the metal rises exponentially in the same energy range until the energy level $E_F - F\,\Delta\psi + 2.3RT$, where the number of electrons in the metal ceases to rise exponentially, as a result of the Fermi distribution (cf. Figure 4). For this reason,† one may neglect the contribution to the rate from electrons below $E_F - F\,\Delta\psi$.

Gurney considers the basic difference between charge transfer at the electrode–solution interface and electron emission into vacuum to be that, in the former case, the rate of charge transfer is strongly dependent on the existence of electron-acceptor levels, *without which no reaction is possible.*

†As will be shown later, these conclusions are not applicable to the final stage in the development of Gurney's theory.

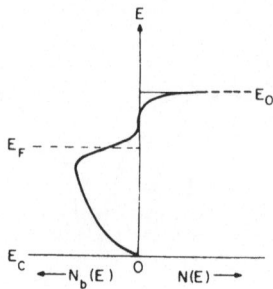

Figure 4. Electron distribution
functions for the metal $N_b(E)$
and for the solvated ion $N(E)$.
Notation as in Figure 3.

By analogy, at the electrode–solution interface, the cathode is
the emitter electrode and the ionic layer in the Helmholtz plane of
the metal–solution double layer is the collector electrode. In the
absence of such an ionic layer, a "space charge" would build up at
the cathode surface. The double layer corresponds to the condition
of very high field in the metal–vacuum system, but there is an
important difference. The ionic layer has a limited capacity for
electrons. For this reason, it is necessary to include a term in the
rate expression to allow for the probability of finding a vacant
electron level of the required energy.

The next stage in the development of Gurney's theory is to
consider the spatial distribution of the distances out from the
electrode at which neutralization occurs. If $N(E, x)$ is the number of
ions per unit area at a distance x from the electrode, then the rate is
proportional to

$$\int \int N_b(E)N(E, x)W_e(E, x)\, dE\, dx \tag{29}$$

In evaluating $W_e(E, x)$ it is assumed that all ions are neutralized at
a mean distance $\bar{x}$ and that the barrier to electron transfer may be
approximated by a rectangular barrier. The Wentzel–Kramers–
Brillouin (WKB) approximation[6] for a rectangular barrier is
employed. This is justifiable since, at the energy levels E of interest,
the potential energy of the electron is a slowly varying function of
distance and rather independent of the shape of the barrier (i.e.,

parabolic, triangular, square, etc.). Substituting equations (25) and (27) and a WKB expression for $\overline{W}_e(E)$ into (29), and utilizing the same assumptions as in (28), one obtains that the rate is proportional to

$$\int_{E_F - F\Delta\psi}^{E_0} \exp[(E - E_0)/RT] \exp[-(E - E_F)/RT]$$

$$\times \exp[-(4\pi x/h)2m_e(E_m - E)^{1/2}] \, dE \qquad (30)$$

where E_m is the electron energy at the barrier maximum (Figure 3).

4. Solvent–Atom Interactions

Up to this point, Gurney was taking the elementary view that the mutual potential energy of the ion and adjacent water molecules is equal to $-L$ before neutralization and is zero after neutralization. This, as Gurney pointed out, is not correct, for, if neutralization occurs with stationary nuclei (Born–Oppenheimer approximation), then the mutual potential energy immediately after neutralization would have some value R. For example, in Figure 5, curve NBC represents the potential energy of the hydrogen-atom–water-molecule interaction and curve $DAFG$ the potential energy with

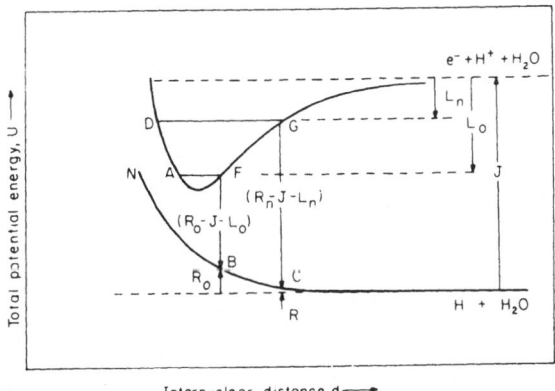

Figure 5. Variation of potential energy of H^+–H_2O + e^- and of H–H_2O with internuclear separation; R_0 and R_n are the H H$_2$O interaction energies produced by adiabatic electron transfer to H^+–H_2O with solvation energies L_0 and L_n, respectively. No allowance has been made for M–H interaction. (After Butler.[7])

distance between H^+ and H_2O. The curves for $e^- + H^+ + H_2O$ and $H + H_2O$, when the nuclei are infinitely separated *in vacuo*, are separated by an amount of energy J, where J is the ionization potential of the hydrogen atom. As the respective nuclei are brought together, the separation of the two curves changes in a way determined by the relative potential energies for $e^- - H^+-H_2O$ and $H-H_2O$. The change in the energy of the electron for neutralization of an ion in its lowest rotation-vibration state is represented in Figure 5 by FB. The energy of the ground-state level AF of H^+-H_2O is L_0 and the energy of $H-H_2O$ is represented by B. Similarly, the energy of an excited level DG of H^+-H_2O is L_n and the energy of $H-H_2O$ is represented by C. From Figure 5, it may be seen that an increase in energy of H^+-H_2O leads to a decrease in energy of $H-H_2O$. In Figure 5, Gurney omits the fact that the electron comes from the metal, and hence energy changes shown in Figure 5 do not include the work function.† This omission makes no difference to the form the rate expression. Inclusion of Φ raises the curve NBC with respect to the curve $DAFG$ (Figure 5) and the resultant figure then has the appearance of the familiar Horiuti–Polanyi potential-energy profile,[8] the curves NBC and $DAFG$ intersecting at a point corresponding to the activated state (Figure 6). Inclusion of the term R into the energy of the electron in the atom means that the electron tunneling condition is‡

$$\Phi + F\,\Delta\psi \leqslant J + L - R \qquad (32)$$

From Figure 5, it is seen that, when L increases (from AF to DG), R decreases (from B to C) so that the change in $L - R$ is greater than the change in L. In the absence of definite information concerning the dependence of R and L on internuclear separation, Gurney uses the empirical relation

$$dL = \beta\,d(L - R) \qquad (33)$$

---

†Perhaps Φ was not included by Gurney in Figure 5 since it is invariant with respect to nuclear separation.
‡The quantity R is defined by

$$H-H_2O \rightarrow H + H_2O; \qquad \Delta H_0 = -R \qquad (31)$$

According to this definition, R is positive.

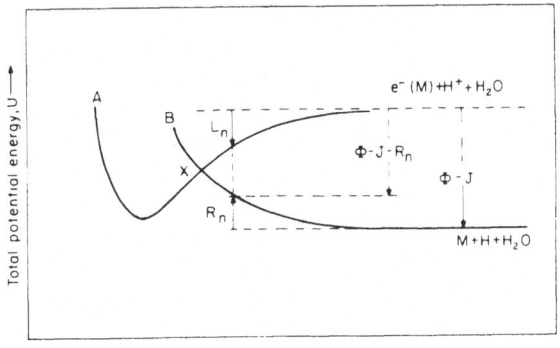

Figure 6. Variation of potential energy of H^+–H_2O + e^-(M) and of M + H–H_2O with internuclear separation, no allowance being made for M–H interaction. (After Butler.[7])

where $0 < \beta < 1$, i.e., the change in L is some fraction of the change in $L - R$. Equation (33) may be rewritten as

$$L - L_0 = \beta[(L - R) - (L_0 - R_0)] \quad (34)$$

Thus, whereas previously the number of atoms with electron energy E was given by equation (27), now, each level E corresponds to a given value of $L - R$ instead of just L. The number of ions with solvation energy L is given by

$$N(L) = N_0 \exp[-(L - L_0)/RT] \quad (35)$$

where N_0 is the number of ions in the ground state ($L = L_0$). From equations (35) and (24), the number of H_3O^+ ions with energy L that form H–H_2O with repulsive energy R is given by

$$N(L, R) = N(L_0, R_0) \exp\{-\beta[(L - R) - (L_0 - R_0)]/RT\} \quad (36)$$

where $N(L_0, R_0)$ is the number of H_3O^+ ions in the ground state ($L = L_0$) that form H–H_2O with repulsive energy R_0. In terms of total electron energies E (Figure 3), we have the number of atoms with the electron energy E given by

$$N(E) = N(E_0) \exp[+\beta(E - E_0)/RT] \quad (37)$$

where E and E_0 are the values of E corresponding to $L - R$ and $L_0 - R_0$, respectively.

The value of β, as defined by equation (33), depends on the relative slopes of the curves FG and BC (Figure 5) and, to a first approximation, is taken as a constant for the energy range under consideration. Gurney finds that the electron tunneling probability $W_e(E)$ is relatively independent of E so that it may be taken as a constant W_e.

By substituting equations (37) and (25) into (28), one obtains that, for large $|E - E_F|$, the rate is proportional to

$$\int_{E_F - F\Delta\psi}^{E_0} \exp[\beta(E - E_0)/RT] \exp[-(E - E_F)/RT]\, dE \qquad (38)$$

which is readily evaluated to give

$$\log i = [\beta(E_F - F\,\Delta\psi - E_0)/RT] + \log T = \text{const} \qquad (39)$$

where i is the current density.

The approximations used in deriving equation (39), although valid in the derivation of equation (30), are no longer valid. The integrand of (38) has a maximum value at $E = E_F - F\,\Delta\psi$ and decreases exponentially on either side. The lower limit of integration in (38) is thus too high. For the same reason, the approximation $|E_F - E| \gg RT$ is not valid. The two errors, however, largely cancel, so that the resulting error in the rate calculation is small.

The symmetry factor β has its essential origin with the theory of Gurney. It is the coefficient relating the electron energy difference in the metal and in the H atom at the electrode–solution interface to the thermal energy required by the ion in order to fulfill the electron-tunneling condition (32),

$$\Phi + F\,\Delta\psi \leq J + L - R$$

i.e., to achieve the activated-state configuration. This was implicit in Gurney's paper. It was not brought out or discussed until very much later.[12,21,81]

The Gurney theory predicted, in agreement with experiment,[4] the following facts:

(i) $\quad d(\log i)/d(\Delta\psi) = C_1/T$ $\hfill (40)$

(ii) $\quad d(\log i)/dt = C_2$

over a small temperature range; $\hfill (41)$

(iii) $\quad C_2$ increases with increasing $\Delta\psi$;

where C_1 and C_2 are independent of temperature. Hence, the qualitative details of the work of Bowden[4] on the h.e.r. were entirely accounted for, and the value of C_1 was given a rational explanation.

5. Summary

Because of the stress it placed on the role played by electron tunneling, Gurney's theory became known as the electron–transfer theory of hydrogen overvoltage. Thus, Butler[7] states that, "In Gurney's theory the potential determining process was the transfer of electrons from the metal to the hydrogen ions in the solution." In contrast, Butler says, in reference to a later theory of Horiuti and Polanyi,[8] that, "Horiuti and Polanyi have suggested an alternative mechanism in which the primary process is the transference of hydrogen ions to adsorption positions at the surface of the metal, in the course of which neutralization occurs." Actually, the stretching of the H^+-OH_2 bond, as manifested by changes in L, is just as important in Gurney's theory as in the later theory of Horiuti and Polanyi.[8] The principal difference in the theories of Gurney's and of Horiuti and Polanyi's is that Gurney was very explicit about when and why neutralization occurs, whereas Horiuti and Polanyi smooth over the distinction between electron transfer and proton transfer by the statement, "in the course of which, neutralization occurs."

The theory of Gurney received little attention because it was criticized by Butler on the grounds that it would give rise to heats of activation which were far too high compared with the experimental values. Butler himself showed how this difficulty might be overcome (see Section III, 1), but he did not clarify the fact that his views were an addendum, not an alternative, to those of Gurney. The fact that Butler pointed out that Gurney's theory gave quite impossible numerical results while his (Butler's) modified Gurney theory did give plausible results caused a hiatus in the theory of electrode kinetics, which, instead of developing along quantum-mechanical lines, continued phenomenologically until the 1950's.

The dependence of rate on the work function Φ of the metal has also been raised in connection with Gurney's theory.[9] It has been argued that, as Φ is increased, the theory predicts that the current density at a given overpotential should decrease. This was considered contrary to the then-known facts.[10] It is now known[11,12,17] that the predicted dependence is observed for the high-hydrogen-over-potential ("irreversible") metals mercury, thallium, gallium, tin, lead, manganese, and indium. This qualitative agreement for the high-overpotential group is, however, fortuitous. The work function Φ may enter into the rate of proton discharge in several ways. It may enter directly as a term obtained in considering the difference in heat contents of the initial and final states for the proton-discharge reaction, or it may enter indirectly by influencing the heat of adsorption of hydrogen on the metal. The former way is the one concerned in the above discussion. The above discussion has not, however, considered the relation between the metal–solution potential difference $\Delta\psi$ and the measured cell potential. When this is done, the contribution of the metal–metal potential difference cancels the direct dependence of rate on the work function of the test electrode. This leaves the dependence of rate on the heat of adsorption of atomic hydrogen as the source of the dependence of rate (at a given cell potential) on work function. This conclusion was reached by Conway and Bockris[11] and recent data[12] support this viewpoint.

III. DEVELOPMENTS OF THE QUANTUM MECHANICAL THEORY

1. Butler's Theory

In 1940, Butler[13] wrote: "It could be supposed that the discharge process" (of H_3O^- ions at electrodes) "takes place in either of the following ways:

(a) Electrons escape from the metal and become attached to nearby hydrogen ions in the solution:
(b) the protons of the hydrogen ions (H_3O^+) come into contact with, or are adsorbed on, the metal and are then neutralized."

The former theory was attributed to Gurney[3] and the latter theory to Horiuti and Polanyi.[8] The view presented by Butler has been carried over to recent times.[9,14–17] The theories due to Horiuti and Polanyi,[8] of Butler,[7] and to Parsons and Bockris[19] will be presented in due course. Marked similarities between these theories and that due to Gurney[3] will become obvious.

Butler allows for the adsorption of atomic hydrogen on the electrode, in keeping with the suggestion of Horiuti and Polanyi,[8] and he applies this additional factor as a correction to the Gurney theory. According to the Gurney theory, the condition for transfer by tunneling of electrons from the Fermi level of the metal to a proton is given by equation (32),

$$\Phi + F \Delta\psi \leq J + L - R$$

Butler replots Gurney's representation of the relation of dL to dR (Figure 5) so as to include the electronic work function Φ. The result is given in Figure 6. Rearranging equation (32), the electron-tunneling condition is

$$\Phi + F \Delta\psi - J + R \leq L \qquad (42)$$

From Figure 6, it is seen that condition (42) is fulfilled at the intersection point x and at points lying to the left of x. This then provides an exact correlation between Gurney's theory and the Horiuti–Polanyi theory except for the allowance of the M–H interaction. Allowance for the metal–atom interaction energy† A changes the electron energy in the atom of reaction (14) so that the electron tunneling condition becomes

$$\Phi + F \Delta\psi \leq J + L - R + A \qquad (43)$$

From equation (43), it is seen that inclusion of the term A makes it easier to fulfill the electron tunneling condition. Since A is dependent on the rotation-vibration level of the ion, for the same reasons that R was dependent on this variable, then the variation of A with distance must also be included in Figure 6. This is shown in Figure 7, where it is shown that allowance for the M–H interaction results in a lowering of the intersection point and consequently to a lower energy of activation.

In the calculations of Butler's, an improbably small value (0.3 Å) is used for the proton-transfer distance.‡ Neither Gurney,

† A is defined by

$$MII \rightarrow M + II; \qquad \Delta II_0 - A \qquad (44)$$

‡ Parsons and Bockris[19] consider the most probable value to be 0.55 Å, while Bockris and Matthews,[21] taking into account evidence from measurements of the capacity of

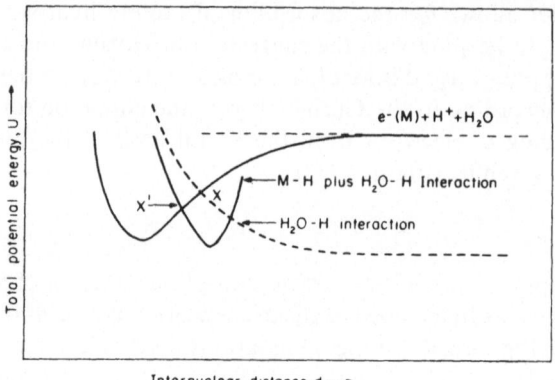

Figure 7. Variation of potential energy of H^+–H_2O + e^-(M)
and of M + H–H_2O with internuclear separation, including
the M–H interaction. (After Butler.[7])

Horiuti and Polyani, nor Butler calculated the symmetry factor β
from their potential-energy–distance plots. Essin and Kozheurov[20]
used the Gurney–Butler theory with larger proton-transfer distances
and obtained $\beta \simeq 0.5$ for the discharge step on Hg and Ni.

2. Gerischer's Theory

The mechanism of redox reactions at metal and semiconductor
electrodes has been treated by Gerischer[22–24] as an extension of the
Gurney theory of charge transfer.

Gerischer considers redox reactions of the type

$$M_{solv}^{z+1} + e^-(M) \rightleftharpoons M_{solv}^z \tag{45}$$

where both ions remain dissolved in the electrolyte and the chemical
interaction with the surface is much less than in the case of, say,
proton discharge to form adsorbed H atoms.

For a redox electrolyte of the type M_{aq}^{z+1}/M_{aq}^z, the M_{aq}^{z+1} represent empty terms and the M_{aq}^z full terms. The position of the empty
terms on the energy scale is given by the work done when an electron
is brought from infinity into the electron state of the ion, which is

---

the electrode–solution double layer and of the dependence of the separation factor on
potential, consider it to be about 2.4 Å.

then changed into an M_{aq}^z ion. Each of the ions has a solvation sheath, in which the structure of the solvent molecules is different from that of the pure solvent and where strong interaction forces between the ion and its surroundings are present. The solvation sheaths of the two ions are normally quite different on a time average, and remain in constant fluctuation because of thermal energy. Correspondingly, the energies of the two kinds of ions M^z and M^{z+1} form a distribution of electron energy states in the redox electrolyte which is analogous to the distribution of electrons in the metal.

The result obtained is that electron exchange takes place in the immediate region of the Fermi level, which indicates that the solvation sheath of the ions favored for electron transfer has a structure between the most probable states of the individual components. Electron transfer occurs only when the ions have reached a favorable configuration.

The definition of β derived in Gerischer's theory is fundamentally the same as that derived in Gurney's theory.

More recently, Gerischer has made considerable progress by extending his theory to problems of catalysis at semiconductor electrodes.[116] Mechanisms of these charge-transfer reactions are discussed under two categories, "weakly interacting" and "strongly interacting" with the semiconductor electrode surface.

Gerischer, in a more recent paper, relates the electron transfer theory, developed earlier, to the general situation in catalysis. Heterogeneous catalysis is usually concerned with electroneutral reactions. Partial reactions within the overall electroneutral reaction may indeed be concerned with charge transfer to the substrate but because there must be an opposite reaction to compensate for the transfer charge, the mechanism of the catalysis is difficult to analyze. In an electrochemical reaction, the individual reactions can be better studied because a net transfer rate can be measured. Thus, chemical catalysis may sometimes be regarded as a limiting case of electrochemical catalysis in which there are equal rates of partial electrochemical reactions.

Gerischer[116] goes on to develop the relation of charge transfer reactions to photocatalytic effects and to the possibility of the initiation of reactions by illumination of electrodes. The mechanism of photosensitization is analyzed in connection with the fundamental processes of charge transfer.

3. Theory Due to Bockris and Matthews

An examination of the theory of charge transfer in which several possible mechanisms are compared has been given by Bockris and Matthews.[21] The theory was developed for the discharge step of the hydrogen-evolution reaction but has more general applicability, e.g., to metal deposition.

(i) Classical Electron Transfer

In the so-called "proton-transfer theory" of the discharge step,

$$H_3O^+ + Me^- \rightarrow MH + H_2O \qquad (46)$$

of the hydrogen-evolution reaction, the role of the electron is not defined. To understand this, it is necessary to consider the potential-energy profile for the electron transfer.

The potential-energy profile for the electron transfer is compounded from two potential-energy curves and for the change in potential energy as the electron moves away from the metal and the change in potential energy as an electron approaches the H_3O^+ ion. The states to be considered are

Charged state (state I): $(H_3O^+)_{dl} + e^-(M)$
Neutralized state (state II): $(M\text{-}\text{-}\text{-}H\text{-}\text{-}\text{-}H_2O)_{dl}$

The neutralized state describes the state immediately after the electron transfer, i.e., before the various atoms and molecules relax into their ground vibration-rotation states.

The forces responsible for the H-atom transfer are the M–H attractive force A and the H–H$_2$O repulsive force R. Following Parsons and Bockris,[19] the variation of A with internuclear separation may be given by the Morse function,

$$A = A_e[2e^{-a(d-d_e)} - e^{-2a(d-d_e)}] \qquad (47)$$

where A_e is the value of A for the internuclear distance d_e corresponding to the minimum of potential energy and a is the Morse constant, given by

$$a = 1.2177 \times 10^7 \omega(\mu/A_e)^{1/2} \quad \text{cm}^{-1} \qquad (48)$$

where μ is the reduced mass and ω the vibration frequency for M–H. The variation of R with internuclear separation d is approximated by the expression

$$R = [0.567 \exp(-24.9d^2) + 0.215d \exp(-2.40d^2)] \times 10^{-10} \quad \text{ergs} \tag{49}$$

The standard enthalpy change $\Delta H_0(e)$ for electron transfer from the Fermi level of the metal electrode to the electron level of the proton, when the proton–solvent system is in its ground state, will be designated by $_0\Delta H_0(e)$. Calculation of $_0\Delta H_0(e)$ is carried out using the following thermodynamic cycle:

$$(H_3O^+)_{dl} + Me^- \xrightarrow{\;\;_0\Delta H_0(e)\;\;} (M\text{- - -}H\text{- - -}OH_2)_{dl}$$

$$\uparrow L_0 \qquad\qquad\qquad\qquad\qquad\qquad \downarrow$$

$$Me^- + H^+ + H_2O(aq) \qquad\qquad M\text{- - -}H + H_2O(aq)$$

$$\uparrow _{-\Phi} \qquad\qquad\qquad\qquad\qquad\qquad \downarrow A$$

$$M + H^+ + e^- + H_2O(aq) \xleftarrow{\qquad J \qquad} M + H + H_2O(aq)$$

The absence of specific adsorption and of a diffuse double layer are assumed.† The values of J, A_e, and Φ are 313,[33] 59,[19] and [34]104 kcal/mole, respectively. In order to obtain the value of $\Delta H_0(e)$ at the p.z.c., the real proton-solvation energy ρ is required.‡ According to Randles,[35] this is -260.5 kcal/mole. R_0 is calculated for an internuclear separation d of 1.05 Å, i.e., the equilibrium $H^+\text{–}OH_2$ bond distance.[29] The calculated value of $_0\Delta H_0(e)$ is $75 - A$ kcal/mole. Calculation of A involves assumptions concerning the double-layer structure, which is assumed to be that proposed by Bockris et al.[30] Thus, cations such as the H_3O^+ ions are thought to be adsorbed in a hydrated state onto the metal electrode, the surface of which is covered by a layer of water dipoles with coverage $\theta_{H_2O} \simeq 0.9$. At potentials sufficiently negative with respect to the potential of zero charge (p.z.c.),§ this layer of water molecules is strongly oriented with the positive end of the water dipole toward the electrode. The model is represented in Figure 8.

From the model of Figure 8, the double-layer width δ_{dl} is found to be about 5.2 Å. The H_3O^+ ion at the outer Hemholtz plane

---

† It is assumed that L_0 is the same for a proton at the electrode–solution interface as in the solution, and that M–H$^+$ interaction is negligible.
‡ So long as $\Delta x \simeq \Delta g$. In the original paper,[20] the *chemical* solvation energy was used.
§ This is the potential range of interest in studies of the h.e.r. on mercury.

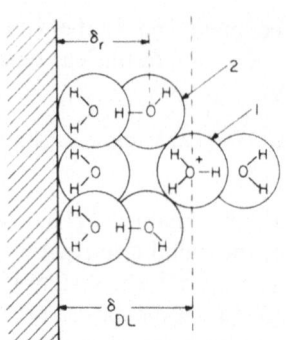

Figure 8. Model of the electrode–
solution double layer for negative
electrode charges, according to
Bockris and Matthews[21]; δ_{dl} is the
double-layer width, and δ_r is the
reaction distance, or the distance
between the electrode and the center
of the discharging hydronium ion.
Proton transfer from the H_3O^+ ion
(1) to the hydrogen-bonded water
molecule (2) brings the proton to its
reaction position. The strongly ori-
entating effect of the negatively
charged electrode on the immediately adjacent water molecules prevents further proton transfer to
these molecules from water molecule (2).

is hydrogen-bonded to its hydration sheath. Transfer of the proton
through a hydrogen bond leads to the formation of a H_3O^+ ion at a
distance δ_r from the electrode surface. Breakdown of the normal water
structure at the electrode surface prevents further transfer of the
proton through hydrogen bonds by the Grotthuss mechanism.
Proton discharge must therefore occur with the H_3O^+ ion at a
distance δ_r from the electrode surface. The model gives $\delta_r = 3.8$ Å,
which is consistent with the proton-transfer distance deduced by
Bockris and Matthews[32] from proton tunneling calculations. The
bond extension $(d - d_e)$ in equation (47) is given by (see Figure 9)

$$d - d_e = \delta_r - d_{O-H}(H_3O^+) - r_H = 2.4 \text{ Å} \qquad (50)$$

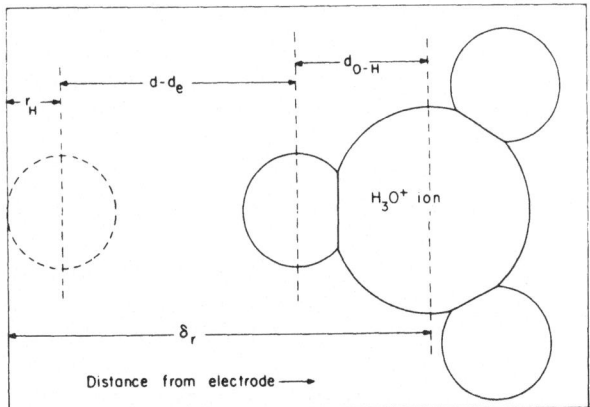

Figure 9. Interrelation between the reaction distance δ_r and the M–H bond distance, $d - d_e$; r_H is the radius of a hydrogen atom, d_{O-H} is the O–H bond length in an H_3O^+ ion.

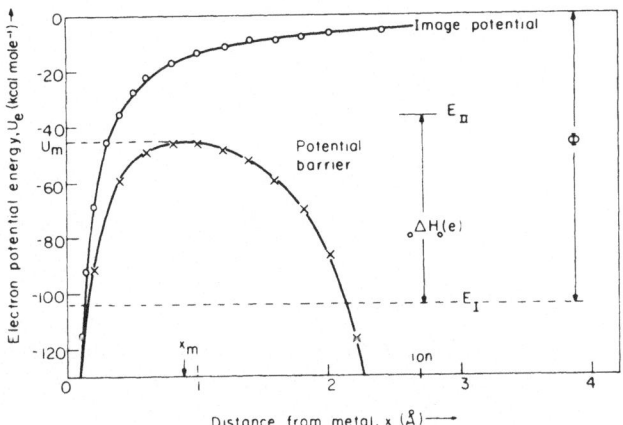

Figure 10. Potential-energy–distance profile for electron transfer from a metal electrode to a proton in solution at $\Delta\psi = 0$. E_I and E_{II} are energies of the states I and II, respectively (see text); $_0\Delta H_0(e)$ is the standard enthalpy of reaction for electron transfer from the Fermi level of the metal to an H_3O^+ ion in its ground rotation-vibration state; U_m is the potential energy of the barrier maximum.

From equation (47), using $d - d_e = 2.4\,\text{Å}$, A is $7\,\text{kcal/mole}$. Therefore $_0\Delta H_0(e)$ is $68\,\text{kcal/mole}$. This quantity is to be identified with the standard real-enthalpy change, or the standard enthalpy change at the potential of zero charge in the absence of specific adsorption and a diffuse double layer.

The potential-energy–distance curve for the electron in going from the metal to the H_3O^+ ion is shown in Figure 10. From a graphic construction, using image and Coulombic potentials with a dielectric constant (ε) of 6, the barrier height for electron transfer† is found to be $58\,\text{kcal/mole}$, i.e., less than $_0\Delta H_0(e)$. The activation energy $\Delta H_0(e)$ for classical electron transfer is thus equal to $_0\Delta H_0(e)$, which is $68\,\text{kcal/mole}$.

One may also analytically arrive at the barrier height by finding (cf. Ref. 11) the position of the barrier maximum. Thus

$$U(x) = U_1(x) + U_2(x) \tag{51}$$

$$= -(e_0^2/4\varepsilon x) - \{e_0^2/[\varepsilon(d - x)]\} \tag{52}$$

At the barrier maximum, $\partial U/\partial x = 0$; therefore,

$$\partial U/\partial x = (e_0^2/4\varepsilon x_m^2) - \{e_0^2/[\varepsilon(d - x_m)]^2\}$$

$$= 0 \tag{53}$$

where x_m is the value of x corresponding to $U = U_{\max}$. From equation (53),

$$x_m = d/3 \tag{54}$$

For $d = 2.7\,\text{Å}$, then, $x_m = 0.9\,\text{Å}$. Substituting the value of x_m into equation (52) for U gives

$$U_m = -9e_0^2/4\varepsilon d \tag{55}$$

and since the barrier height, as measured from E_1, equals $\Phi + U_m$, then the barrier height is $104 - (9e_0^2/4\varepsilon d)$. For $\varepsilon = 6$ and $d = 2.7\,\text{Å}$, the barrier height for electron transfer is $58\,\text{kcal/mole}$.

Because of the large value of $\Delta H_0^{\ddagger}(e)$ for classical electron transfer compared to the measured heat of activation for the h.e.r. on

---

†In an earlier publication,[21] a dielectric constant of unity was assumed, which is inconsistent with the model due to Bockris et al.,[30] and the image and Coulombic potentials were summed incorrectly.

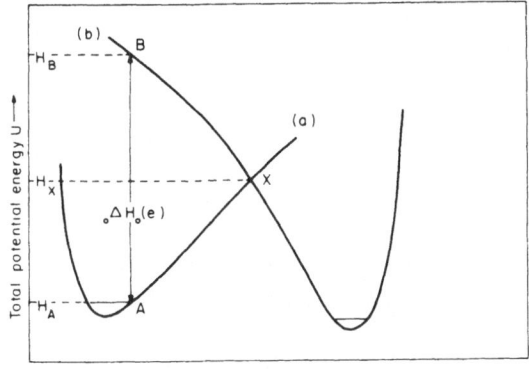

Figure 11. Potential-energy–distance profile for the proton-discharge reaction; H_A is the ground-state energy level of the reactants; curves (a) and (b) show the variation of potential energy with internuclear separation for $Me^- - H^+ - OH_2$ and $M-H-OH_2$, respectively; $_0\Delta H_0(e)$ is the standard enthalpy or reaction for electron transfer with reactants in their ground states.

mercury at the p.z.c., it is concluded that classical electron transfer makes a negligible contribution to the rate of proton discharge.

(ii) Electron Tunneling to a Proton in Its Ground State

The electron has a nonzero probability of tunneling from the metal to the proton only if there is an empty electron level of suitable energy to accept the electron. The probability of electron tunneling is zero for all energy levels below E_{II} (see Figure 10). The standard enthalpy of the reaction, $_0\Delta H_0(e)$, for electron transfer thus becomes the effective standard enthalpy of activation for electron transfer by quantum mechanical tunneling. The value of $_0\Delta H_0(e)$ is so large (68 kcal/mole) that electron tunnel transfer to a proton *in its ground vibrational level* makes a negligible contribution to the rate of proton discharge.

(iii) Classical Proton Transfer

Referring to Figure 10, we note that, if the electron energy E_{II} of the vacant electron level of the ion is decreased to become equal to E_1, then neutralization may occur by electron tunneling from the Fermi level of the metal to the proton. This may be achieved by exciting the H_3O^+ ion to a higher rotation-vibration level and stretching the H^+-OH_2 bond (see Figure 11).

The vertical transition AB corresponds to the process

$$e^-(M) + (H_3O^+)_{dl} \rightarrow (M\text{-}\text{-}\text{-}H\text{-}\text{-}\text{-}OH_2)_{dl} \qquad (56)$$

i.e., to taking an electron from the Fermi level of the metal to an H_3O^+ ion in its ground rotation-vibration state, with *no change* in the proton coordinate d. The energy change for this process is $_0\Delta H_0(e)$. Such a vertical transition for an H_3O^+ ion *not in its ground state* will be accompanied by an energy change $\Delta H_0(e)$. Stretching the H^+-OH_2 bond results in a decrease in $\Delta H_0(e)$. Eventually, at the intersection point x of curves a and b in Figure 11, $\Delta H_0(e) = 0$ and electron tunneling from the metal to the proton becomes possible (cf. Figure 10). Figure 11 also shows that the change $d[\Delta H_0(e)]$ brought about by a change dH in the total energy of the system is greater than dH. Changes in energy H may be related to changes in energy E (Figure 13) by

$$dH = -\beta \, dE \qquad (57)$$

where $0 \le \beta \le 1$.

Neutralization occurs in the activated state for the proton-discharge step of the hydrogen-evolution reaction. It is noteworthy that Figure 11 is identical to that used in socalled "proton-transfer theories" of the discharge reaction In effect, therefore, the model of proton discharge which arises from consideration of the quantum-mechanical tunneling of electrons to protons turns out to be, with respect to the calculation of heats of activation, the same as that which would arise from a proton-transfer model. This is because the factor which controls electron tunneling is the $O-H$ stretching. It is probable that such a view can be generalized: Charge transfer at interfaces occurs by tunneling of electrons between particles in solution and the metal, but the rate depends on the displacement of the ion–solvent sheath such that empty electron energy levels, equal in energy to full electron levels in the metal (for reduction), become available.

(iv) Combined Classical Proton Transfer and Electron Tunneling

The rate of proton discharge depends on: (a) the number of electrons $v(E) \, dE$ with energy between E and $E \, dE$ that strike a unit area of the surface in unit time from within the metal† ; (b) the number $N(\varepsilon)$ of protons with energy H that on neutralization yield hydrogen atoms with electron energy E ; (c) the probability $W_e(E)$ of electron tunneling at the energy level E ; (d) the probability of collision between an electron and a proton.

Following Gurney, and including the proportionality constant as obtained from thermionic-emission theory, the rate of proton discharge is given by

$$i = [H^+](\pi r_H^2/3)\overline{W}_e[4\pi m_e e_0(kT)^2/h^3(1 - \bar{\beta})]\exp[-\bar{\beta}(E_{II} - E_I)/RT]$$
(58)

where $[H^+]$ is the number of protons per unit area of the interface, r_H is the radius of an H atom, and $\bar{\beta}$ is the average value of the symmetry factor.

The rate expression (58) contains only one activation energy term, $\bar{\beta}(E_{II} - E_I)$. The *rate-controlling* process in the proton-discharge reaction is thus the stretching of the H^+-OH_2 bond. The reaction mechanism may be termed "proton transfer."

The crossing point x (Figure 11) thus refers to that excited state which satisfies the electron tunneling condition $\Delta H_0(e) = 0$. In general, there are many such crossing points, one for each electron energy level in the metal electrode, but the crossing point corresponding to the electron in the Fermi level of the metal gives the lowest activation energy. Hence, this latter reaction path is most probable. This type of model is general, being applicable to metal deposition and to redox reactions.[22,23] Excitations other than stretching of the H^+-OH_2 bond lead to higher activation energies. This is a direct result of the appearance of the symmetry factor β in the rate expression. Consider, for example, an excitation process which leads merely to a vertical shift curve of a with respect to curve b (Figure 11) such that curve a intersects curve b at point B. The energy expended is $H_B - H_A$, which is clearly greater than $H_x - H_A$. Consulting our definition of β as given by equation (57), we see that, for such a mode of excitation, $\beta = 1$. In general, we can say that, for a given reaction, that path which gives the lowest value of β will be the most probable, since for this path the electron tunneling condition has been fulfilled by the lowest energy expenditure.

---

†This is governed by the supply function of thermionic- and field-emission theory.

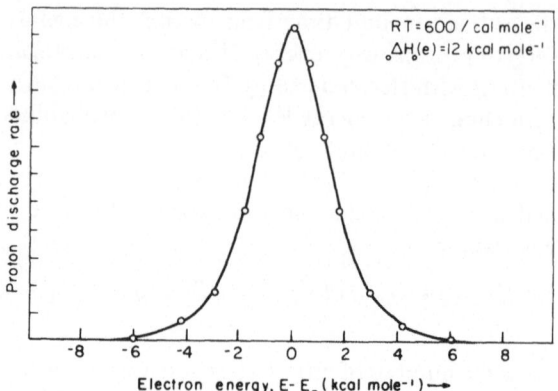

Figure 12. Variation of the rate of proton discharge with electron energy E; $_0\Delta H_0(e)$ is the standard enthalpy of reaction for electron transfer with reactants in their ground states.

In deriving equation (58), use is made of the Richardson thermionic-emission equation, which assumes that $E \gg RT$. This approximation does not invalidate the above conclusions, but does introduce an error of about a factor of about two in equation (58). Furthermore, the limits of integration did not allow for the contribution to the rate of proton discharge by electrons from energy levels below the Fermi level. This leads to an error of a factor of about 1/2 in the rate.

A more exact derivation of the rate expression does not approximate the Fermi to the Boltzmann distribution of electrons among energy levels of the metal, and the corresponding rate expression is

$$i = [H^+](\pi r_H^2/3)(4\pi m_e e_0 kT/h^3)\overline{W}_e \int_{-\infty}^{\infty} [1 + \exp\{(E - E_1)/RT\}]^{-1}$$

$$\times \exp[-\beta(E_0 - E)/RT]\, dE \tag{59}$$

Equation (59) has the form shown in Figure 12. For sufficiently large $E - E_1$, the equation is symmetrical about $E = E_1$, and the rate is a maximum at $E = E_1$. The result of the exact treatment is that the major contribution to the rate is from tunneling of electrons at the Fermi level to excited vibration-rotation states of the proton. There is a small contribution from electrons both above and below the

Fermi level, which diminishes rapidly with change of electron energy from the Fermi level.

(v) Charge Distribution in the Activated State

The discharge step may, according to the above theory, be written

$$H_3O^+ + e^-(M) \rightleftharpoons Me^- \text{-}\text{-}\text{-}H^+ \text{-}\text{-}\text{-}OH_2 \rightleftharpoons M\text{-}\text{-}\text{-}H\text{-}\text{-}\text{-}OH_2$$
$$\rightarrow MH + H_2O \qquad (60)$$

The activated state consists of two resonant states. The transition from one activated state to the other is accomplished by electron tunnel transfer between the metal and the proton. According to the WKB approximation, the probability of electron tunneling is a function of the barrier height as measured from E_I or E_{II} (depending upon the direction of electron transfer). Since electron transfer becomes possible when $E_I = E_{II}$, then the tunneling probability for an electron between metal and the ion is the same in either direction under these conditions. The two resonant states are thus equally probable and the effective charges (i.e., the time-average charge) on the H nucleus and on the metal are $+e_0/2$ and $-e_0/2$, respectively.

(vi) The Symmetry Factor

The symmetry factor β relates the electrode potential to the change in activation energy which it produces. Thus, the standard electrochemical free energy of activation for the discharge reaction, involving unit charge transfer, is written

$$\Delta\bar{G}_0^{\ddagger} = \Delta G_0^{\ddagger} + \beta F \, \Delta\phi \qquad (61)$$

where $\Delta G_0^{\ddagger}$ is the standard free energy of activation at $\Delta\phi = 0$; $\Delta G_0^{\ddagger}$ *should not* be identified with the so-called "chemical free energy of activation," because this may depend on $\Delta\phi$.

According to the method due to Gurney,[3] the dependence of activation energy on electrode potential is obtained by vertically shifting the initial-state curve with respect to the final-state curve by an amount equal to the electrode potential. This has the effect of changing the activation energy by some fraction of the electrode potential. This change in activation energy is a measure of the

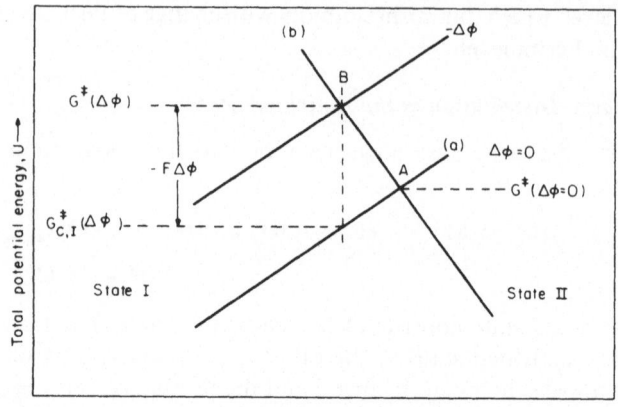

Figure 13. Variation of the free energy, $G^{\ddagger}$ of the activated state with electrode potential $\Delta\phi$. Curves (a) and (b) are the same potential-energy–distance curves shown in Figure 11, but enlarged in the vicinity of the crossing point x; $G^{\ddagger}(\Delta\phi)$ and $G^{\ddagger}(\Delta\phi = 0)$ are the free energies of the activated state at electrode potentials $\Delta\phi$ and zero, respectively; and $G^{\ddagger}_{c,I}(\Delta\phi)$ is the chemical part of the free energy of resonant activated state I at an electrode potential $\Delta\phi$.

symmetry factor β. Horiuti and Polanyi's concept of the discharge step of the h.e.r. involved a fully charged proton in the activated state in the act of being transferred to the electrode. A similar model was used by Essin and Kuzheurov[20] and by Parsons and Bockris[19] to show that $\beta = 1/2$. However, as shown above, the average charge on the H nucleus in the activated state is $e_0/2$, which may lead one to expect $\beta \simeq 3/4$ instead of $1/2$ (*cf.* p. 224, ref. 81, and footnote to p. 337, ref. 80). The present model, however, does not lead to this conclusion. The crossing point of the potential-energy curves defines the activated state and the energy at the crossing point as the *electrochemical* energy of the activated state. The method of vertical shift thus leads directly to the dependence of activation energy on potential without any need to inquire as to the charge distribution[38,39] in the activated state.

When the energy of the activated state is analyzed in terms of chemical and electrical parts, it is found that both parts are functions of the electrode potential. This is a result of the change in configuration of the activated state with electrode potential, which itself is a consequence of the dependence on $\Delta\phi$ of the H^+-OH_2 stretching required to produce the condition $\Delta H_0(e) = 0$.

Consider the region of intersection of the potential-energy curves in Figure 11 shown magnified in Figure 13 at $\Delta\phi = 0$ and at a given electrode potential $-\Delta\phi$. At $\Delta\phi = 0$, the curves for the two states $MH + H_2O$ and $e^-(M) + H_3O^+$ intersect at A. The free energy of the activated complex, $G^{\ddagger}$, is the same for the two states, and, since $\Delta\phi = 0$, then

$$G^{\ddagger}(\Delta\phi = 0) = G_c^{\ddagger}(\Delta\phi = 0) = G_{c,I}(\Delta\phi = 0) = G_{c,II}^{\ddagger}(\Delta\phi = 0) \quad (62)$$

where $G_{c,I}^{\ddagger}$ and $G_{c,II}^{\ddagger}$ refer to the chemical part of the free energy of activated resonant states I and II respectively. For the electrode potential $-\Delta\phi$, the curve for state I is shifted vertically by an amount $-F\,\Delta\phi$, producing the new intersection point B. The energy of the new activated complex is now part chemical and part electrical and the ratio is different for the resonant states I and II. For the charged state I, the chemical energy of the activated state is $G_{c,I}^{\ddagger}(\Delta\phi)$ and the electrical energy is $-F\,\Delta\phi$; thus,

$$G_I^{\ddagger}(\Delta\phi) = G_{c,I}^{\ddagger}(\Delta\phi) - F\,\Delta\phi \quad (63)$$

For the uncharged state II,

$$G_{II}^{\ddagger}(\Delta\phi) = G_{c,II}^{\ddagger}(\Delta\phi) \quad (64)$$

The states I and II make equal contributions to the energy of the activated state; hence, at $\Delta\phi = 0$,

$$G^{\ddagger}(\Delta\phi = 0) = [G_{c,I}^{\ddagger}(\Delta\phi = 0) + G_{c,II}^{\ddagger}(\Delta\phi = 0)]/2 \quad (65)$$

$$= G_{c,I}^{\ddagger}(\Delta\phi = 0) \quad (66)$$

and at $\Delta\phi$,

$$G^{\ddagger}(\Delta\phi) = [G_I^{\ddagger}(\Delta\phi) + G_{II}^{\ddagger}(\Delta\phi)]/2 \quad (67)$$

$$= [G_{c,I}^{\ddagger}(\Delta\phi) + G_{c,II}^{\ddagger}(\Delta\phi) - F\,\Delta\phi]/2 \quad (68)$$

Subtracting equation (66) from equation (68), we obtain

$$\delta G^{\ddagger} = G^{\ddagger}(\Delta\phi) - G^{\ddagger}(\Delta\phi = 0)$$

$$= [G_{c,I}^{\ddagger}(\Delta\phi) + G_{c,II}^{\ddagger}(\Delta\phi) - 2G_{c,I}^{\ddagger}(\Delta\phi = 0) - F\,\Delta\phi]/2 \quad (69)$$

or

$$\delta G^{\ddagger} = \delta G_c^{\ddagger} - F\,\Delta\phi/2 \quad (70)$$

where

$$\delta G_c^{\ddagger} = G_c^{\ddagger}(\Delta\phi) - G_c(\Delta\phi = 0) \tag{71}$$

$$\delta G_c^{\ddagger} = [G_{c,\mathrm{I}}^{\ddagger}(\Delta\phi) + G_{c,\mathrm{II}}^{\ddagger}(\Delta\phi) - 2G_{c,\mathrm{I}}^{\ddagger}(\Delta\phi = 0)]/2 \tag{72}$$

From equation (72), it is seen that only if the decrease in $G_{c,\mathrm{I}}^{\ddagger}$ equals the increase in $G_{c,\mathrm{II}}^{\ddagger}$ will $\delta G_c^{\ddagger}$ be independent of potential, and the condition for this to occur is for the curves for the states I and II to have equal slopes in the vicinity of their crossing points.

The free energy of activation $\Delta G^{\ddagger}$ at an electrode potential $-\Delta\phi$ may be obtained as

$$\Delta G^{\ddagger}(\Delta\phi) = G^{\ddagger}(\Delta\phi) - G_{\mathrm{I}}(\Delta\phi)$$

$$= [G_{c,\mathrm{I}}^{\ddagger}(\Delta\phi) + G_{c,\mathrm{II}}^{\ddagger}(\Delta\phi)]/2 - G_{\mathrm{I},c} + F\,\Delta\phi/2 \tag{73}$$

at $\Delta\phi = 0$ as

$$\Delta G^{\ddagger}(\Delta\phi = 0) = G_{c,\mathrm{I}}^{\ddagger}(\Delta\phi = 0) - G_{\mathrm{I},c} \tag{74}$$

Solving equation (74) for $G_{\mathrm{I},c}$ and substituting the result in equation (73) gives

$$\Delta G^{\ddagger} = \Delta G^{\ddagger}(\Delta\phi = 0) + \beta F\,\Delta\phi \tag{75}$$

where the symmetry factor β is given by

$$\beta = d[\Delta G^{\ddagger}(\Delta\phi)]/dF\,\Delta\phi) \tag{76}$$

$$= d(\delta G_c^{\ddagger})/d(F\,\Delta\phi) + 1/2 \tag{77}$$

By contrast, it is often written that

$$\Delta G^{\ddagger}(\Delta\phi) = \Delta G_c^{\ddagger} + \beta F\,\Delta\phi \tag{78}$$

with the presumption that $\Delta G_c^{\ddagger}$ is independent of potential. Such an assumption is valid only for $\delta G_c^{\ddagger} = 0$, i.e., the curves I and II are of equal slope in the vicinity of their crossing point, in which case, from equation (77), $\beta = 1/2$. If one ignores the dependence of $G_c^{\ddagger}$ on electrode potential, then one ignores the dependence of β on the relative slopes of the potential-energy curves, taking it always equal to $1/2$.

The expression for the dependence of the activation energy on electrode potential may be made more general by allowing for the case of unequal probabilities of the resonant states in the activated complex. In this case, one obtains

$$\Delta G^{\ddagger}(\Delta\phi) = \Delta G^{\ddagger}(\Delta\phi = 0) + \delta G_c^{\ddagger} + (1 - g_{\mathrm{I}})F\,\Delta\phi \tag{79}$$

where g_I is the fractional contribution of the state I to the activated complex. The above equations may be made more explicit by introducing parameters for the slopes of the potential-energy curves, and the result obtained is that *the symmetry factor is independent of the charge distribution in the activated state.*

(vii) Application of the Theory to Systems Other Than the Proton-Discharge Step of the h.e.r.

The theory can be applied to other charge-transfer processes such as metal deposition and redox reactions. The charge-transfer step in metal deposition is visualized as proceeding by metal-ion–solvent-sheath stretching such that the metal ion moves toward the electrode. Movement of the ion toward the electrode ensures that $\beta < 1$. Since various ion–solvent configurations will have the same energy, it will be necessary to introduce weighting, or degeneracy, coefficients in the manner done by Gerischer.[22,23]

Care must be exercised in applying the theory to fast reactions. For fast reactions, $_0\Delta H_0(e)$ may be small, so that the dependence of rate on energy level E will not have a sharp maximum at $E = E_I$, i.e., electrons no longer come predominantly from the Fermi level of the metal.[113–115] This effect will be coupled with that predicted by Despic and Bockris,[27] namely, that for very fast reactions, $\beta < 1/2$ and β decreases with decrease in electrode potential ("barrierless discharge"). This effect is a consequence of the curvature of the potential-energy curves in the vicinity of their minima.

IV. ABSOLUTE REACTION RATE APPROACH TO CHARGE-TRANSFER REACTIONS

An approach to the theoretical development of the rates of charge-transfer reactions based on the use of potential–energy profiles or surfaces has been widely utilized by workers in the field of charge-transfer theory. This approach, which originates with the work of Horiuti and Polanyi,[8] has sometimes been thought to be different and more realistic than the model of Gurney's.[3] However, it is now believed that the theory of Gurney's is the more inclusive theory and that the approach originated by Horiuti and Polyani and adopted by others is, of itself, more a method for determining

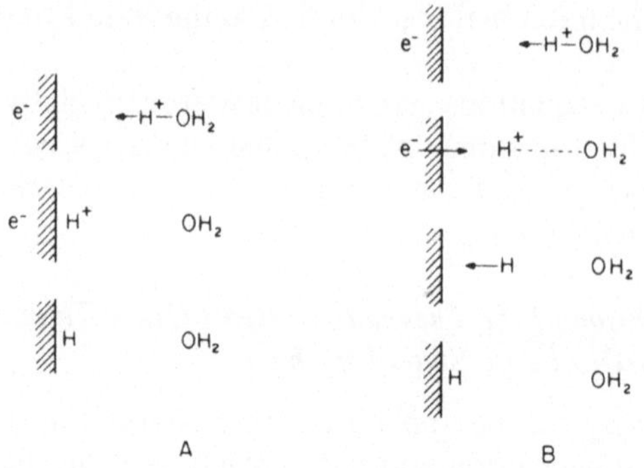

e^- $\leftarrow H^+_- OH_2$

e^- $\leftarrow H^+_- OH_2$

e^- H^+ ..... OH_2

e^- H^+ OH_2

$\leftarrow H$ OH_2

H OH_2

H OH_2

A   B

Figure 14. Comparison between the described (A)
and the implicit (B) proton-discharge model due to
Horiuti and Polanyi.[8]

activation energies as a function of such parameters as M–H bond
strength than a theory of charge transfer. In addition, the potential-
energy profile method, sometimes said to have been introduced by
Horiuti and Planyi in 1936, was first used by Gurney in 1931.

1. Theory Due to Horiuti and Polanyi

The transfer of a proton over an energy barrier was considered
by Horiuti and Polanyi[8] to constitute the rate-determining step
(r.d.s.) of many proton-transfer processes, including the discharge
step of the h.e.r. Potential-energy–distance curves were constructed
for the initial and final states of the reaction using thermodynamic
data to obtain the energy difference between the potential-energy
minima. For the discharge step of the h.e.r., the H atom was assumed
to be adsorbed on the electrode. This was an important advance in
the theory of the discharge step. Horiuti and Polanyi were the first
authors to point out that the state of adsorption of the H atom would
affect the rate of proton discharge, so that the experimental connection between the properties of the metal surface and overpotential
became understandable without necessarily assuming that a surface-
catalyzed reaction was rate-determining. It is of importance to note
that, by using an initial-state curve corresponding to H^+–OH_2
stretching plus an electron in the metal, and a final-state curve
corresponding to M–H stretching plus a water molecule, Horiuti
and Polanyi have implicitly assumed that neutralization occurs in

the activated state, i.e., at the crossing point of the curves, in accord with neutralization by electron transfer to a proton in an excited rotation-vibration level of the H_3O^+ ion, as proposed by Gurney.†
The distinction between the described and the implicit mechanisms is shown in Figure 14.

2. Calculations of Parsons and Bockris

An absolute calculation of the rate of the discharge step of the h.e.r. was carried out by Parsons and Bockris.[19] The standard electrochemical potential of the initial state is calculated in such a way as to avoid the difficulty of surface potentials, which in turn permits the calculation of the standard enthalpy of activation at the potential of zero charge.

At equilibrium, we have for the reaction

$$H^+(\text{solution}) + e^-(\text{metal}) = (1/2)H_2 \qquad (80)$$

that

$$\bar{\mu}_{H^+} + \mu_e = (1/2)\mu_{H_2} \qquad (81)$$

Rewriting in terms of standard potentials and using

$$\bar{\mu}_i = \mu_0 + z_i(\psi_i + \chi_i)F \qquad (82)$$

one obtains

$$(\mu_0)_l - F\,\Delta\psi + RT \ln a_{H^+} = (1/2)(\mu_0)_{H_2} + (RT/2)\ln p_{H_2} \qquad (83)$$

where

$$(\mu_0)_l = (\mu_0)_{H^+} + \mu_e - F\,\Delta\chi \qquad (84)$$

At the electrocapillary maximum, $\Delta\psi = 0$. By assigning to a_{H^+} the value at which a hydrogen electrode under atmospheric pressure would be in equilibrium at the potential of the electrocapillary maximum in dilute solution, assuming absence of specific adsorption, $(\mu_0)_l$ may be calculated from equation (83); $(\mu_0)_l - F\,\Delta\psi$ is then the standard electrochemical potential of the initial state.

The enthalpies of the final and initial states are assumed to approximate the potential energies of these states. Using Morse

†For this reason [cf. Section III, 4(iii)], the treatments of charge transfer that involve implicitly *ion transfer* should in fact give rates and dependences of rates upon, e.g., electrocatalytic parameters, which are the same as those calculated from a purely quantum mechanical (Gurney-type) model.

functions, the variation of the potential energy of these states with displacement of the proton is calculated. Both the standard enthalpy of activation and the symmetry factor (from the relative slopes of the potential-energy–distance curves) are calculated as a function of the parameters employed in the calculations. Standard entropies of activation are also estimated. The dependence of the transfer coefficient α on temperature, which had been observed by Bockris and Parsons,[40] is ascribed to a change in the double-layer thickness with temperature.

The conclusion reached by Parsons and Bockris is that, "Although no completely satisfactory quantitative agreement between theory and experiment can be expected using the available data for the various parameters, this description of the discharge process is not inconsistent with the observed facts." This statement sums up the uncertainty of the potential-energy profile method as far as quantitative calculations are concerned.

3. Calculations of Conway and Bockris

The potential-energy profile method was used by Conway and Bockris[41] to calculate relative activation energies for the electro-chemical desorption step,

$$H_3O^+ + e^-(M) + H_{ads} \rightarrow M + H_2 + H_2O \qquad (85)$$

of the h.e.r. on Cu, Ni, W, and Hg. The activation energy is shown to increase with increasing heat of adsorption of H on the metal. The converse is true for the proton-discharge step on the h.e.r. The method is shown to be of value in calculating kinetic-isotope effects.[42,43]

In the case of metal deposition, Conway and Bockris[44] picture the reaction as occurring by the stretching of ion–solvent bonds during transfer to various metal sites. Calculations of the relative activation energies for various processes are described and the probable r.d.s. for the cases of Cu^{2+}, Ni^{2+}, and Ag^+ ion deposition on the respective metals are deduced. Direct transfer of ions to sites other than a planar site is found to be relatively improbable owing to the large heat of activation resulting from large heats of dehydration at sites of high coordination number.

Potential-energy profile diagrams for Cu^{2+} and Ag^+ ions with simultaneous neutralization to form *adatoms* indicate heats of activation between 30 and 80 kcal/mole, depending on the ion and the type of site to which transfer occurs. Such values are inconsistent with the observed i_0 values. The symmetry factor for direct formation of adatoms is calculated as 0.8–0.9, whereas experimentally a value nearer to 0.5 is found. For these reasons, neutral "adatom" formation is considered improbable by Conway and Bockris, i.e., adions are formed. This conclusion is consistent with the degree of ionic character (30–40%) of the surface entity in Ag^+ deposition determined by Gerischer.[45]

4. Calculations of Despic and Bockris

The potential-energy profile method for calculating the symmetry factor was extended by Despic and Bockris[27] to include the electrode-potential dependence of β. According to this method,[46]

$$\beta_c = (\tan \gamma)/(\tan \gamma + \tan \theta) \qquad (86)$$

where θ and γ are the slopes of the initial- and final-state potential energy-distance curves. According to the symmetry view, which relates β to the position of the barrier maximum with respect to the potential-energy minima (see Figure 15),

$$\beta_c = (x_2 - x^{\ddagger})/(x_2 - x_1) \qquad (87)$$

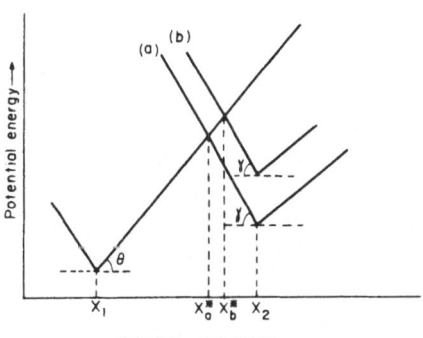

Figure 15. The effect of electrode potential on the potential-energy profile for a charge-transfer reaction according to the linear analog approximation. Curves (a) and (b) are at two different electrode potentials. (After Bockris.[49])

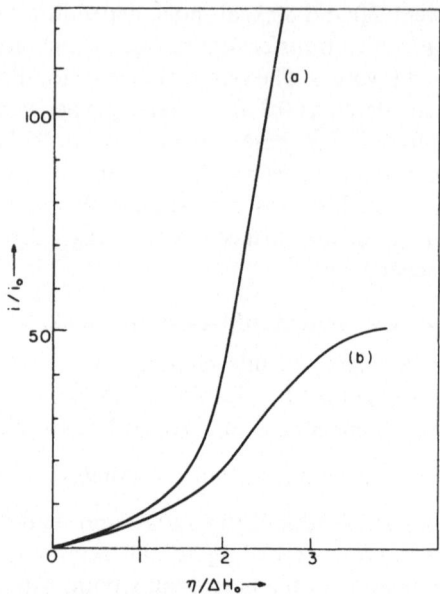

Figure 16. The dependence of the current density i on the overpotential η assuming in (a) the symmetry factor β is constant, and in (b) that it varies with potential. i_0 is the exchange current density and ΔH_0 is the standard enthalpy of reaction. (After Despic and Bockris.[27])

where $x^{\ddagger}$, x_1, and x_2 are the distance coordinates of the activated state, the initial state, and the final state, respectively. Despic and Bockris found, using linear analogs of the potential-energy curves, that the above two definitions of β are equivalent provided that the potential-energy curves do not have large radii of curvature at their minima. This, however, refers only to a special case, when the standard enthalpy of reaction is approximately zero. At electrode potentials where $|\Delta \overline{H}_0| > 0$, the two definitions (90) and (91) are no longer equivalent even in a linear analog to a potential-energy profile. For example, in Figure 15, the potential-energy curves are drawn such that θ and γ are constant, so that, by definition (86), β is constant. However, $x^{\ddagger}$ is not constant, and hence, according to definition (87), β is a function of electrode potential. A Tafel relation is only to be expected if, over the range of the electrode-potential measurement, tan γ and tan θ are constant. For $\beta = 0.5$, the condition for a Tafel relation is that tan $\theta = $ tan γ over the given electrode-potential range.

For silver deposition, the dependence of β_c on overpotential is theoretically obtained by calculating the slopes of the Morse curves as a function of electrode potential (see Figure 16). In this way, the (non-Tafelian) experimental dependence of log i on overpotential for silver deposition is explained.

As can be seen from Figure 16, as the overpotential increases one will reach a potential at which further increase does not increase the reaction rate. The thermal heat of activation has been completely wiped out by the electrical potential difference in the double layer (Despic and Bockris[27]). Such "barrierless discharge" has been developed and discussed particularly by Kristallik.

An objection to the calculations so far reviewed in this section concerns the use of potential-energy profiles rather than surfaces to represent the reaction. The method is a good approximation for electron and proton transfer, but may involve substantial approximations when applied to metal deposition. Nevertheless, as a comparative technique for indicating reaction paths and for determining the dependence of reaction rates on variables such as electronic work function or adsorption energy, the method has utility even, e.g., for metal deposition.

V. ELECTROSTATIC TREATMENT OF THE RATE OF REDOX REACTIONS

1. Hush's Theory for Redox Reactions

A nonmechanistic treatment of the energetics of redox reactions has been given by Hush.[38,39] The unique aspect of this theory is the description of the redox reaction in terms of the variation of electron charge density λ along the reaction coordinate.

The type of process under consideration can be expressed as

$$A^z + e^-(M) \rightarrow A^{z-1} + M \tag{88}$$

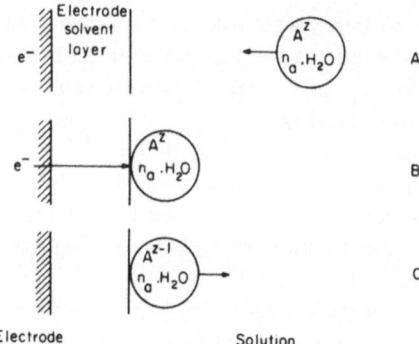

Figure 17. Model for the redox reaction at an electrode according to Hush.[39] (A) Diffusion of the reactant to the electrode; (B) the electron transfer; (C) diffusion of the product from the electrode.

i.e., one electron transfer from the metal electrode to the kernel A (atomic or molecular). A typical example is

$$Fe^{3+} + e^-(M) \rightarrow Fe^{2+} + M \tag{89}$$

Hush, like Marcus[48] and Levich,[47] considered only redox reactions.

The resonance energy of the activated state is assumed to be sufficiently small (say 0.03 eV) to give an overlap integral equal to zero and therefore to allow the reaction to proceed with a Zener transmission coefficient[50] of about unity (see Appendix II).

For convenience in calculating the energies of the various states, the energy of the system M- - -e^-- - -A is partitioned into (i) the binding energy of the electron; and (ii) the environmental energy of the complex M- - -e^-- - -A with respect to its environment. For any state s, characterized by the value λ^s of the probability density parameter, we write

$$\mu^s = \mu^s_{env} + \mu^s_{bind} \tag{90}$$

and

$$\mu^s_{bind} = \lambda^s \mu^f_{bind} + (1 - \lambda^s)\mu^i_{bind} \tag{91}$$

where μ denotes the standard free energy and the superscripts f and i denote the final and initial states, respectively. In states i and f, the M–A separation is infinite. The states immediately preceding

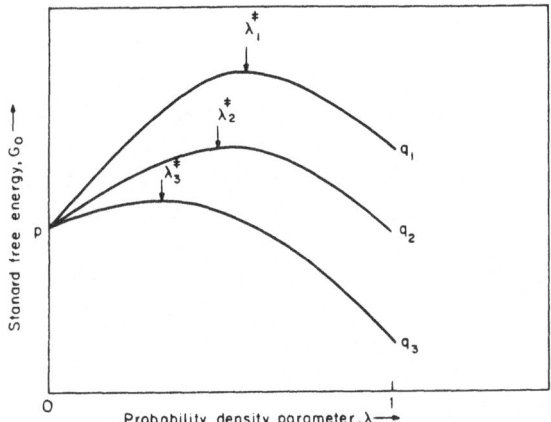

Figure 18. Standard free energy as a function of the probability-density parameter λ for reactions with overall free-energy change: (1) positive, (2) zero, (3) negative. $\lambda^{\ddagger}$ indicates the value of λ in the transition state, p denotes the free energy of the state immediately before the transition state, and q denotes the free energy of the state immediately after the transition state. (After Hush.[39])

and succeeding the activated state t will be denoted by p and q, respectively. The reaction is shown schematically in Figure 17. It is assumed that the M–A separation is constant over the range of states p through t to q.

The standard free energy of activation for the redox reaction is given by

$$\Delta G_0^{\ddagger} = {}^i\Delta^p\mu_0 + {}^p\Delta^t\mu_{\text{env}} + {}^p\Delta^t\mu_{\text{bind}} \tag{92}$$

The first term on the r.h.s. of equation (92) represents the change in standard free energy when the reactants are brought together in solution to the M–A separation of state p. The remaining terms represent the change of environmental and binding energies on passing from state p to state t at the top of the barrier.

From equations (91) and (92), we obtain (Figure 18)

$$\Delta G_0^{\ddagger} = {}^i\Delta^p\mu_0 + ({}^p\Delta^t\mu_{\text{env}} - \lambda^{\ddagger}\,{}^i\Delta^r\mu_{\text{env}}) + \lambda^{\ddagger}\,{}^i\Delta^r\mu_0 \tag{93}$$

The environmental energy terms in equation (93) constitute essentially the free energy of solvation of the ions. The assumption that the size of the solvated ions is constant makes the calculation simpler. However, it is a crude assumption.[49] The dielectric charging energy $\theta(\lambda^\ddagger)_{diel}$ and the cohesive energy of the solvated ion $\theta(\lambda^\ddagger)_{cpx}$ constituted the environmental energy. The contribution due to the former is calculated using the familiar Born expression and the latter by a model of ion–water dipole interaction. Hush considers the species to have a radius of $r_a + 2r_w$ (with r_a the ionic radius and r_w the radius of the water molecule), with charges of ze_0 units and $(z - 1)e_0$ units for fractions $(1 - \lambda)$ and λ of the time. Therefore,

$$\theta(\lambda^\ddagger)_{diel} + \theta(\lambda^\ddagger)_{cpx} = {}^p\Delta^t\mu_{env} - \lambda^\ddagger\,{}^i\Delta^t\mu_{env}$$

Now, $\theta(\lambda^\ddagger)_{diel}$ is obtained as

$$\theta(\lambda^\ddagger)_{diel} = \lambda^\ddagger(1 - \lambda^\ddagger)e^2[(1/\varepsilon_0) - (1/\varepsilon)]/2(r_a + 2r_w) \qquad (94)$$

The other energy term $\theta(\lambda^\ddagger)_{cpx}$ is obtained by minimizing the distance of separation between the ion and the solvating water molecules for a charge of $(z - \lambda)$ units. This turns out to be

$$\theta(\lambda^\ddagger)_{cpx} = \lambda^\ddagger(1 - \lambda^\ddagger)n_a e_0 vn/d_{e,z}^2 z(n - 2) \qquad (95)$$

where n_a is the number of solvating water molecules and v the dipole moment of a water molecule. The value of the constant is defined by the ion–water interaction energy $U(d, q)$ as

$$U(d, q) = -(n_a ze_0 v/d^2) + (\xi/d_n) \qquad (96)$$

In the above equation, d is the ion–oxygen distance. When n, ξ, v, and n_a are independent of charge,

$$d_{e,z-\lambda} = d_{e,z}[z/(z - \lambda)]^{1/n-2} \qquad (97)$$

Equation (95) differs by a factor of $n/(n - 2)$ from that given by Hush, which appears to be in error. Now, from equations (93), (94), and (96),

$$\Delta G_0^\ddagger = {}^i\Delta^p\mu_0 + \lambda^\ddagger\,{}^i\Delta^t\mu_0 + c\lambda^\ddagger(1 - \lambda^\ddagger) \qquad (98)$$

where

$$c = \frac{e^2[(1/\varepsilon_0) - (1/\varepsilon)]}{2(r_a + 2r_w)} + \frac{n_a e_0 vn}{d_{e,z}^2 z(n - 2)^2} \qquad (99)$$

The value of $\lambda^\ddagger$ that minimizes $\Delta G_0^\ddagger$ is obtained as

$$\lambda^\ddagger = (i\Delta^t\mu_0 + c)/2c \qquad (100)$$

Therefore, from equation (98),

$$\Delta G_{0,\min}^{\ddagger} = {}^{i}\Delta^{p}\mu_{0} + \tfrac{1}{2}({}^{i}\Delta^{r}\mu_{0}) + [({}^{i}\Delta^{r}\mu_{0})^{2}/4c] + \tfrac{1}{4}c \qquad (101)$$

Now, substitution for $\Delta G_{0,\min}^{\ddagger}$ ($= \Delta G_{0}^{\ddagger}$) in the expression for rate constant,

$$k_{r} = \kappa\tau(kT/h)\exp(-\Delta G_{0}^{\ddagger}/RT) \qquad (102)$$

enables one to calculate the reaction rate.
At equilibrium, for equal concentrations of reactants and products, ${}^{i}\Delta^{r}\mu = 0$. Hence,

$$\Delta G_{0}^{\ddagger}(\eta = 0) = {}^{i}\Delta^{p}\mu_{0} + \tfrac{1}{4}c \qquad (103)$$

Differentiating the above equation with respect to ${}^{i}\Delta^{r}\mu_{0}$ gives the symmetry factor as†

$$\beta = (1/2) + [\eta F/8\,\Delta G_{0}^{\ddagger}(\eta = 0)] \qquad (104)$$

The expressions for $\Delta G_{0}^{\ddagger}$ and β obtained from the theory of Hush bear a strong resemblance to those of Marcus (cf. Section V, 2). The methods used by Marcus and Hush, however, are superficially quite different; in particular, Hush assumed electrostatic equilibrium between the average charge $(z - \lambda^{\ddagger})e_{0}$ on the ion in the activated complex and the environment, whereas Marcus considered that the atomic and dipole orientation polarizations are not in equilibrium with the charge in the activated complex. This difference in approach results in a simpler derivation in the case of Hush compared to that of Marcus.

Later, Hush attempted to unify the theory of redox reactions in the high-temperature limit (thermal electron transfer) with the theory of optical electron transfer.[117] This theory is developed in terms of the polaron. The electrons are supposed to be essentially localized in the fields of individual ions, and electron plus accompanying lattice vibration constitutes a "small polaron." The theoretical considerations have been developed to correlate the thermal and optical electron transfer at a metal–solution interface. The peak frequencies of optical transitions depend on the potential.

†The symmetry factor is equal to $\lambda^{\ddagger}$, but it is not easy to trace back from equation (97) through equations (98), (94), etc., what the physical significance of the important term $1/2$ in equation (104) is. Harris and Weir have concluded that the view that $\lambda^{\ddagger}$ is representative of the actual charge density of the reactant species in the activated state is invalid.[107]

Photocatalysis of the interfacial electron transfer is predicted. The relationship between the transmission coefficient in adiabatic thermal exchange and the transition probability in the corresponding optical transfer is discussed.

The description of redox reactions at electrodes given by Hush makes no mention of the properties of the electrode, but implicit in the theory is the result that the electronic work function of the electrode metal does not influence the rate of outer-sphere adiabatic redox reactions. Models due to Gurney,[3] Butler,[7] and, more recently, Gerischer[22,23] and Bockris and Matthews[21] are definitive about the role of the electron. In metals, the electrons come predominantly from the Fermi level.

2. Marcus' Theory for Redox Reactions[51–54,56]

R. A. Marcus[45,51] has developed a theory for homogeneous electron-transfer reactions which derives from the work of Weiss,[110] Libby,[111] and R. J. Marcus et al.[112] The theory concerns reactions such as

$$Fe^{2+} + Fe^{3+} \rightarrow Fe^{3+} + Fe^{2+} \qquad (105)$$

It uses the assumption of small overlap of the electronic orbitals of the reactants in the activated state. Subsequently, the theory was applied to electrode redox reactions.[48,52] Later,[53] the theory of electrode redox reactions was approached in terms of potential-energy profile diagrams, apparently taking the ion–solvent bond-stretching energy into account.

The overall homogeneous electron-transfer reaction was envisaged by Marcus as occurring in the following way: When the reactants are near each other, a suitable solvent fluctuation can result in the formation of the state $X^{\ddagger}$. The atomic configuration in this state is that of the activated complex and the electronic configuration is that of the reactants. This state either re-forms the reactants by relaxation of some of the perturbed solvent molecules, or it forms the state X by an electron transfer. The atomic configuration of X is the same as that of $X^{\ddagger}$ but the electronic configuration is that of the products. The state X either re-forms the state $X^{\ddagger}$ by electron transfer, or it may form the products by solvent relaxation. The pair of states X and $X^{\ddagger}$ constitutes the activated complex. The reaction scheme is

$$A + B \rightleftharpoons X^{\ddagger} \tag{106}$$

$$X^{\ddagger} \rightleftharpoons X \tag{107}$$

$$X \rightarrow \text{products} \tag{108}$$

The energy requirement for the formation of the activated complex can also be derived in terms of the Franck–Condon principle (see Appendix II). When one electron configuration is formed from the other by an electronic transition, the electronic motion is so rapid that the solvent molecules do not have time to move during the electron jump. Conservation of energy leads to the requirement that, in the absence of radiant-energy flux, the total energy of these two states must be the same.

In order to calculate the energy of the activated states $X^{\ddagger}$ and X, it is necessary to use expressions which do not assume that the solvent molecules are oriented toward the ions in an equilibrium manner. The electrical polarization of the solvent is not in electrostatic equilibrium with the electrical field produced by ionic charges.

Of the infinite number of pairs of $X^{\ddagger}$ and X, the most probable pair will constitute the activated complex. The most probable pair is determined with the aid of the calculus of variations by minimizing the free energy of formation of $X^{\ddagger}$ from the reactants subject to the restriction that X and $X^{\ddagger}$ have the same total energy.

The free energy of a nonequilibrium state of a system can be calculated if a reversible path for reaching that state can be found. The difference in free energy arising from electrostatic interactions between any two states that do not differ in the number of each species of charged particles present is given by the difference in work in charging up each of the states. Accordingly, in any state, the contribution to the free energy arising from the electrostatic interaction of charges with each other and with the polarized medium may be calculated by subtracting from the electrostatic free energy G the work W_{iso} required to charge up the state in vacuum when the charges are infinitely distant from each other.

Nonequilibrium states are considered whose polarization consists of two types:

(a) The U-type polarization, which is not in equilibrium with the given charge distribution, denoted by a vector $\mathbf{P}_u(\mathbf{r})$.

(b) The E-type polarization, denoted by $P_e(r)$, which is in equilibrium with the electric field arising from the charge distribution and from the U-type polarization.

In most nonequilibrium systems of physical interest, the E-type polarization will be electronic and the U-type will correspond to atomic and orientation polarization.

The reversible charging process used to create the desired nonequilibrium state is as follows:

(a) In the first stage, the final value of the U-type polarization, $P_u^0(r)$ say, is produced in a reversible manner by the formation of some appropriate charge distribution. Meanwhile, the E-type polarization attains some value, $P_e^0(r)$ say.

(b) In the second stage, the polarization $P_u^0(r)$ is held fixed, but the charge distribution is reversibly altered until the final charge distribution is achieved. During this process, the E-type polarization changes from $P_e^0(r)$ to $P_e(r)$, say, whose value is dictated by the charge distribution and by $P_u^0(r)$.

This charging scheme enables the free energy to be calculated for each of the activated states X and $X^\ddagger$ using the Born model for calculations of the solvation-energy change due to change in charge. Minimization of the free energy of $X^\ddagger$, subject to the restraint of equal energies for $X^\ddagger$ and X, using Lagrange's method of undetermined multipliers gives

$$P_u^0 = \alpha_u[E^\ddagger + (E^\ddagger - E)m_1] \qquad (109)$$

where E is the electric-field strength, α_u is the polarizability, defined by

$$4\pi\alpha_u = \varepsilon - \varepsilon_0 \qquad (110)$$

and m_1 is the Lagrangian multiplier.

Equation (109) for P_u^0 gives some insight into the physical meaning of the Lagrangian multiplier m_1, which appears to be a measure of the relative contributions of the states $X^\ddagger$ and X to the electric-field strength produced by the charged activated complex. We note that, if $m_1 = -1/2$, then $P_u^0 = \alpha E/2$ and, for reasons of symmetry, $P_u^0 = \alpha_u E^\ddagger/2$; hence, $E = E^\ddagger$. The free energy of activation $\Delta G^\ddagger$ is finally obtained as

$$\Delta G^\ddagger = (q_1^\ddagger q_2^\ddagger/RD) + m_1^2(\Delta q)^2[(1/2a_1) + (1/2a_2) - (1/d)]$$
$$\times [(1/\varepsilon_0) - (1/\varepsilon)] \qquad (111)$$

where

$$\Delta q = q_1 - q_1^\ddagger = q_2^\ddagger - q_2 \qquad (112)$$

q_1 and $q_1^\ddagger$ are the charges on the first reactant in the states X and X‡, respectively; q_2 and $q_2^\ddagger$ are the corresponding charges for the second reactant; and d is the distance between the centers of the two reactants.

The above theory of homogeneous redox reactions was extended by Marcus to include electrode redox reactions such as

$$Fe^{3+} + e^-(M) \rightarrow Fe^{2+} + M \qquad (113)$$

The rate constant for the electrode process is given by†

$$k_r = A \exp(-\Delta \bar{G}_0^\ddagger / RT) \qquad (114)$$

$$\Delta \bar{G}_0^\ddagger = {}^f\Delta^q\bar{G}_0 + \tfrac{1}{2}m_1^2 B \qquad (115)$$

$$B = e_0^2[(1/a) - (1/\delta_r)][(1/\varepsilon_0) - (1/\varepsilon)] \qquad (116)$$

$$(2m_1 + 1)B/2 = -{}^p\Delta^q\bar{G}_0 - {}^i\Delta^p\bar{G}_0 + {}^f\Delta^q\bar{G}_0 \qquad (117)$$

$$^p\Delta^q\bar{G}_0 = \eta F \qquad (118)$$

The term A is the collision frequency, ${}^i\Delta^p\bar{G}_0$ is the work needed to bring reactants together, ${}^f\Delta^q\bar{G}_0$ is the corresponding term for the products, ${}^p\Delta^q\bar{G}_0$ is the free-energy change of the elementary electron-transfer step in the prevailing medium, a is the average radius of the hydrated ion, and δ_r is the distance between the electrode and the center of the ion. The central hydrated ion is treated as a sphere within which no changes in interatomic distances occur during the reaction and outside of which the solvent is treated as a dielectrically unsaturated continuum.‡ Equal concentrations of reactants and products are assumed.

From equations (117) and (118), we obtain

$$-m_1 = (1/2) + \{[\eta F - \Delta(\Delta \bar{G}_0)]/B\} \qquad (119)$$

$$-m^2\lambda/2 = \{[\eta F - \Delta(\Delta \bar{G}_0)]^2/2B\}$$
$$+ \{\eta F - \Delta(\Delta \bar{G}_0) + [(B/4)/2]\} \qquad (120)$$

---

†These expressions are somewhat modified in more recent papers of Marcus, the factors of 1/2 in the expressions for $\Delta \bar{G}_0^\ddagger$ and for m_1 being replaced by unity.
‡Both of these assumptions are gross approximations.

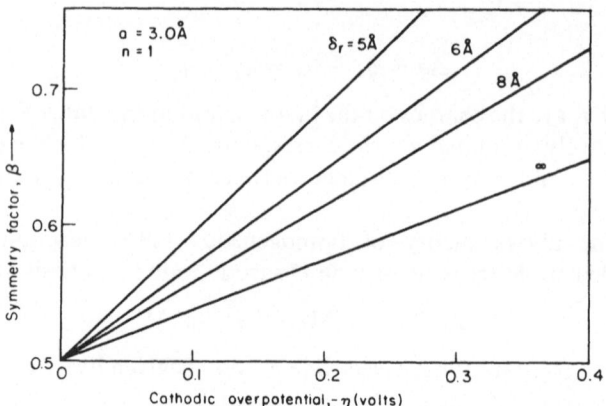

Figure 19. The symmetry factor β as a function of overpotential for various ion–electrode distances δ_r, according to the theory of Marcus. a is the radius of the solvated ion, and n is the number of electrons transferred.

Substituting equation (120) into equation (115) gives

$$\Delta G_0^{\ddagger} = {}^f\Delta^q\bar{G}_0 + \{[\eta F - \Delta(\Delta\bar{G}_0)]^2/2B\}$$
$$+ \{\eta F - \Delta(\Delta\bar{G}_0) + [(B/4)/2]\} \qquad (121)$$

where

$$\Delta(\Delta\bar{G}_0) = {}^f\Delta^p\bar{G}_0 - {}^i\Delta^q\bar{G}_0 \qquad (122)$$

From equation (121) for $\Delta(\Delta\bar{G}_0) = 0$ is obtained

$$\beta = (1/2) + [\eta F/8\Delta\bar{G}_0^{\ddagger}(\eta =)] \qquad (123)$$

which is identical with the result obtained by Hush [Section V,1, equation (104)]. The predicted dependence of β on η is shown in Figure 19.

That the theories due to Hush and to Marcus should give such similar results is surprising in view of the difference in models. Whereas Hush considers a highly nonadiabatic reaction leading to fast electron transfer and an average ionic charge in equilibrium with the environment, Marcus considers a more or less adiabatic reaction leading to a partitioned ionic charge which is *not* in equilibrium with the environment. Both authors claim to be considering weak orbital overlap or interaction.

A recent study[55] of the Cr(II)–Cr(III) redox reaction at a mercury electrode in acidified $0.5\ M$ $NaClO_4$ shows the transfer coefficient α to be a linear function of overpotential, in agreement with the predictions by Marcus and of Hush. Experimentally determined values of $\Delta\bar{G}_0^{\ddagger}(\eta = 0)$, when substituted into equation (123), gave very good agreement with the experimental dependence of α on overpotential. No attempt was made to calculate $\Delta\bar{G}_0^{\ddagger}(\eta = 0)$ by, say, equation (121). The value of $\alpha(\eta = 0)$ was found to be 0.4, compared to the predicted value of 0.5. This latter discrepancy cannot be attributed to the choice of parameters.

A different approach to that described above has been utilized more recently by Marcus[53] to describe homogeneous and electrode redox reactions. This more recent approach is in many ways similar to that utilized by the earliest workers in the field of charge transfer at electrodes.[3,7,8,19,41]

The potential-energy surface for reacting species with electronic structure of the reactants is denoted by PS^p and that for the products is denoted by PS^q. The *"reaction coordinate"* is a complex combination of translation, rotation, and vibration coordinates. The surfaces PS^p and PS^q intersect, and the set of configurations describing this intersection form a hypersurface PHS in configuration space. We need to calculate the probability that the vibration-rotation-translation coordinates of the entire system are such that the system is in the vicinity of the multidimensional intersection hypersurface PHS. It is assumed that the distribution of systems in the vicinity of the intersection region is an equilibrium one. The rate constant for a heterogeneous reaction in this system is given by

$$k_{het} = (kT/2\pi)^{1/2} \int_{PHS} [e^{-U/RT}(\mu^*)^{-1/2}/Q]\, ds \qquad (124)$$

where μ^* is the effective mass for motion normal to the hypersurface PHS; Q is the configurational integral for the reactants; and dS is the area element in a multidimensional internal coordinate space. In equation (124), integration has already been performed over the coordinates parallel to the solid–solution interface. For a metal electrode, the electron transfer is assumed to occur to and from the Fermi level. The electrode reaction will be assumed to be adiabatic, with Zener-transmission coefficient κ approximately equal to unity.

The parameter m is introduced. If U^p, U^q, and $U^{\ddagger}$ are the potential-energy functions for reactants, products, and activated complex, then, for a given value of the reaction coordinate r, m is defined by

$$U^{\ddagger} = U^p + m(U^p - U^q) \tag{125}$$

m is a function of R, and, on the reaction hypersurface, $U^p = U^q = U^{\ddagger}$.

The transmission coefficient κ is included in the integrand of (124), and the reaction coordinate r is separated by writing dS as $dS'\,dr$. The factor κ depends primarily on r, so that the integral of equation (124) is written as

$$\int_r \bar{\kappa} \left[\int_S e^{-U/RT}(u^*)^{-1/2}\,dS' \right] dr \tag{126}$$

where $\bar{\kappa}$ is equal to the transmission coefficient averaged over S'. Solution of the integral over S', making use of the fact that the resulting integrand has a maximum at some value of r (when r is large, $\kappa \to 0$, and when r is small, the van der Waals repulsion leads to high energies of activation), one obtains from (126)

$$\kappa\sigma(m^{\ddagger})^{-1/2} \exp[-G^{\ddagger}(r)/RT] \tag{127}$$

where $G^{\ddagger}(r)$ is the configurational free energy for the optimum value of r and σ is a factor which contains differences resulting from the use of an optimum value of r rather than integrating over r.

Let G^i be the configurational free energy associated with Q

$$G^i = kT \ln Q \tag{128}$$

G^i pertains to infinite separation of the reactants; let $G^p(r)$ be the corresponding quantity when the reactants are separated by r. We then have

$$^i\Delta^p G = G^i(r) - G^p(r) \tag{129}$$

Similarly,

$$\Delta G^{\ddagger}(r) = G^{\ddagger}(r) - G^f(r) \tag{130}$$

Equation (124) for the rate constant may thus be written

$$k_{\text{het}} = \kappa\sigma Z_{\text{het}} \exp[-(^i\Delta^p G/RT)] \exp[-\Delta G^{\ddagger}(r)/RT] \tag{131}$$

where

$$Z_{het} = (kT/2\pi\mu^{\ddagger})^{1/2} \tag{132}$$

is the collision number of an uncharged species with unit area of the electrode, when it has unit concentration and mass $\mu^{\ddagger}$.

In obtaining an expression for the configurational free energy $G^{\ddagger}(r)$, it is convenient to divide the internal coordinates at a given r into two groups: V_i coordinates describing the positions of the atoms in the inner coordination shells of the reactants, and V_0 coordinates describing the positions of the atoms of the medium relative to each other and to those in the inner coordination shells. We write

$$G^{\ddagger}(r) = G_0^{\ddagger}(r) + G_i^{\ddagger}(r) \tag{133}$$

Each free-energy term may be written as the sum of a potential-energy term and a distribution-function term,

$$G^{\ddagger}(r) = \langle u^p \rangle + kT \langle \ln f^{\ddagger} \rangle \tag{134}$$

where

$$\langle u^p \rangle = \int u^p f^{\ddagger} \, dV' \tag{135}$$

$$\langle \ln f^{\ddagger} \rangle = \int (\ln f^{\ddagger}) f^{\ddagger} \, dV' \tag{136}$$

V' is the volume element at fixed r and $-k \langle \ln f^{\ddagger} \rangle$ is the configurational entropy of a system having the distribution function $f^{\ddagger}$.

Using a harmonic approximation for the vibrational potential energy of the inner coordination shell and assuming that the force constants are approximately the same for reactants and products gives a simple expression for $\Delta G_i^{\ddagger}(r)$,

$$\Delta G_i^{\ddagger}(r) = (1/2)m^2 \sum k_s (\Delta d_e)^2 \tag{137}$$

where the k_s are the reduced force constants in the case where only the diagonal stretching contributions to the vibrational potential energy are considered, i.e.,

$$k_s = 2k_s^p k_s^q / (k_s^p + k_s^q) \tag{138}$$

and the Δd_e are the differences in equilibrium values of bond coordinates for the "diagonal stretching only" case. The result for

$k_s^p \neq k_s^q$ has also been treated by Marcus. Cases other than that for "diagonal stretching only" have not been treated. Equation (137) may be rewritten as

$$\Delta G_i^{\ddagger}(r) = m^2 g_i \tag{139}$$

where

$$g_i = (1/2) \sum_s k_s (\Delta d_s)^2 \tag{140}$$

The quantity $\Delta G_0^{\ddagger}(r)$ is given by

$$\Delta G_0^{\ddagger}(r) = \Delta G^{op} - \Delta G \tag{141}$$

where ΔG^{op} and ΔG denote the polar contributions to the free energies of two hypothetical equilibrium and dielectrically unsaturated systems each having the charge density $m(\rho^p - \rho^q)$ on each reactant. The first system is an "*optical-polarization*" system, which responds to the charge density via the electronic polarization. The second system responds via all polarization terms. The use here of a charge-distribution description is peculiar to the 1965 paper[53] of Marcus and affords a meeting point with the work of Hush. The charge-distribution description is achieved by subdividing the potential energy U of the system into four terms and considering the charge-density dependence of each term. Thus, we write

$$U = U_i + U_0 \tag{142}$$

where

$$U_0 = U(0) + U(1) + U(2) \tag{143}$$

In equation (142), U_i is the intraparticle term for the reactants; U_0 is the sum of the intraparticle term for the medium and of the interparticle term; and $U(0)$, $U(1)$, and $U(2)$ depend on the zeroth, first, and second powers of the charge density ρ_i of the reactants and on the second, first, and zeroth powers of the charge density ρ_0 of the medium. U_i and ρ_a depend only on the intraparticle coordinates V_i of the reactants, and ρ_m depends only on the medium coordinates V_0. For each term in equations (142) and (143), we may write an equation similar to equation (125) with the approximate superscripts. One such equation is

$$U^{\ddagger}(1) = U^p(1) + m[U^p(1) - U^q(1)] \tag{144}$$

Since $U(1)$ is linearly dependent on ρ_a, we may write

$$\rho_1^{\ddagger} = \rho_i^p + m(\rho_i^p - \rho_i^q) \tag{145}$$

Both G^{op} and G in equation (141) are quadratic functions of $\rho_i^{\ddagger} - \rho_1^p$, and hence, of $m(\rho_i^p - \rho_i^q)$. We may then describe the dependence of $\Delta G_0^{\ddagger}$ on $u(1)$ by

$$\Delta G_0^{\ddagger} = m^2 g_0 \tag{146}$$

where

$$g_0 = \langle [U^p(1) - U^q(1) - \langle U^p(1) - U^q(1)\rangle]^2\rangle/RT \tag{147}$$

Using equation (146) and the relation

$$\eta F = \Delta G_0^p - \Delta G_0^q + \Delta G^{\ddagger}(r) - \Delta G^{\ddagger}(r) \tag{148}$$

where η is the overpotential, ΔG_0^p and ΔG_0^q are the free-energy changes when reactants and products, respectively, are brought together,

$$\Delta G_i^{\ddagger}(r) = m^2 g_i \tag{149}$$

and

$$\Delta G^{\ddagger} = \Delta G_0^{\ddagger} + \Delta G_i^{\ddagger} \tag{150}$$

we obtain

$$\Delta G^{\ddagger} = \tfrac{1}{2}(\Delta G_0^p + \Delta G_0^q) + \tfrac{1}{4}g + \tfrac{1}{2}\eta F + [(\eta F + \Delta G_0^q - \Delta G_0^p)^2/4g] \tag{151}$$

where

$$g = g_0 + g_i \tag{152}$$

g_0 and g_i are given by equations (147) and (148), respectively.

For constant ΔG_0^q and ΔG_0^p, differentiation of equation (151) gives

$$\beta = (1/2) + (\eta F/2g) \tag{153}$$

for the case of approximately equal force constants for reactant and product ion–solvent vibrations. Equation (153) may be modified to allow for differences in k^p and k^q. For $\eta F/2g \ll 1/2, \beta = 1/2 = -m$.

From equation (145) for $m = -1/2$,

$$\rho_i^{\ddagger} = (\rho_i^p + \rho_i^q)/2 \qquad (154)$$

which is consistent with the theory of Hush.[38,39]

There is an advantage in the thoroughgoing simplicity of the purely electrostatic treatment, which avoids the complexities and inaccuracies of the treatment of potential-energy surfaces (also treated by the same author). Some disadvantages must be weighed against these positive aspects. Thus, the coefficient β, which has a clear physical meaning in the Gurney theory, arises in the electrostatic treatment with a minimum of physical identity. The effect of the substrate, often considered important in treatments of heterogeneous reactions, is not explicitly treated. The use of a Born-like model to calculate solvation energies may be considered to be too much of a compromise on the side of simplicity, in view of the more sophisticated models available. Lastly, the model is not explicit with respect to the physical process of electron transfer, departing, perhaps to some disadvantage, from the earlier models.[3,22]

Marcus' treatment of 1965 represents a relatively sophisticated approach to the theory of redox reactions and one which shows marked advantages with respect to his earlier work. The advantages must be compared to the inevitable difficulties arising from the complexity of the treatment for calculating rates of reaction.

3. Theory Due to Levich, Dogonadze, and Chizmadzhev

Subsequent to the work by Hush[38] and by Marcus,[36,37] Levich and Dogonadze[47,57] developed a theory of homogeneous redox reactions employing models and methods similar to those of Marcus. On the other hand, a later paper of Marcus[53] contains certain features first presented by Levich. Application of the Levich–Dogonadze theory to electrode redox reactions has been carried out by Dogonadze and Chizmadzhev.[58]

The Levich–Dogonadze–Chizmadzhev theory has recently been extensively reviewed by Levich.[118]

A one-dimensional coordinate r is used to represent the change from the state p, immediately preceding the activated state, to the state q, immediately following the activated state. The value of this "solvent coordinate" for the equilibrium state of p is written as r^p, and the analogous value for q as r^q. The variation of potential energy

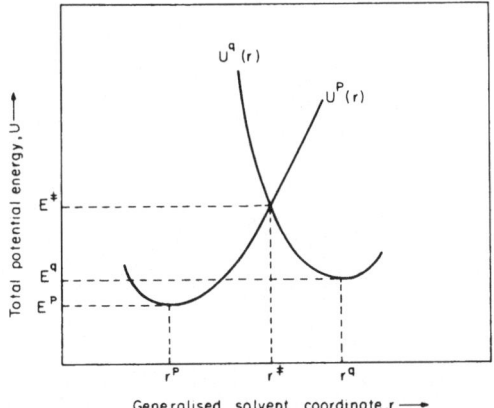

Figure 20. Total potential energy as a function of the solvent coordinate for an electron-transfer reaction. (After Levich.[47])

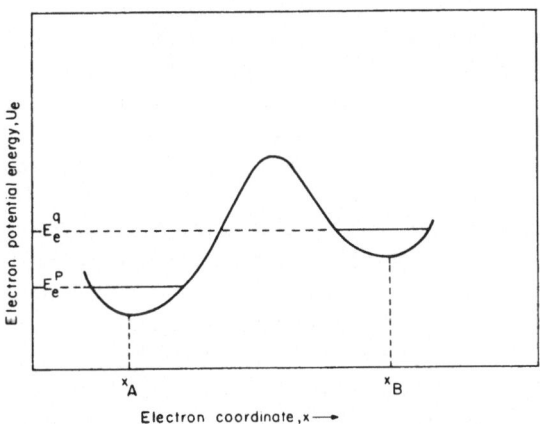

Figure 21. Electron potential energy as a function of the distance between the nuclei of two ions A^n and B^m (After Levich.[47]) E_e^p and E_e^q are the electron energies of p and q, respectively.

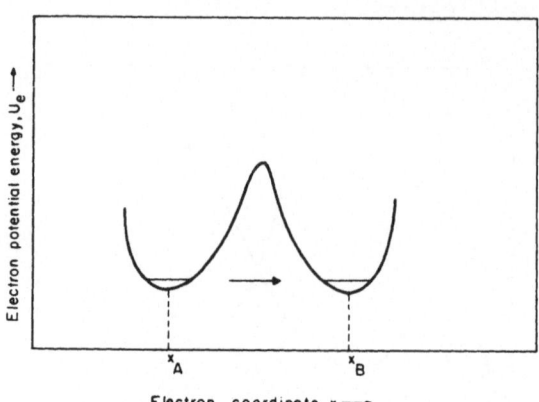

Figure 22. Electron potential energy as a function of the
distance between the nuclei of the two ions A^n and B^m after
suitable solvent fluctuation. (After Levich.[47])

U as a function of r is given by $U^p(r)$ and $U^q(r)$ for p and q, respectively (Figure 20). The energy given by the intersection of $U^p(r)$ and $U^q(r)$ at $r = r^{\ddagger}$ is designated as $E^{\ddagger}$. The energy of activation is supplied by thermal fluctuation of the solvent.

The behavior of the electron in the process of transition is depicted by a plot of the electron potential U_e as a function of the electron coordinate x. U_e depends on the polarization state of the solvent and hence on the solvent coordinate r.

Figure 21 shows the dependence of U_e on x for two ions A^n and B^m in a polar solvent. E_e^p and E_e^q correspond to the electron energies of states on the curves $U^p(r)$ and $U^q(r)$ respectively (Figure 20) for a given value of r. Solvent molecules and ion nuclei remain fixed during such an electron transition. Distinction is made between the process of Figure 21 and an electron tunnel transition. For the latter process, it is necessary, as recognized by Gurney,[3] that the electron energy levels in the left- and right-hand wells be coincident as is shown in Figure 22. The motion of the system along the curve $U^p(r)$ (Figure 20) from r_e^p to $r^{\ddagger}$ corresponds to a continuous transition from curves of the type shown in Figure 21 to the curve shown in Figure 22. The polarization at $r = r^{\ddagger}$ corresponds to equal values of $U^p(r)$ and $U^q(r)$ and to equal electron energies E_e^p and E_e^q. An electron tunnel transition may take place only in a situation corresponding to $r = r^{\ddagger}$ (Figure 20). Up to this time, the distinction between the *electron* potential-energy curve and the *total* potential-energy curve had not been clearly pointed out for redox reactions, although it had been recognized very early by Gurney[3] for electrode reactions. Marcus[48] has also stressed the above distinction. The Schrödinger equation for the complete system is solved using the Born–Oppenheimer approximation, in which the total wave function is written as the product of two wave functions, one for the fast subsystem (electron) and one for the slow subsystem (ions). In the zeroth approximation, the overlap between the reactants is taken as small. This allows a solution of the Schrödinger equation for the case of weak interactions.

The probability of the solvent passing the intersection point $r^{\ddagger}$ of the potential-energy curves $U^p(r)$ and $U^q(r)$ (Figure 20) is deduced, using the laws of classical mechanics, to be

$$W_s = (h/4\pi^2 kT) \exp(-\Delta E^{\ddagger}/kT) \exp[-h(r^p)^2/4\pi\omega_0 kT]\dot{r}^p \qquad (155)$$

where $\dot{r}^p$ is the velocity with which the system passes through the intersection point $r^{\ddagger}$.

A semiclassical calculation of the electron-transition probability at $r = r^{\ddagger}$ using Zener's formula[50] yields

$$W_e = 1 - \exp[-(8\pi^2\Delta^p\Delta^q/h^2\omega_0\delta_r\dot{r}^p)] \qquad (156)$$

where Δ is the exchange integral.

The probability of transition of the complete system from the state p to the state q in unit time is determined by

$$^pW^q = \int_0^\infty W_s W_e \, d\dot{r}^p \qquad (157)$$

Integration is carried out over all positive values of the velocity $\dot{r}^p$. Two limiting cases are considered:

(a) Let the exchange integrals Δ be small, then

$$^pW^q = (4\pi^3/h^2 kTE_s)^{1/2}\Delta^p\Delta^q \exp[-(\Delta E^{\ddagger}/kT)] \qquad (158)$$

where the solvent rearrangement energy E_s is given by

$$E_s = h\omega_0^2\delta_r^2/4\pi \qquad (159)$$

and δ_r is the distance between potential-energy minima corresponding to states p and q. This corresponds to a nonadiabatic transition.

(b) Let the exchange integral Δ be large, then $W_e \simeq 1$, and

$$^pW^q = (\omega_0/2\pi)\exp[-(\Delta E^{\ddagger}/kT)] \tag{160}$$

This corresponds to an adiabatic transition.

A rough value of Δ^q is obtained by using corrected Slater functions[59] for the wave function $\Psi(d)$. The result is

$$\Delta^q = (2\sigma^2/\varepsilon_{eff})e^{-\sigma d} \tag{161}$$

where

$$\sigma = z_{eff}e_0^2/a_0\varepsilon_{eff} \tag{162}$$

where ε_{eff} is the effective dielectric constant of the medium, $z_{eff}e_0$ is the effective charge, and a_0 is the radius of the first Bohr orbit of hydrogen.

For electrode redox reactions, an optimum distance δ_r of the ions from the electrode is assumed, and δ_r is equated to δ_{dl}. As pointed out elsewhere, this is a gross approximation and is not realistic. The theory leads to the result that a Tafel relation between i and η is not predicted except for small η.

For strongly hydrated ions, a linear relation between $\log i$ and η may be obtained with a symmetry factor β of $1/2$. It is rightly pointed out by Levich[47] that, although the symmetry factor is of a universal character in electrode kinetics, the above assumption is not, and in reality, appreciable deviations from it may be observed. The result $\beta = 1/2$ is dependent on the use of the harmonic approximation for the solvent–ion interaction. Anharmonicity, however, is frequently present, so that $\beta = 1/2$ should be the exception rather than the rule.

In a recent paper by Dogonadze et al.,[115] the above theory has been extended to cover the proton-discharge step of the h.e.r. The main feature of the paper is the exact treatment of transition probabilities. Stress is also put on the role of the solvent in producing a continuum of energy levels, a feature only lightly touched upon by earlier workers. A difficulty for this work exists, however, in the prediction that slow discharge mechanisms for the h.e.r. will be accompanied by a region in which the Tafel slope will be RT/F, although this has not yet been observed. The use of a continuum treatment for the solvent right up to the contact with the metal is unrealistic. The largest difficulty is that the paper does not predict the dependence of the rate of the h.e.r. upon the metal, although this dependence exists clearly, even for the high-overvoltage metals for which the theory is intended. The theory (being adiabatic in model) is not consistent with the facts concerning the isotopic rate ratio.

German *et al.*,[119] have recently attempted to evaluate the ratio of the isotopic reaction rates for the discharge of tritium and protons. They neglect all previous work, with its explicit experimental values, detailed models, and agreement with experiment both with respect to the absolute values and dependence on potential. They utilize the Levich model for the electron transfer rate, in which a major point is that the heat of activation is supposed to be associated primarily with the solvent reorganization. However, now they state that the heat of activation is given by:

$$E_A = \frac{(E_S + \Delta I)^2}{4E_S}$$

Assuming that E_S equals 2.0 and $\Delta I = 1.5$ eV, roughly correct for mercury, that the E_A is only about 60% solvent reorganization energy. (The ΔI term is the heat of the proton discharge reaction and presumably includes the heat of bonding of H to the metal.) The absolute value comes to some 33 kcal/mole with the above values.

The contribution from the zero-point energy, calculated by present workers, shows isotopic rate constant ratio which is about 2.3, and much too small. They, therefore, abandoned the previously used concept that charge transfer reactions at interfaces are adiabatic. They formulated a nonadiabatic version of the reaction and obtained an explicit expression for the separation factor, which now depends upon potential. They did not evaluate their expression numerically.

Kuznetsov[120] has recently discussed the work function in semiconductors. There is no effect of the Fermi level of an intrinsic semiconductor on the rate of a redox electrode process. However, if one is dealing with a doped semiconductor, there is an effect of the work function, because the concentration of electrons at the interface depends on its value.

The theory which Kuznetsov uses is that due to Levich *et al.* The predicted change of rate compared with the corresponding metal should be, according to Kuznetsov, $e^{E_G/RT}$, where E_G is the gap energy. With a value of E_G of about 1 eV, the acceleration should be approximately 10^7 times the electron transfer velocity of a corresponding metal. There is no evidence at the present time for such an increase (in Kuznetsov discussion, diffusional limitations of the electrons in the semiconductor are neglected).

There is a major difference in the model suggested by Hush, Marcus, and Levich on the one hand, and Gurney, Gerischer, and Bockris and Matthews on the other. In the former group of authors, the origin of the activation, which allows acceptance by the ion of an electron is electrostatic energy, physically the change of the Born solvation energy, In the latter, the activation is primarily thermal, i.e., it is the energy distribution among bonds which decides how many receptor molecules will have electronic states which are free to receive electrons. A number of tests can be devised to distinguish between these models, for example, the dependence on temperature in which the first model does not predict a linear $\log(i/1/T)$ relation. Calculations of these distinguishing criteria are not complete.

VI. ION TUNNELING

Charge transfer by quantum mechanical tunneling of ions, particularly the hydrogen ion, has been discussed by Christov,[60,62] Conway,[63-65] and Bockris and Matthews.[32] Ion tunneling is restricted to ions of small mass and only hydrogen-ion tunneling will be considered here.

The possibility of proton tunneling contributing to the mechanism of h.e.r. was referred to by Topley and Eyring.[66] According to these authors, "The relative rates of discharge of the proton and of the deuteron will depend upon two factors associated with the height and width of the (potential energy) barrier, namely the rates of leakage through the barrier and the difference between heats of activation arising from the difference in zero-point energies of the bonds H–O and D–O in the hydrated ions." An evaluation of the importance of the tunnel effect based on the use of a parabolic barrier 1.5 Å wide and 5 kcal high led to the conclusion that the tunnel effect is of secondary importance. The use of barrier models,

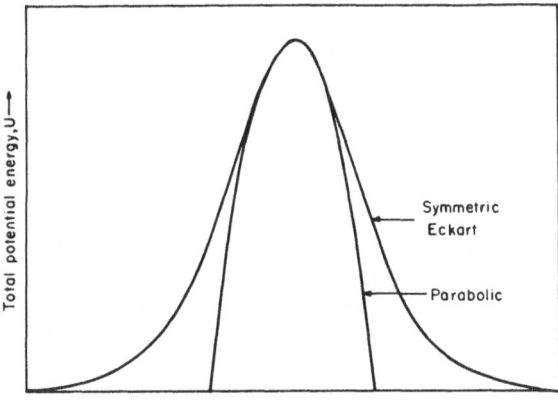

Figure 23. Relation of shapes of parabolic and symmetrical Eckart barriers having the same curvature in the vicinity of their maximums.

such as rectangular, triangular, parabolic, and that due to Eckart,[67] however, raises the question of the relation of the width of the actual, or real, potential-energy barrier to that of the model barrier. This question has been discussed by Bockris and co-workers,[32,68] who propose, as a rough guide, that the barrier widths possess the following proportionality (Figure 23)

$$(2d)_{\text{Eckart}}:(2d)_{\text{real}}:(2d)_{\text{parabolic}} = 6:3:2 \tag{163}$$

The barrier model of Topley and Eyring[66] thus corresponds to a proton-transfer distance of about 2.3 Å and to an overpotential for mercury in acid solution of about -1.5 V.† In terms of more recent views, there would indeed be minor proton tunneling under these conditions.

On the other hand, Bawn and Odgen[77] conclude that proton tunneling should play a major role in the h.e.r. These conclusions

†The activation energy at $\eta = 0$ is 21 kcal/mole for Hg in acid solution.[69–76] The barrier height is >21 kcal, say 22 kcal. For $\beta = 0.5$, the barrier height is 5 kcal at an overpotential of about -1.5 V.

are based on an Eckart-barrier model with $2d = 1.5$ Å and a height of 15 kcal, corresponding to a proton-transfer distance of about 0.8 Å and an overpotential on mercury in acid solution of about -0.6 V. Bawn and Odgen predicted a value of the protium-deuterium separation factor of 74 at 273°K, excluding zero-point energy-difference contributions.

The models chosen by Topley and Eyring and by Bawn and Ogden correspond to widely different physical conditions. Indeed, this is a major problem in this field and several authors, rather than choosing a single model as the best representation of physical reality, have considered several models with variation of parameters in order to obtain a fit for the experimental data.

1. Christov's Work

Various barrier models are investigated by Christov,[60] such as the symmetrical Eckart[67] plus linear-field model given by

$$U(x) = [Be^{2\pi x/d}/(1 + e^{2\pi x/d})^2] + (Ax/2d) + (A/2) \qquad (164)$$

where $A = e_0\eta$, $2d$ is the barrier width, and B is a constant. The linear term $(Ax/2d) + (A/2)$ accounts for the effect of electrode potential. The barrier is shown in Figure 24.

To determine the rate, use is made of the equation

$$i_q = k_1 e_0 c(1 - \theta) \int_0^\infty W_p(E, \eta) w(E, T) \, dE \qquad (165)$$

where i_q is the quantum mechanically calculated current density for the forward reaction, c is the number of protons per cm^2 of the double layer, $W_p(E, \eta)$ is the probability of proton transfer at the energy level E, and $w(E, T)$ is the probability of the initial state $(H_3O^+ + Me^-)$ possessing energy between E and $E + dE$. Tunneling of the system has been approximated to tunneling of a proton, so that the generalized reaction coordinate is reduced to a proton coordinate.

Use of the WKB approximation for $W_p(E, \eta)$ leads to a linear dependence of $\log i$ on overpotential with Tafel slope of 106 mV

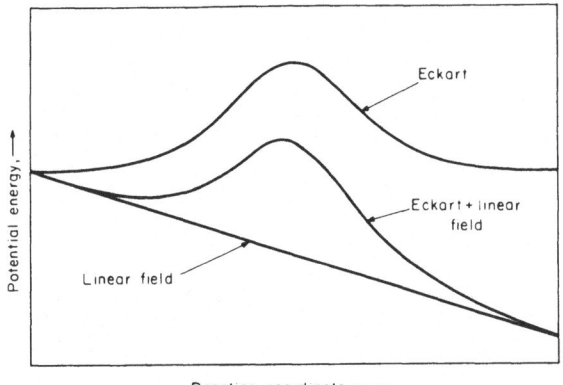

Figure 24. The symmetrical Eckart–linear-field barrier model.

for $\eta = 0$ to -0.31 V, $E^{\ddagger}(\eta = 0) = 1 \times 10^{-12}$ erg (14.4 kcal/mole), $2d = 2$ Å, and $T = 273°$K. Under these conditions, the ratio of currents for tunneling through (i_p) and over (i_r) the barrier [equation (164)] is 300:1. Thus, the degree of tunneling through the barrier can be quite high while little or no deviation from linearity of the Tafel plot is exhibited.

Later work of Christov's[61,62] reexamines the conditions under which the Tafel behavior is obtained and also considers the temperature dependence of the current density.

The total, quantum mechanically calculated current density given by equation (165) is divided into two parts,

$$i_q = i_p + i_r \tag{166}$$

where i_p corresponds to tunneling through the barrier and i_r corresponds to that over the barrier. For simplicity, i_r is equated to the classical current density i_{cl}, thus ignoring the quantum mechanical phenomenon of reflection ($W < 1$ for $E > E^{\ddagger}$). The graphical solution of equation (165) as a function of $\gamma = E^{\ddagger}/kT$ and $\delta = 2\pi d(2mE^{\ddagger})^{1/2}$ using the WKB approximation for W and an asymmetrical Eckart barrier model demonstrates that, as δ increases, the extent of the Tafel region increases. Increase of either mass or barrier width leads to a decrease of tunneling and a closer approach

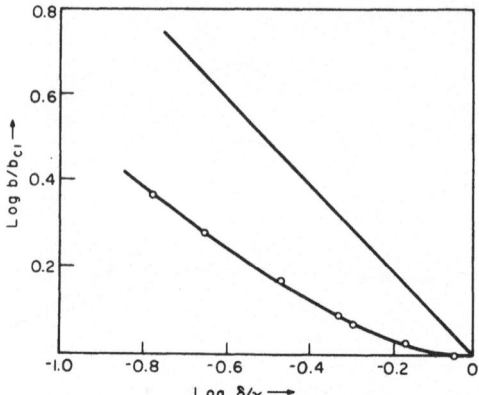

Figure 25. Dependence of Tafel slope b on the barrier parameters δ and γ; (A) asymmetrical Eckart barrier; parabolic + linear-field barrier. (After Christov.[62])

to the classical Tafel behavior (Figure 25). For $\gamma = 26.5$, the Tafel slope b equals the classical value when $\delta = 24$; for $T = 273°K$ and $E^{\ddagger} = 1 \times 10^{-12}$ ergs (14.4 kcal/mole), a value of δ equal to 24 corresponds to an Eckart barrier width of 2.7 Å. From the dependence of Tafel slope on γ at constant δ, it is found that increase in temperature results in b tending to b_{cl}. The dependence of b on δ and γ obeys a simple law according to which b/b_{cl} is determined by δ/γ.

For a model consisting of a symmetrical parabolic barrier plus a linear electric field,

$$U(x) = E^{\ddagger}[1 - (x^2/d^2)] + [(x/2d) + (1/2)]e_0\eta \qquad (167)$$

it is also found that b/b_{cl} depends solely on δ/γ. As with the Eckart barrier, increase of mass, barrier width, or temperature causes b/b_{cl} to tend to unity (Figure 25).

Christov's method of accounting for the change in barrier shape as a function of electrode potential is, at best, a very rough approximation. The main changes brought about in barrier shape occur in the vicinity of the activated state (Figure 26). A better approximation of the changes in barrier shape may be accounted for by the use of a nonlinear potential-energy function such as $U_A - U_B$ in Figure 26, but it should be kept in mind that this

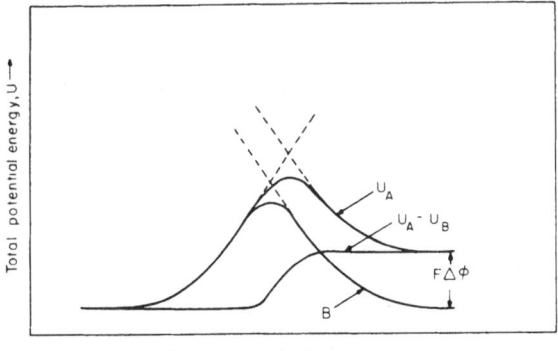

Figure 26. The vertical-shift method of allowing for the effect of electrode potential on the potential-energy barrier.

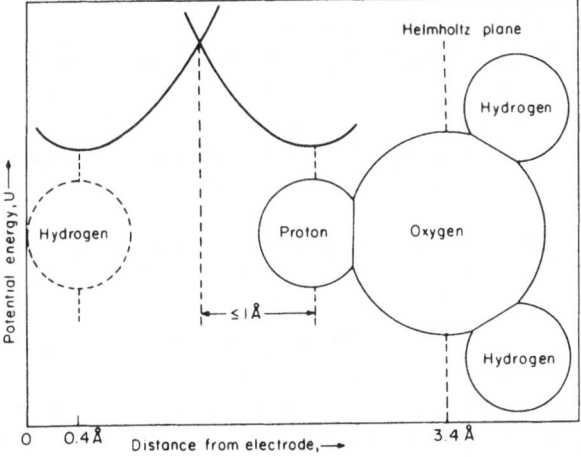

Figure 27. The relation between the metal-electrolyte double-layer structure and the proton-transfer barrier.

function has no simple significance or relation to physical reality. That portion of the proton-transfer barrier influenced by the field is, however, only a fraction of the double-layer width, as shown in Figure 27.

In order to obtain the barrier parameters, use is made of the results of Post and Hiskey.[72,78] *By ignoring contributions made by the zero-point energy of initial and activated states*, equations for apparent activation energies, preexponential factors, and separation factors are solved simultaneously for the barrier height and width (the barrier parameter A being fixed in these calculations). The separation factor calculated is the so-called "imaginary separation factor" $S_{H,D}$ defined by

$$S'_{H,D} = i_{0,H}/i_{0,D}; \qquad C_{H_3O^+} = C_{D_3O^+} \tag{168}$$

Calculation of i_0 is carried out according to formula (165) for $\eta = 0$. However, it was not recognized by Christov that these calculated values of i_0 are not directly comparable with the i_0 values obtained experimentally by extrapolation from high cathodic overpotentials. Dependence of proton tunneling on barrier height, and hence on overpotential, may lead to large differences between the extrapolated and calculated values of i_0. Bockris and Matthews[32] have overcome this problem by calculating i as a function of η and extrapolating from the experimentally accessible η region to $\eta = 0$ in order to obtain a calculated i_0 comparable with the experimental values. Calculations by other workers show that zero-point energy differences make a major contribution to isotope effects. Hence, the neglect of these invalidates the quantitative conclusions of Christov, and results in the assignment of an abnormally large role to proton tunneling.

Christov's calculations give $E^{\ddagger}(\eta = 0) = 1.6 \times 10^{-12}$ erg (23 kcal/mole), $2d_{Eck} = 4.7$ Å, and $2d_{par} = 1.65$ Å for an assumed value of $A(\eta = 0) = 0$. Using these barrier parameters, it is found that an Arrhenius relation between the rate of proton discharge and temperature is obtained in the temperature range 300–220°K.

Fortuitous agreement between $2d_{par}$ and the then-accepted double-layer width ($\simeq 1.75$ Å) led Christov to erroneously conclude that the parabolic barrier was a better model than the Eckart barrier. The model barrier widths deduced by Christov correspond to a proton-transfer distance of about 2.4 Å or a double-layer width of about 3.5 Å.

Christov concluded that, "The tunnel effect is important for both (protium and deuterium) isotopes; it is however comparable with reflection of particles over the barrier." It is understood that by "tunneling effect" Christov is referring to the contribution of i_p to the total current density. At 273°K, it is found that $i_p/i_q = 0.82$ for protium discharge and $i_p/i_q = 0.50$ for deuterium discharge.

Christov proposed several criteria to determine the occurrence and degree of proton tunneling: (a) the measurement of Tafel slopes at low temperatures, (b) the measurement of the rate as a function of temperature at low temperatures, and (c) the measurement of the isotopic separation factor (or kinetic isotope effect) at low temperatures.

The work of Christov on proton tunneling in the h.e.r. was the first systematic work in this area. Christov was the first to recognize the uncertainty that existed as to the barrier height and width and he made large advances over the earlier work of Topley and Eyring[66] and Bawn and Ogden[77] when he used the experimental results of Post and Hiskey[72,78] to obtain an estimate of the barrier parameters for the h.e.r. on mercury.

Despite the detailed work by Christov, neglect of zero-point energies and other errors still leaves doubts as to the role of proton tunneling in the h.e.r. There are several reasons for the inconclusive nature of Christov's work; some of these reasons stem from conceptual difficulties, such as the lack of precise understanding of the charge transfer mechanism, and others are the result of invalid assumptions or approximations. However, it may be said that the work by Christov added considerable impetus and interest to the problem of proton tunneling in the h.e.r., but that it did not give decisive or quantitative answers to this problem.

2. Conway's Work

An asymmetrical Eckart barrier of width $2d = 0.5$ Å and height $E^{\ddagger} = 19$ kcal is used by Conway[63] to calculate the protium–deuterium separation factor $S_{H,D}$ for the h.e.r. taking into account the zero-point energy difference. The Eckart barrier width of 0.5 Å corresponds to a real barrier width of about 0.3 Å, which is unrealistically low. One might expect such a barrier to produce large separation factors by virtue of high degrees of proton tunneling. In actual fact, the calculated separation factors are quite low (0.7–12.40) due to the additional existence of high degrees of *deuteron* tunneling for such a thin barrier. As a consequence of tunneling, the calculated Tafel plots have a high initial slope (250 mV for protium, 170 mV for deuterium) and become nonlinear at potentials cathodic to -0.5 V. The calculations are carried out using the approximation $B = 4E^{\ddagger}$, which is valid only for $A = 0$.

Conway proposes the criterion of high Tafel slope to detect proton tunneling.

Conway and Salomon[15,64,65,80,81] have concluded that proton tunneling plays a negligible role in the h.e.r. They find, in agreement with the earlier work of Rome and Hiskey,[79] that the protium–deuterium separation factor $S_{H,D}$ decreases with an increase of cathodic overpotential. Unlike Bockris and Matthews,[32] Conway and Salomon *exclude* proton tunneling as the explanation of the potential dependence of the separation factor (although they do take into account the difference of OH and OD zero-point energy). The effect is initially given qualitative interpretation in terms of the potential dependence of the symmetrical vibration mode of the activated state.† In later publications,[80,81] Conway and Salomon supplement their original explanation by considering three more factors: (a) electrostriction may result in discharge at a dual site; (b) electrostriction may affect the force constant of the laterial bending modes of the activated state; and (c) the electrode field may affect the libration behavior of neighboring solvent molecules in the double layer.

Conway and Salomon[65,81] have preferred five qualitative explanations to compensate for the lack of assumed proton tunneling. The evidence quoted by Conway and Salomon[65] against a significant degree of proton tunneling in the h.e.r. is, however, open to criticism. Exclusion of proton tunneling on the basis of Tafel slopes or on the basis of the temperature dependence of current density has been shown by Bockris and Matthews[32] to be useful only in the case of very large degrees of proton tunneling, i.e., these two tests for proton tunneling are insensitive. Conway and Salomon[65,81] and Conway and MacKinnon[65] concluded, from results on the h.e.r. and d.e.r. down to $-110°C$, that tunneling was not appreciable. In terms of H/D effects on Tafel slopes, these results are complicated by the temperature dependence of b discussed by Conway and Mac-Kinnon.[65]

3. Work of Bockris and Matthews

Bockris and Matthews[32] reexamined the question of proton tunneling in the h.e.r. The quantum mechanical current density i_q

---

†The importance of this mode to the calculation of the separation factor was first recognized by Bockris and Srinivasan.[42]

is calculated for protium, deuterium, and tritium for a large number of combinations of the asymmetrical Eckart-barrier parameters $2d$, $E^{\ddagger}$, and A, at various temperatures and electrode potentials. From these computations, Tafel parameters, Arrhenius parameters, quantum mechanical transmission coefficients τ, quantum mechanical corrections Γ to the separation factor, and the dependence of Γ on electrode potential are calculated.

The quantum mechanical transmission coefficient is given by

$$\tau = i_q/cT \tag{169}$$

$$= (1/kT)\exp[E^{\ddagger} - E_0)/kT]J \tag{170}$$

where

$$i_q = k_q cJ \tag{171}$$

$$J = \int_{E_0}^{\infty} \exp[-(E - E_0)/kT]W_p(E)\,dE \tag{172}$$

and k_q is a frequency factor, c is the proton concentration, E_0 is the zero-point energy level for H^+–OH_2 stretching, and $W_p(E)$ is the probability of proton tunneling at the energy level E.

The quantum mechanical correction to the separation factor is given by

$$\Gamma_{H,T} = \tau_H/\tau_T \tag{173}$$

$$= \{\exp[(E_{0,T} - E_{0,H})/kT]\}J_H/J_T \tag{174}$$

and the separation factor is

$$S_{H,T} = \Gamma_{H,T}(S_{H,T})_{cl} \tag{175}$$

where $(S_{H,T})_{cl}$ is the classical separation factor including zero-point energy differences but excluding proton tunneling contributions. The dependence of $S_{H,T}$ on overpotential is given by

$$dS_{H,T}/d\eta = (S_{H,T})_{cl}(d\Gamma_{H,T}/d\eta) \tag{176}$$

Tafel slopes are obtained by computation of i_q according to equations (171) and (172) as a function of overpotential.

The exchange current density i_0 was not computed directly, but was obtained by graphic extrapolation from regions of moderate (0.5–1.0 V) cathodic overpotential, thus obtaining a value of i_0

suitable for comparison with experiment. A plot of $\log i_0$ against $1/T$ at temperatures above $0°C$ yielded the activation energy excluding zero-point energy differences. Zero-point differences were obtained from the work by Bockris and Srinivasan.[42] The activation energy thus obtained could be compared directly to experiment.

The barrier model employed is the asymmetrical Eckart barrier[67]

$$U(x) = \frac{A \exp(2\pi x/d)}{1 + \exp(2\pi x/d)} + \frac{B \exp(2\pi x/d)}{[1 + \exp(2\pi x/d)]^2} \tag{177}$$

where

$$A = A_0 + e_0\eta \tag{178}$$

$$B = 2E^{\ddagger} - A = 2[E^{\ddagger}(E^{\ddagger} - A)]^{1/2} \tag{179}$$

$$E^{\ddagger} = E_0^{\ddagger} - \beta e_0\eta \tag{180}$$

with A_0, $E_0^{\ddagger}$, and $2d$ the barrier parameters corresponding to energy of reaction at $\eta = 0$, energy of activation at $\eta = 0$, and proton-transfer distance, respectively. The symmetry factor β is taken as 0.5. Inclusion of the zero-point energy level E_0 in the model led not only to a better correspondence between real and model barriers, but also simplified the tunneling probability expression.

For the range of barrier parameters considered, the Tafel plot is found to be linear from 0 to -1.5 V. It is found that the criteria, the dependence of Tafel slope b on mass and temperature, and the dependence of rate on temperature, are not sensitive tests for the existence of proton tunneling.

The computed dependence of $\Gamma_{H,T}$ on overpotential for various values of $A(\eta = 0)$ and temperature are shown in Figures 28 and 29, respectively. The dependence of Γ on overpotential is found to be linear in the range $1.2–2.0 \times 10^{-12}$ ergs (0.75–1.25 V). *Both Γ and $d\Gamma/d\eta$ (the latter especially) are found to be sensitive tests for proton tunneling.* The dependence of S on temperature is shown to be complex and often experimentally unattainable, so that dS/dT is usually not a satisfactory test for proton tunneling.

Comparison of the experimentally determined[75] values of $S_{H,T}$ as a function of overpotential at $27°$ and $50°C$ with computed values leads to an assignment of the most probable barrier parameters as

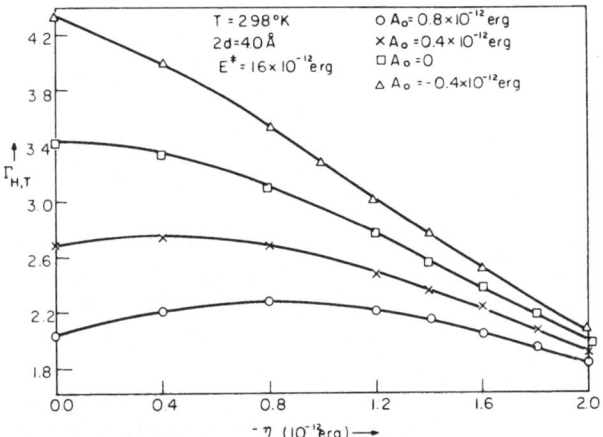

Figure 28. The dependence of the quantum mechanical correction $\Gamma_{H,T}$ to the separation factor on overpotential η for various values of A_0. (After Bockris and Matthews.[32])

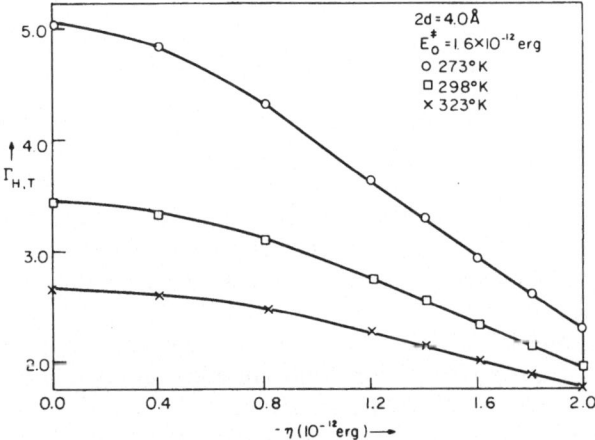

Figure 29. The dependence of the quantum mechanical correction $\Gamma_{H,T}$ to the separation factor on overpotential η for various temperatures. (After Bockris and Matthews.[32])

$2d = 4 \text{ Å}, \quad E^{\ddagger}(\eta=0) = 1.5 \times 10^{-12}$ ergs, and $A(\eta=0) = -0.6 \times 10^{-2}$ ergs. The degree of tunneling, defined by,

$$Q = 100(i_q - i_c)/i_q \tag{181}$$

$$= 100[1 - (1/\tau)] \tag{182}$$

is found to be 68.7% at 27°C and $\eta = -1$ V for the most probable barrier. At $\eta = 0$ and 27°C, $Q_H = 85.9\%$, compared to 70% obtained by Christov.[62]

The value of $2d = 4.0$ Å has important consequences for double-layer theory since it corresponds to a proton-transfer distance of about 2 Å. The result is consistent with the presence of a layer of electrode-oriented water molecules between the electrode and the discharging proton, in agreement with the model of Bockris *et al.*[30]

It has been shown that, even at room temperature and for high cathodic overvoltages, proton tunneling makes large contributions to the rate of proton discharge on mercury, leading to a trebling of the rate. At lower temperatures and at lower cathodic overpotentials, the contribution of proton tunneling increases. The potential dependence of the kinetic isotope effect has proved to be the most sensitive test for proton tunneling.

VII. CONCLUDING REMARKS

The central aspect in electrode kinetics, i.e., in reactions at interfaces between an electronic conductor and an ionic conductor, is that of metal–electrolyte charge transfer. The basic theory of this phenomenon was given in 1931 by Gurney in one of the first applications of quantum mechanics to chemistry. Gurney's 1931 paper was the origin of two main lines of attack on the theory of interfacial charge transfer.

A quite different approach to the theory of charge transfer originated qualitatively with Libby in 1952 and was formulated mathematically by Weiss in 1954, although the considerations were restricted to charge transfer between ions in solution. It has been developed more recently by R. A. Marcus, and also by Levich, Dogonadze, and Chizmadzev.

In a third line of work, the quantum mechanical tunneling aspects of ion transfer, particularly proton transfer, have been investigated, the work being originated by Topley and Eyring, and developed extensively by Christov and by Bockris and Matthews. Such work sheds light on the mechanism of proton transfer.

The advantage of the approach originated by Gurney is that it is a direct quantum mechanical treatment of the reaction rate; it deals with the actual act of electron transfer and concentrates on consideration of the state of the electron both in the metal and in the electrolyte. It was originated for charge transfer *at electrodes* and hence the electrical quantities such as interfacial potential difference are foci of attention from the beginning. However, exact calculations are complex and lengthy. Introduction of simplifying potential-energy profile diagrams is associated with insufficiently defined degrees of approximation.

In the approach arising from the theory of electron transfer in solution, the great advantage is simplicity of thought process and model. The energy changes taken into account (and proposed as being the origin of activation) are simply those of the change of solvation energy with ionic charge. These changes are determined largely in terms of changes in the Born term of the solvation energy. Continuum dielectric theory is used and therefrom arises much of the mathematical simplicity of the formulation. Corresponding disadvantages arise as the consequences of the simple model used. Quantum mechanical aspects of the electron transfer are only considered at the boundaries of the treatment, to discuss the adiabatic or nonadiabatic character of the change. The approach is less well applied to *bonding* reactions.

Although there are such great differences in these two approaches, they both treat basically the same model: charge transfer by quantum-mechanical electron tunneling when the reactant molecules are perturbed in such a way as to offer vacant energy states to electrons, which optimize their rate of transfer. The approaches begin to overlap in the recent work of Marcus (1965).[53]

An advance has been the clarification of the significance of β, earlier regarded by electrochemists as an empirical factor. The significance of β was essentially stated in the Gurney quantum-mechanical theory; it is a proportionality factor which relates changes in the electron energy difference between metal and electrolyte to changes in the total interaction energy of the reactants. In the electrostatic model, the significance of β is less easy to state simply in physical terms.

Four final remarks will be made:

(1) One of the first applications of quantum mechanics in chemistry was to the basic electrochemical act. This theory was then almost totally neglected for some 30 years, the most unfortunate event to occur in the history of fundamental electrochemistry.

(2) Both quantum-mechanical and electrostatic approaches to charge-transfer theory agree in predicting *no direct effect of the electronic properties of the metal upon the reaction velocity.* In fact, this conclusion is independent of the model of charge transfer. This fact is stressed because it is contrary to the intuitive appreciation of most workers.

(3) It follows from (2) that electrocatalytic properties of metals depend principally on bonding of the reactants, intermediates, and products to the surface and in the electrolyte. Of course, such bonding involves the electronic states of the metal, but these effects are not the direct ones which many at first expected when considering the electronic work function and electrocatalysis.

(4) The questions of the adiabatic and nonadiabatic properties of charge transfer have to be resolved. The principal contribution has been due to Levich, Dogonadze, and Chizmadzev, most other workers having assumed an adiabatic reaction.

The present state of this field is a very challenging one. There is one extremely clear objective, namely, to determine the relative degree of contribution of thermal and electrostatic activation in the theory of the fundamental act. When this problem is decided, the electrocatalytic problem will clearly be approachable.

ACKNOWLEDGMENTS

The authors thank the National Science Foundation for financial support of their work; the U.S. Air Force for the support of certain antecedent work from which the present work grew; and the Materials Science Department of the University of Virginia for its cooperation. One of the authors (D.B.M.) wishes to acknowledge discussions and encouragement from Dr. Mino Green during the problem formulation stage, and to thank Drs. M. A. V. Devanathan, K. Müller, H. Wroblowa, A. K. N. Reddy, and S. Srinivasan for discussions in which the ideas presented here were clarified. Thanks are due to Dr. Klaus Müller for his thorough checking of the manuscript, and for his helpful remarks thereon.

APPENDIX I. THE TRANSFER COEFFICIENT

In order to relate the rate of the overall reaction to the rate of the rate-determining step (r.d.s.), it is necessary to consider the relative numbers of charges transferred in the respective processes. Let the r.d.s. occur v times when the overall reaction occurs once. If n is the total number of electrons transferred in one act of the overall reaction, then the number of electrons transferred in the overall act of formation of one activated complex is $n/v = \lambda$. The quantity v is called the *stoichiometric number* and λ is called the *electron number*.† Let r be the number of electrons transferred in one act of the r.d.s., and let s and t be the numbers of electrons transferred in steps preceding and succeeding, respectively, the r.d.s. The total number of electrons transferred per act of the overall reaction is thus

$$n = s + t + vr \tag{183}$$

Consider a generalized reaction sequence,

$$aA = bB \rightleftharpoons cC + xX \tag{184}$$

$$c'C + dD \xrightarrow{\text{r.d.s.}} fF + yY \tag{185}$$

$$f'F + gG \rightleftharpoons mM + zZ \tag{186}$$

After some reflection on the matter, it is realized that, besides the parameter v relating f to f', we need a parameter relating c either to c' or to f'. Let v^p be the number of times reaction (184) must occur in order for the reaction (186) to occur once. One act of the overall

---

†The stoichiometric number was first devised as a concept by Horiuti and Ikusima,[82] who applied it to the consideration of equilibrium at electrodes. The analogous quantity, the electron number, was introduced into electrode *kinetics* by Bockris and Potter.[83] Application of the stoichiometric number to mechanism analysis in electrode kinetics was subsequently carried out by Horiuti and Nakamura,[84] Parsons,[85] Oldham,[86] Makrides,[87] Mauser,[88] Riddiford,[89] Mohilner,[90] Barnartt,[91] and Despic.[92]

reaction is thus comprised of the following steps:

$$v^p a A + v^p b B \rightleftharpoons v^p c C + v^p x X \tag{187}$$

$$v^p c C + v d D \xrightarrow{\text{r.d.s.}} v f F + v y Y \tag{188}$$

$$v f F + g G \rightleftharpoons m M + z Z \tag{189}$$

Consider now a generalized sequence of charge-transfer reactions at an electrode,[†]

$$v^p a A^{z_1} + s e^- \rightleftharpoons v^p c C^{z_2} + v^p x X \tag{190}$$

$$v^p c C^{z_2} + v r e^- \xrightarrow{\text{r.d.s.}} v f F^{z_3} + v y Y \tag{191}$$

$$v f F^{z_3} + t e^- \rightleftharpoons \text{products} \tag{192}$$

where the z's are the ion valences. It is not necessary, for the present purposes, to discuss the nature of the "products." Conservation of charge requires that

$$c z^2 = a z_1 - (s/v^p) \tag{193}$$

$$v f z_3 = v^p c z_2 - v r \tag{194}$$

The symbols i, p, t, q, and f will be used to denote the initial state [left-hand side of equation (190)], the state immediately preceding the activated state [left-hand side of equation (195)], the activated state $A^{\ddagger}$, the state immediately following the activated state [right-hand side of equation (195)], and the final state [right-hand side of equation (192)], respectively.

One act of the r.d.s. is given by

$$(v^p/v) c C^{z_2} + r e^- \rightleftharpoons A^{\ddagger} \rightarrow f F^{z_3} + y Y \tag{195}$$

where $A^{\ddagger}$ denotes the activated state. We define v^i by

$$v^i = v^p/v \tag{196}$$

The standard electrochemical free energy of activation for the r.d.s. (195) is

$$\Delta \bar{G}_0^{\ddagger}(p \rightarrow t) = \bar{\mu}_0^{\ddagger} - v^i c \bar{\mu}_0(C) - r \bar{\mu}_0(e^-) \tag{197}$$

[†]More general reaction schemes have been treated by Riddiford[89] and Despic.[92]

In order to write $\Delta \bar{G}_0^{\ddagger}$ as a function of the electrode–solution potential difference $\Delta \psi$, we utilize the relation

$$\bar{\mu}_0 = zF(\psi + \chi) + \mu_0 \tag{198}$$

This relation may be applied to each species in each of the states except the activated state $A^{\ddagger}$. Equation (198) has not been successfully applied to the activated state and the following approach is routine. The standard electrochemical free energy of activation is written as

$$\Delta \bar{G}_0^{\ddagger}(p \to t) = \beta \, \Delta \bar{G}_0(p \to q) + \text{const} \tag{199}$$

where $\Delta \bar{G}_0$ is the standard electrochemical free-energy change of the r.d.s. (195),

$$\Delta \bar{G}_0(p \to q) = \Delta G_0(p \to q) + rF(\Delta \psi + \Delta \chi) \tag{200}$$

and ΔG_0 is the standard chemical free-energy change of the r.d.s. (195),

$$\Delta G_0(p \to q) = f\mu_0(F) + y\mu_0(Y) - v^i c\mu_0(C) - r\mu_0(e) \tag{201}$$

Equation (200) assumes no change of the dipole potential χ when phases are brought together to form an interface. Allowance for such changes are expressed by Parsons.[2] Substituting equation (200) for $\Delta \bar{G}_0(p \to q)$ into equation (199) for $\Delta \bar{G}_0^{\ddagger}(p \to t)$ and assuming $\Delta \chi$ to be independent of $\Delta \psi$ gives

$$\Delta \bar{G}_0^{\ddagger}(p \to t) = \beta rF \, \Delta \psi + \Delta \bar{G}_0^{\ddagger}(p \to t, \Delta \psi = 0) \tag{202}$$

Equation (202) is a more familiar, but less general, definition of the symmetry factor β than is equation (199).

Combining equations (197) and (198) for $\Delta \bar{G}_0^{\ddagger}(p \to t)$ gives the required expression for the standard electrochemical potential of the activated state:

$$\bar{\mu}_0^{\ddagger} = \Delta \bar{G}_0^{\ddagger}(p \to t) + v^i c \bar{\mu}_0(C) + r\bar{\mu}_0(e^-) \tag{203}$$

$$= \Delta \bar{G}_0^{\ddagger}(p \to t, \Delta \psi = 0) + \beta rF \, \Delta \psi + v^i c \bar{\mu}_0(C) + r\bar{\mu}_0(e) \tag{204}$$

The expression for $\bar{\mu}_0^{\ddagger}$ may now be used in deriving the standard electrochemical free-energy change $\Delta \bar{G}_0^{\ddagger}(i \rightarrow t)$ for the overall reaction,

$$v^i a A^z + s e^- + r e^- \rightleftharpoons A^{\ddagger} + v^i x X \rightarrow \text{products} \qquad (205)$$

For the overall reaction (205),

$$\Delta \bar{G}_0^{\ddagger}(i \rightarrow t) = \bar{\mu}_0^{\ddagger} - v^i a \bar{\mu}_0(A) - s \bar{\mu}_0(e) - v r \bar{\mu}_0(e) + v^i x \bar{\mu}_0(X) \quad (206)$$

Substituting equation (204) for $\bar{\mu}_0^{\ddagger}$ into equation (206) gives

$$\Delta \bar{G}_0^{\ddagger}(i \rightarrow t) = \Delta \bar{G}_0^{\ddagger}(p \rightarrow t, \Delta \psi = 0) + \beta r F \, \Delta \psi + v^i c \bar{\mu}_0(C)$$
$$- v^i a \bar{\mu}_0(A) - s \bar{\mu}_0(e) + v^i x \bar{\mu}_0(X) \qquad (207)$$

Substituting for $\bar{\mu}_0$ according to equation (208) and utilizing equations (193) and (208),

$$c z_2 = a z_1 - (s/v^p)$$

$$\rho = \mu_0 + z F \chi \qquad (208)$$

where ρ is the real potential, we obtain

$$\Delta \bar{G}_0^{\ddagger}(i \rightarrow t) = \Delta \bar{G}_0^{\ddagger}(p \rightarrow t, \Delta \psi = 0) + (s + \beta v r) F \, \Delta \psi$$
$$+ v^i [c \rho(C) + x \bar{\mu}_0(X) - a p(A)] - [s \rho(e)/v] \quad (209)$$
$$= \alpha F \Delta \psi + \text{const} \qquad (210)$$

where the transfer coefficient α is defined by

$$\alpha = (s + \beta v r)/v \qquad (211)$$

The above derivation assumes the absence of double-layer effects and limitingly low coverage of the electrode with reactants and products.

APPENDIX II. THE ADIABATIC PRINCIPLE AND RELATED APPROXIMATIONS

Some misunderstanding has been shown in the past concerning the significance of the adiabatic principle, the Born–Oppenheimer approximation, and the Franck–Condon principle with respect to charge-transfer reactions.

The Adiabatic Principle

The adiabatic principle, formulated by Ehrenfest[96] in 1916, states that a system will always remain in a definite quantum state (eigenstate) if its surrounding are changed sufficiently slowly. According to this principle, if it is possible to go from one system [e.g., $Fe^{3+} + e^- M$)] to another (e.g., $Fe^{2+} + M$) by applying a continuous perturbation, then each eigenstate of one system can be related to a definite eigenstate of the other. The adiabatic approximation has been described as follows:[97] "If the Hamiltonian changes very slowly with time then the solutions of the Schrödinger equation may be approximated by means of stationary energy eigenfunctions of the instantaneous (time independent) Hamiltonian, so that a particular eigenfunction at one time goes over continuously into an eigenfunction at a later time."

An example of adiabatic behavior is seen in the elastic collision between molecules in a gas.[98] As two molecules move into each other's fields of force, their electronic energy levels are continuously distorted and one might expect electronic transitions to occur. But since electrons move much faster than whole atoms, the distortion is essentially adiabatic, so that, when the molecules separate, their rotational and vibrational states may have been altered but their electronic states, which determine the molecular binding, remain unchanged.

Two illustrations of the adiabatic principle have been given by Fong.[99] In the first example, an oscillator in an electric field is considered. If the electric field is slowly turned up, then the quantum state after the field has reached its full value is an eigenstate and has a definite energy (eigenvalue). In the adiabatic approximation, if the field is turned down gradually, the quantum state (eigenstate) will gradually change to the initial quantum state, remaining an eigenstate all the time. Thus, when a field is switched on and off very slowly, the quantum state returns to the initial state. As a result, no net energy exchange between the oscillator and the field occurs (hence the term "adiabatic"). The second example is a vibrating, rotating diatomic molecule AB. Electrons in atoms A and B are arranged in "orbitals" described by wave functions. These wave functions are dependent on the interatomic distances in the molecule. As the molecule vibrates and rotates, the electronic wave functions change accordingly. Because the nuclear mass is much greater than the electronic mass, the period of the nuclear motion (i.e., vibration and rotation of the molecule) is much longer than that of the electronic motion. Thus, the disturbance of electronic motion due to the nuclear motion may be treated by the adiabatic approximation. We may speak of the electrons being in certain eigenstates with definite energy eigenvalues while the vibrational and rotational motion of the molecule is going on.

The latter example shows the connection between the adiabatic principle and the Born–Oppenheimer approximation, which states that the nuclear motions in ordinary vibrations are so slow that they do not affect the electronic state of the molecules.

It has been suggested by London[100] that the electronic motions involved in chemical reactions are adiabatic, and hence one may use the time-independent perturbation theory.

In connection with reaction-rate theory, Glasstone et al.[101] have defined an adiabatic change as one in which there is a continuous equilibrium between electrons and nuclei, there being no abrupt electronic rearrangement involving a transition from one electronic energy level to another; the entire process takes place on a single potential-energy surface. For an adiabatic reaction, a single eigenfunction can be used to represent the state of an electron throughout the course of the reaction.

The degree to which a reaction is adiabatic is commonly expressed in terms of a transition probability (between potential-energy surfaces) or in terms of a transmission coefficient κ. The latter gives the probability that, having achieved the activated-state configuration, the system will remain on the lower potential-energy surface and form the final state (adiabatic change) as opposed to making a transition to the upper potential-energy surface (nonadiabatic change).

To further elucidate the term "adiabatic reaction," let us consider what happens when two potential-energy surfaces (each surface calculated using the adiabatic approximation) intersect as in Figure 20. As the reactant nuclei are brought together, the potential energy of the system changes along the adiabatic surface $U^p(r)$, and as the product nuclei are brought together, the potential energy of the system changes along the adiabatic surface $U^q(r)$. At specific nuclear configurations, the surfaces intersect. That intersection point X with the lowest energy with respect to the initial state specifies the activated state of the system. The activated state has a unique configuration. In the vicinity of the activated state, the wave function of the system is a hybrid function of the reactants and the products. Throughout the reaction, the wave function of the system can be represented by

$$\Psi = c_1\psi_1 + c_2\psi_2 \tag{212}$$

where ψ_1 and ψ_2 are wave funtions for the reactants and products, respectively. The quantities c_1 and c_2 are functions of the internuclear separation. In the initial state, $c_1 = 1$ and $c_2 = 0$, whereas in the final state, $c_1 = 0$ and $c_2 = 1$. As a consequence of the resonance nature of the system at the crossing point, the energy of the system is not equal to that of the intersection point, but is either lower or greater than it by an amount of energy E_{res} called the resonance energy. The single energy level is split into two, leading to the formation of upper and lower potential-energy surfaces as in Figure 30.

When a point representative of the reacting nuclei approaches the activated-state configuration, it may either remain on the lower potential-energy surface or make a transition to the upper potential-energy surface depending on the amount of resonance splitting

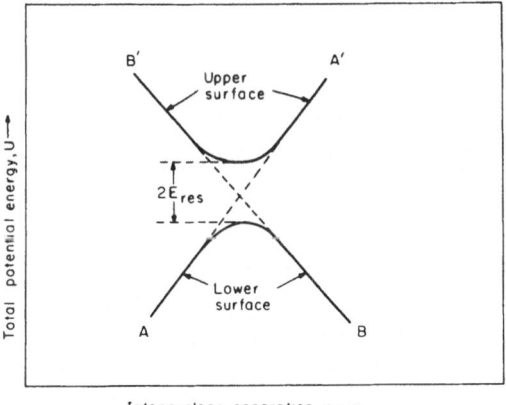

Figure 30. The region of intersection of two potential-energy surfaces AA' and BB'; E_{res} is the resonance energy.

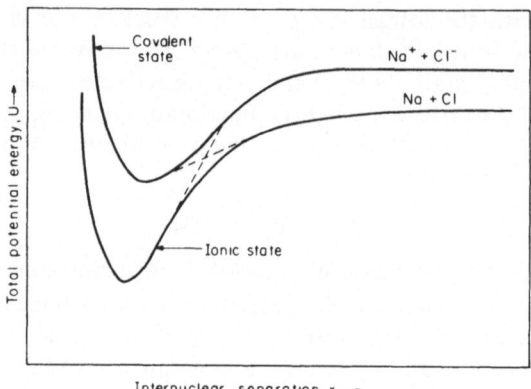

Figure 31. Variation of potential energy with internuclear
separation for Na^+ and Cl^- ions, and for Na and Cl atoms.
(After Kauzmann.[102])

$2E_{res}$, the reduced mass of the system, and the velocity of the representative point. If the representative point remains on the lower
surface AB, then the change is adiabatic. If the mass point makes a
transition to the upper surface $B'A'$, the change is nonadiabatic,
since it results from conditions incompatible with the adiabatic
principle, such as small resonance energies, high velocities, and
large differences in slopes of the intersecting surfaces.

The crossing of potential-energy curves has special significance
for charge-transfer reactions, e.g., charge-transfer adsorption, field
ionization, homogeneous redox reactions, and electrode reactions.
The charge-transfer reaction $Na^+ + Cl^- \rightarrow Na + Cl$ has been
considered by Kauzmann.[102] The potential-energy curves for the
states $(Na^+ + Cl^-)$ and $(Na + Cl)$ are shown in Figure 31. Resonance splitting leads to the formation of upper and lower surfaces.
From Figure 31, it is seen that, for large internuclear separation, the
most stable state (lowest energy) is $(Na + Cl)$, while at small internuclear separation, the ionic state $(Na^+ + Cl^-)$ is the more stable.
Charge transfer between Na and Cl occurs in the vicinity of the
"crossing point" at an internuclear separation of about 10 Å. If a Na
atom and a Cl atom are slowly brought together, the atoms will
remain approximately neutral until they are some 10 Å apart.

The Franck–Condon principle states[103] that an electron transition in a molecule takes place so rapidly in comparison to the vibrational motion that the nuclei have nearly the same relative positions after the electron transition as before it. Electron transitions are hence commonly referred to as "vertical" transitions. In the present context, we are sometimes concerned with the Franck–Condon principle for perturbations as stated by Herzberg,[103] namely two vibrational states belonging to two different electronic states and lying at approximately the same energy level will influence each other strongly only if, classically, the system can go from one state to the other without a large change in position and configuration. This condition is best fulfilled in the vicinity of the crossing point of the two potential-energy surfaces, where, as we have seen earlier, conditions for electron transfer are most favorable.

REFERENCES

[1] R. Parsons in *Modern Aspects of Electrochemistry*, Vol. 1, Chapter 3, Ed., J. O'M. Bockris, Butterworths, London (1954).

[2] R. Parsons, *Surface Sci.* **2** (1964) 418.

[3] R. W. Gurney, *Proc. Roy. Soc. (London)* **A134** (1931) 137.

[4] F. P. Bowden, *Proc. Roy. Soc. (London)* **A125** (1929) 446.

[5] M. L. E. Oliphant and P. B. Moon, *Proc. Roy. Soc. (London)* **A127** (1930) 388.

[6] G. Wentzel, *Z. Physik* **39** (1926) 518.
 H. A. Kramers, *Z. Physik* **39** (1926) 828.
 L. Brillouin, *Compt. Rend.* **183** (1926) 24.

[7] J. A. V. Butler, *Proc. Roy. Soc. (London)* **A157** (1936) 423.

[8] J. Horiuti and M. Polanyi, *Acta Physicochim. URSS* **16** (1942) 169.

[9] St. G. Christov, *Z. Electrochem.* **62** (1958) 567.

[10] J. O'M. Bockris, *Nature* **159** (1947) 539.

[11] B. E. Conway and J. O'M. Bockris, *J. Chem. Phys.* **26** (1957) 532.

[12] D. B. Matthews, Ph.D. Thesis, University of Pennsylvania (1965).

[13] J. A. V. Butler, *Electrocapillarity*, Methuen and Co., London (1940).

[14] P. Delahay, *Double Layer and Electrode Kinetics*, Interscience Publishers, New York (1965), p. 173.

[15] B. E. Conway, *Progress in Reaction Kinetics*, Vol. 4, Ed., G. Porter, Pergamon Press, New York (1967), Chapter 10.

[16] J. O'M. Bockris, *Trans. Faraday Soc.* **43** (1947) 417.

[17] D. B. Matthews, *J. Electrochem. Soc.* **113** (1966) 1109.

[18] R. Parsons, *Trans. Faraday Soc.* **54** (1958) 1053.

[19] R. Parsons and J. O'M. Bockris, *Trans. Faraday Soc.* **47** (1951) 914.

[20] O. A. Essin and V. Kozheurov, *Acta Physicochem.* **16** (1942) 169.

[21] J. O'M. Bockris and D. E. Matthews, *Proc. Roy. Soc. (London)* **A292** (1966) 479.

[22] H. Gerischer, *Z. Physik. Chem. (Frankfurt)* **26** (1960) 223.

[23] H. Gerischer, *Z. Physik. Chem. (Frankfurt)* **26** (1960) 325.

[24] H. Gerischer, *Z. Physik. Chem. (Frankfurt)* **27** (1961) 48.

[25] F. Seitz, *Modern Theory of Solids*, McGraw-Hill Book Co., New York (1940).
[26] J. E. B. Randles, *Trans. Faraday Soc.* **48** (1952) 828.
[27] A. R. Despic and J. O'M. Bockris, *J. Chem. Phys.* **32** (1960) 389.
[28] L. Amdur, *J. Chem. Phys.* **17** (1949) 844.
[29] F. Hund, *Z. Physik.* **32** (1925) 1.
[30] J. O'M. Bockris, M. A. V. Devanathan, and K. Muller, *Proc. Roy. Soc. (London)* **A274** (1963) 55.
[31] T. N. Andersen and J. O'M. Bockris, *Electrochim. Acta* **9** (1964) 347.
[32] J. O'M. Bockris and D. B. Matthews, *J. Chem. Phys.* **44** (1966) 298.
[33] E. A. Moelwyn-Hughes, *Physical Chemistry*, Pergamon Press, London and New York (1961).
[34] O. Klein and E. Lange, *Z. Elektrochem.* **43** (1937) 570.
[35] J. E. B. Randles, *Trans. Faraday Soc.* **52** (1956) 1573.
[36] R. A. Marcus, *J. Chem. Phys.* **24** (1956) 966.
[37] R. A. Marcus, *J. Chem. Phys.* **26** (1957) 867.
[38] N. S. Hush, *J. Chem. Phys.* **28** (1958) 962.
[39] N. S. Hush, *Trans. Faraday Soc.* **57** (1961) 557.
[40] J. O'M. Bockris and R. Parsons, *Trans. Faraday Soc.* **45** (1949) 916.
[41] B. E. Conway and J. O'M. Bockris, *Can. J. Chem.* **35** (1957) 1124.
[42] J. O'M. Bockris and S. Srinivasan, *J. Electrochem. Soc.* **111** (1964) 844.
[43] J. O'M. Bockris and S. Srinivasan, *J. Electrochem. Soc.* **111** (1964) 853.
[44] B. E. Conway and J. O'M. Bockris, *Electrochim. Acta* **3** (1961) 340.
[45] H. Gerischer, *Z. Elektrochem.* **62** (1958) 256.
[46] J. O'M. Bockris, *Modern Aspects of Electrochemistry*, Vol. 1, Ed., J. O'M. Bockris, Butterworths Scientific Publications, London (1954), Chapter 4.
[47] V. G. Levich, *Advances in Electrochemistry and Electrochemical Engineering*, Vol. 4, Ed., P. Delahay and C. W. Tobias, Interscience Publishers, New York, (1966), p. 249.
[48] R. A. Marcus, *Annual Rev. Phys.-Chem.* **15** (1964) 155.
[49] J. O'M. Bockris, *Modern Aspects of Electrochemisty*, Vol. 1, Ed., J. O'M. Bockris, Butterworths Scientific Publications, London (1954), p. 65.
[50] C. Zener, *Proc. Roy. Soc. (London)* **A137** (1932) 696; **A140** (1933) 660.
[51] R. A. Marcus, *J. Chem. Phys.* **24** (1956) 966; *J. Chem. Phys.* **26** (1957); 867; *Trans. N.Y. Acad. Sci.* **19** (1957) 423; *J. Chem. Phys.* **38** (1963) 1858; *J. Chem. Phys.* **39** (1963) 1734.
[52] R. A. Marcus, *J. Chem. Phys.* **38** (1963) 1858; *J. Chem. Phys.* **39** (1963) 1734; *Can. J. Chem.* **37** (1959) 155; *Transactions of the Symposium on Electrode Processes*, Ed., E. Yeager, John Wiley and Sons, New York (1961), p. 239; *J. Phys. Chem.* **67** (1963) 853.
[53] R. A. Marcus, *J. Chem. Phys.* **43** (1965) 679.
[54] R. A. Marcus, *Disc. Faraday Soc.* **29** (1960) 21; *J. Phys. Chem.* **67** (1963) 853.
[55] R. Parsons and E. Passeron, *J. Electroanal. Chem.* **12** (1966) 524.
[56] R. A. Marcus, *J. Chem. Phys.* **41** (1964) 264.
[57] V. G. Levich and R. R. Dogonadze, *Dokl. Akad. Nauk SSSR* **133** (1960) 158; *Dokl. Akad. Nauk SSSR* **133** (1960) 1368; *Coll. Czech. Chem. Commun.* **25** (1961) 193; *Dokl. Akad. Nauk SSSR* **142** (1962) 1108.
[58] R. R. Dogonadze and Y. A. Chizmadzhev, *Dokl. Akad. Nauk. SSSR* **144** (1962) 1077; **145** (1962) 849; **150** (1963) 333.
[59] J. C. Slater, *Phys. Rev.* **36** (1930) 57.
[60] St. G. Christov, *Ann. de l'Univ. Sofia Fac. Phys. Math.* **XLII** (1945/46) 2, 69; *Z. Electrochem.* **62** (1958) 567.

⁶¹St. G. Christov, Z. Phys. Chem. **212** (1959) 40; Z. Elektrochem. **64** (1960) 840; Dokl. Akad. Nauk SSSR **125** (1959) 143; Z. Phys. Chem. **214** (1960) 40.

⁶²St. G. Christov, Electrochim Acta **4** (1961) 194; Electrochimica Acta **4** (1961) 306.

⁶³B. E. Conway, Can. J. Chem. **37** (1959) 178.

⁶⁴B. E. Conway and M. Salomon, J. Phys. Chem. **68** (1964) 2009.

⁶⁵B. E. Conway and M. Salomon, J. Chem. Phys. **41** (1964) 3169. B. E. Conway and D. J. MacKinnon, J. Electrochem. Soc. **116** (1969) 1665.

⁶⁶B. Topley and H. Eyring, J. Am. Chem. Soc. **55** (1933) 5058; Trans. Faraday Soc. **66** (1970) 1203.

⁶⁷C. Eckart, Phys. Rev. **35** (1930) 1303.

⁶⁸J. O'M. Bockris, S. Srinivasan, and D. B. Matthews, Disc. Faraday Soc. **39** (1965) 239.

⁶⁹F. P. Bowden, Proc. Roy. Soc. (London) **A125** (1929) 446; Proc. Roy. Soc. (London) **A126** (1929) 107.

⁷⁰Z. A. Jofa and K. P. Mikulin, Zh. Fiz. Khim. **18** (1944) 137.

⁷¹J. O'M. Bockris and R. Parsons, Trans. Faraday Soc. **45** (1949) 916.

⁷²B. Post and C. F. Hiskey, J. Am. Chem. Soc. **72** (1950) 4203.

⁷³S. Minc and J. Sobkowski, Bull Acad. Polon. Sci., Ser. Sci. Chim. Geol. et Geograph 7 (1959) 29.

⁷⁴Z. A. Jofa and V. Stepanova, Zh. Fiz. Khim. **19** (1945) 125.

⁷⁵J. O'M. Bockris and D. B. Matthews, Electrochim. Acta **11** (1966) 143.

⁷⁶J. N. Butler and M. L. Meehan, Trans. Faraday Soc. **62** (1966) 3524.

⁷⁷C. E. H. Bawn and G. Ogden, Trans. Faraday Soc. **30** (1934) 432.

⁷⁸B. Post and C. F. Hiskey, J. Am. Chem. Soc. **73** (1951) 161.

⁷⁹M. Rome and C. F. Hiskey, J. Am. Chem. Soc. **76** (1954) 5207.

⁸⁰B. E. Conway and M. Salomon, Ber. Bunsenges Physik. Chem. **68** (1964) 331; M. Salomon, Can. J. Chem. **44** (1966) 689.

⁸¹M. Salomon and B. E. Conway, Disc. Faraday Sic. **39** (1965) 223.

⁸²J. Horiuti and M. Ikusima, Proc. Imperial Acad. Tokyo **15** (1939) 39.

⁸³E. C. Potter, Ph.D. Thesis, London (1950); J. O'M. Bockris and E. C. Potter, J. Electrochem. Soc. **99** (1952) 169; J. O'M. Bockris and E. C. Potter, J. Chem. Phys. **20** (1952) 614.

⁸⁴J. Horiuti and T. Nakamura, Z. Physik Chem. (Frankfurt) **11** (1957) 358.

⁸⁵R. Parsons, Trans. Faraday Soc. **47** (1951) 1332.

⁸⁶K. B. Oldham, J. Am. Chem. Soc. **77** (1955) 4697.

⁸⁷A. C. Makrides, J. Electrochem. Soc. **104** (1957) 677; **109** (1962) 256.

⁸⁸H. Mauser, Z. Elektrochem. **62** (1958) 419; J. O'M. Bockris and H. Mauser, Can. J. Chem. **37** (1959) 475.

⁸⁹A. C. Riddiford, J. Chem. Soc. (1960) 1175.

⁹⁰D. M. Mohilner, J. Phys. Chem. **68** (1964) 632.

⁹¹S. Barnartt, J. Phys. Chem. **70** (1966) 412.

⁹²A. R. Despic, Bull. Soc. Chim. (Belgrade) **30** (1955) 293; Parts II to IV in press.

⁹³B. E. Conway, Electrode Processes, Ronald Press Co., New York (1965).

⁹⁴J. O'M. Bockris, Disc. Faraday Soc. **1** (1947) 132; Nature **159** (1947) 539; Trans. Faraday Soc. **43** (1947) 417.

⁹⁵N. K. Adam, The Physics and Chemistry of Surfaces, 3rd ed., Oxford University Press, London (1941), footnote p. 332.

⁹⁶P. Ehrenfest, Ann. Phys. **51** (1916) 327.

⁹⁷L. I. Schiff, Quantum Mechanics, McGraw-Hill Book Co., New York (1949), p. 207.

⁹⁸D. Park, Introduction to Quantum Theory, McGraw-Hill Book Co., New York (1964), p. 241.

[99]P. Fong, *Elementary Quantum Mechanics*, Addison-Wesley Publishing Co., Reading, Mass. (1962), pp. 244, 254.

[100]F. London, *Z. Elektrochem.* **35** (1929) 552.

[101]S. Glasstone, K. J. Laidler, and H. Eyring, *The Theory of Rate Processes*, McGraw-Hill Book Co., New York (1941) p. 87.

[102]W. Kauzmann, *Quantum Chemistry*, Academic Press, New York (1957), p. 536.

[103]G. Herzberg, *Spectra of Diatomic Molecules*, 2nd ed., D. van Nostrand, New York (1950).

[104]P. H. Cutler and D. Nagy, *Surface Sci.* **3** (1965) 71.

[105]K. Müller, *J. Res. Inst. Catal. Hokkaido Univ.* **XIV** (1966) 224.

[106]R. D. Young, *Phys. Rev.* **113** (1959) 110.

[107]R. Harris and W. D. Weir, *J. Chem. Phys.* **47** (1967) 3247.

[108]P. van Rysselberghe, *Electrochim. Acta* **8** (1963) 583, 709.

[109]St. G. Christov, *Ber. Bunsenges. Physik. Chem.* **67** (1963) 117.

[110]J. Weiss, *Proc. Roy. Soc. (London)* **A222** (1954) 128.

[111]W. F. Libby, *J. Phys. Chem.* **56** (1952) 863.

[112]R. J. Marcus, B. J. Zwolinski, H. Eyring, *J. Phys. Chem.* **58** (1954) 432.

[113]D. B. Matthews, *Aust. J. Chem.* **22** (1969) 1349.

[114]J. M. Hale, *J. Electroanal. Chem.* **19** (1968) 315.

[115]R. R. Dogonadze, A. M. Kuznetsov, and V. G. Levich, *Electrochim. Acta* **13** (1968) 1025.

[116]H. Gerischer, *Surface Sci.* **18** (1969) 97.

[117]N. S. Hush, *Electrochim. Acta* **13** (1968) 1005.

[118]V. G. Levich, *Physical Chemistry*, Vol. 9B, Ed., Eyring, Henderson, and Jost, John Wiley and Sons, 1970.

[119]E. D. German, R. R. Dogonadze, A. M. Kuznetsov, V. G. Levich, and Y. I. Kharkats, *Electrorkimia* **6** (1970) 350–354.

[120]A. M. Kuznetsov, *Electrochimica Acta* **13** (1968) 1293.

5

Electrochemical Processes in Glow Discharge at the Gas-Solution Interface

A. Hickling

Department of Inorganic, Physical, and Industrial Chemistry
The Donnan Laboratories
The University of Liverpool, Liverpool, England

I. INTRODUCTION

The term electrolysis is conventionally applied to chemical changes brought about by passing an electric current between conducting electrodes dipping into a liquid phase containing ions, where the changes can be satisfactorily explained by electron transfer between the ions and the electrodes. If, however, the liquid phase is itself made an electrode and an electrical glow-discharge is passed to it from a conductor located in the gas space above the surface, a completely different situation arises in which novel chemical reactions can be brought about in the liquid phase, and this process is referred to as glow-discharge electrolysis (GDE). It is worth emphasizing that it differs fundamentally from chemical decomposition brought about by electrical discharge between metallic electrodes in gases at low pressures, since in GDE, the reactions of interest are initiated in the liquid phase, and the quantity of electricity passed rather than electrical power dissipated is found to be the governing variable, as in conventional electrolysis. In the experimental arrangement most commonly adopted, the electrode above the surface is the anode, while the cathode is immersed in the electrolyte; once the discharge has been initiated, e.g., by momentarily touching the anode to the surface or by applying a high voltage pulse to the system, it can be maintained at voltages of 500 V and upward with the gas or vapor at reduced pressure, and substantial currents can be passed to the liquid surface. The technique first excited interest because it seemed to provide a way of carrying out electrolysis without a solid electrode in contact with the electrolyte, but early work showed that the chemical effects produced are much greater quantitatively than would be expected from Faraday's laws. Recent work suggests that charge transfer is only a minor factor in GDE and that the chemical effects are produced by charged particles which are accelerated in the potential drop near the electrolyte surface and enter the liquid with appreciable energies, probably of the order of 100 eV. These can bring about ionization, excitation, or dissociation of solvent molecules by collision (in addition to charge-transfer reactions), and the chemical effects are due to the reactive species thus produced. The situation, in fact, closely resembles that in radiation chemistry where ionizing radiations produce chemical reactions, and the analogy is particularly close when comparison is made with low-energy alpha particles. Thus, a new type of electrolysis arises in which energy transfer in addition to charge transfer brings about chemical change.

II. HISTORY

The possibility of electrolysis using an electrical discharge was shown by Gubkin[1] in 1887, who carried out electrolysis of aqueous solutions of metallic salts using a glow-discharge cathode and achieved reduction in some cases, and by Klüpfel[2] in 1905, who liberated iodine from solutions containing potassium iodide using a glow-discharge anode. That something over and above ordinary electrolysis was involved was first demonstrated by Makowetsky[3] in 1911. He showed that in the GDE of dilute sulfuric acid solutions, hydrogen peroxide was one of the main products in solution; the yield exceeded 1 equiv F^{-1}, which might have been expected from Faraday's laws, but when the amounts of hydrogen and oxygen evolved were considered, an overall balance was found to exist. Klemenc et al.[4] (1914–54) investigated GDE using a very wide variety of aqueous electrolytes, including sulfuric acid solutions, halide and oxyhalogen salts, and oxidizable substances. The results were often very complicated and no simple pattern emerged, but it was shown that in general two types of reaction occurred, which were designated *polar* and *apolar* reactions respectively. The first corresponded to the expected electrolytic charge-transfer reaction, but the second, which often predominated, corresponded in general to oxidation accompanied by hydrogen evolution from the solution. The origin of the apolar reaction was not clearly established, although in his later publications, Klemenc seems to have considered that it was due to OH radicals produced in the cathode drop in the discharge; these were thought to dissolve in the surface of the solution and to react with oxidizable substrates. Recent fundamental investigations by Hickling and associates[5,5a,b] (1950–69) have suggested a related but somewhat different point of view. The reactions appear to originate in the liquid phase and show many points of similarity to those produced by ionizing radiation. It is believed, therefore, that they have a common origin in the initial decomposition of solvent molecules by an energy-transfer process to produce reactive radicals which may interact among themselves or be scavenged by suitable reactive species in solution. Many of the concepts of radiation chemistry can be fruitfully applied to this situation, and the idea seems capable of further development.

Miscellaneous publications[6,6a–e] on particular aspects and applications of GDE are fairly numerous. While GDE at reduced pressure with an electrode above the surface is a well-defined situation, other discharge processes may arise which seem to give closely related results. Thus, spark electrolysis at atmospheric pressure has been investigated, particularly by de Beco,[7] and, under abnormal conditions, immersed electrodes may sometimes become covered by a vapor film through which a discharge spontaneously occurs if the applied voltage is sufficiently high.[8] In all these cases, we seem to be dealing with particular aspects of the same general phenomenon.

III. EXPERIMENTAL TECHNIQUE

Basic requirements for GDE are relatively simple and a suitable experimental arrangement for carrying it out with an anode above the liquid surface is shown in Figure 1. As source of current, a rectifier of continuously variable voltage between 0 and 1500 V is suitable, and the positive lead is connected to the electrode above the surface through a multirange ammeter and appropriate ballast resistor. The value of this resistor is not critical and it is usually about 3000 ohms; it depends upon the current to be used and it should be such as to give a voltage drop of about 100–200 V. In the negative lead to the cell, which is grounded, a Lingane pattern hydrogen–oxygen coulometer is included to measure the quantity of electricity passed. With a steady dc voltage of about 1000 V applied, some device is necessary to initiate the discharge. This can be done mechanically by touching the anode momentarily to the liquid surface, but it is usually more convenient to use a high-voltage pulse from an induction coil or transformer. Once started, the discharge is very steady and the current can be fixed in the range 0.02–0.2 A by varying the applied dc voltage manually. Recently, a convenient electronic circuit for current control and discharge initiation has been described[9] which is especially useful for short-duration GDE.

Many types of experimental cell are possible and some convenient patterns are shown in Figure 2. Here, A is a simple one-compartment cell which is convenient when the products formed in solution are not reduced at the cathode; B and C are cells in which anode and cathode compartments are separated by a sintered-glass diaphragm. For anode, a platinum wire fixed to a tungsten rod sealed into a glass holder is suitable, the tip of the wire being at a distance of 0.5–1 cm

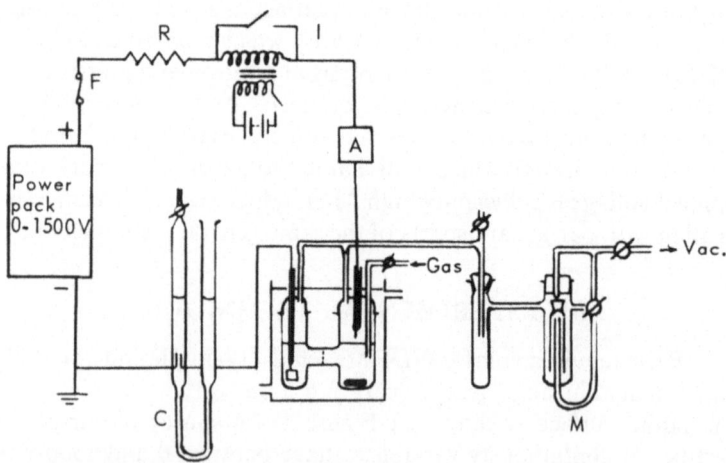

Figure 1. Apparatus for glow-discharge electrolysis. (F) Fuse, (R) ballast resistor, (I) discharge initiator, (A) multirange milliammeter, (C) coulometer, (M) manostat.

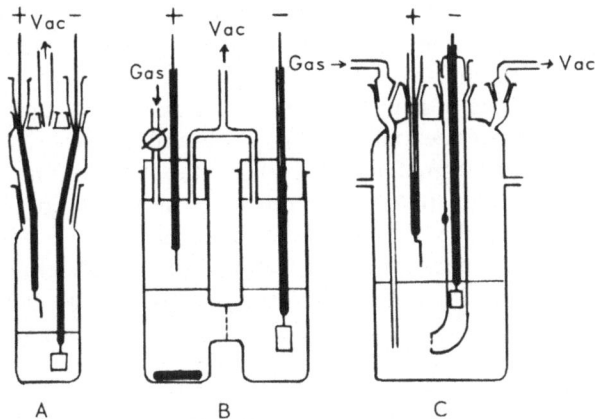

Figure 2. Cells for glow-discharge electrolysis.

from the liquid surface; the nature of the cathode is not usually important, provided it is not attacked by the cell solution. Reduced pressure is maintained in the cell during electrolysis by connecting it to a vacuum line in which a constant pressure can be maintained by a manostat; the minimum pressure that can be used is fixed by the vapor pressure of the solution, and, in general, working pressures in the range 25–250 mm Hg are employed. Usually, it is desirable to control the gaseous atmosphere in the discharge, and this can be done by admitting a slow stream of gas to the cell during electrolysis. An inert atmosphere of nitrogen is commonly satisfactory, and it serves also to prevent the accumulation of hydrogen and oxygen in the gas space when aqueous solutions are used, with its possible risk of explosion. This arrangement is satisfactory if it is only the products of GDE in the solution which are of interest. If gaseous products are to be examined, it must be modified. The cell is then initially evacuated to the vapor pressure of the solution, and any gas evolved on GDE is pumped away at a convenient pressure for measurement and analysis. Stirring of the electrolyte is often desirable and this can be achieved either with a gas stream or by using a magnetic stirrer. Considerable heat is generated when the discharge is passing and the cell must be largely immersed in a suitable cooling bath. With circulated tap water, the bulk electrolyte temperature can normally be maintained roughly constant in the range 20–25°C, although the temperature directly below the glow-spot is probably the boiling point of the solution at the pressure used.

With the kind of setup described, there exists a voltage drop of about 400 V very close to the liquid surface, and it is the existence of this considerable drop of potential adjacent to the solution which is essential to bring about the characteristic chemical effects of GDE. If, instead of making the electrode above the surface positive, it is made negative by reversing the cell connections, quite a different state of affairs results. The main voltage drop is now close to the metal electrode in the gas space, and this electrode may become white hot and melt unless it is of sufficiently massive construction. Near the liquid surface, there is only a small potential drop of about 30 V, and the chemical effects produced in the solution are small. If it is wished to run a glow-discharge in this way, the electrode above the surface may take the form of a platinum cone with its point toward the liquid mounted on a robust metal rod to dissipate heat. GDE can be operated with alternating current, but it is then found, as would be expected, that most of the chemical effect is produced in the half cycles in which the electrode above the surface is anodic relative to the solution.

If the wire anode as used in GDE is brought sufficiently close to the electrolyte so that it touches, conventional electrolysis of course occurs. If, however, the current is sufficiently large and the immersed area of electrode is small, the solvent may be vaporized locally and form a vapor sheath around the electrode through which a glow-discharge passes, producing luminescence. This phenomena, which has been termed contact glow-discharge electrolysis (CGDE), has been utilized[8] in a very simple method of bringing about GDE at atmospheric pressure. The thin wire anode may be immersed for a short distance below the surface, or alternatively, the wire may project from a glass holder which is immersed in the body of the solution, preferably with the wire pointing upward to permit ready evolution of gas bubbles. The voltage applied to the cell is gradually increased from an initially low value, and when the power dissipated near the anode surface becomes sufficiently great, the liquid separates from it and GDE occurs spontaneously through a vapor sheath. This method has the advantage of very great simplicity, and the discharge is automatically started and maintained at a relatively low overall applied voltage, usually about 500 V; the characteristics of this type of discharge are considered later. It is, however, somewhat less controllable than GDE with an electrode above the liquid surface, and the experimental conditions are less precisely defined.

IV. PHYSICAL FEATURES OF GDE

Very little systematic work has been done on the physics of electrical discharges to the surface of solutions (see, however, Sternberg[6a]) and therefore there does not exist any firm theoretical basis for the understanding of the phenomena of GDE. Progress can only be made by taking over some of the concepts arising from the study of glow-discharge between metallic electrodes in gases at low pressures, and making use of such physical observations as have been made on GDE, although these have usually been very much in the context of the chemical effects produced. With the common arrangement of a wire anode above the electrolyte surface, the discharge usually appears as a well-defined cone between the tip of the anode and a luminous disk on the liquid surface; sometimes this glow-spot shows up as a slowly rotating cloverleaf pattern. The color of the discharge depends upon the nature of the vapor present; with aqueous electrolytes, the glow-spot is usually pink and there is a pale blue, diffuse negative glow. Occasionally, the discharge may be colored by substances present in solution.[10] Alteration of the size and shape of the metal anode has very little influence on the form of the discharge, which remains cone-shaped, originating from one point on the anode surface, although this may wander from time to time.

In a conventional glow-discharge between metal electrodes, the total voltage V between the electrodes is made up of contributions from the cathode drop V_c, the anode drop V_a, and the drop over the positive column V_p, so that $V = V_c + V_a + V_p$. Of these, the cathode drop V_c depends upon the nature of the gas and the cathode material, but is independent of most other variables and it is usually large, e.g., 200–300 V with diatomic gases; V_a is much smaller, and V_p depends upon the length of the positive column. With the usual arrangement for GDE, similar considerations apply, although with the electrode spacing usually adopted, the positive column is very short compared to conventional gas discharges. Furthermore, since the cathode is a liquid, which is the same in different experiments, and

Table 1
Voltages between Probe Electrode and Liquid Surface

Surface–probe distance, cm	Probe voltage, V		
	at 27	at 50	at 100 mm Hg
0.10	425	432	443
0.25	447	463	499
0.50	467	501	557

the gas immediately adjacent to it is its own vapor, one might expect the cathode drop V_c to be constant for a given solvent. This has been investigated by using a small wire probe electrode inserted into the discharge parallel to the liquid surface and measuring its potential relative to the electrolyte at various distances using a valve voltmeter. Although the results obtained with such probe electrodes are not necessarily accurate, in this case they are quite illuminating and some values for a neutral aqueous phosphate electrolyte at 0.075 A with the anode 0.75 cm from the surface in a hydrogen atmosphere are given in Table 1.

The voltage falls toward an apparently constant value as the probe approaches the liquid surface. It is not practicable to bring it nearer than 0.1 cm, but if these values are plotted as in Figure 3, the three curves intersect at a distance of about 0.05 cm from the surface when the voltage is about 415 V. Changing the nature of the electrolyte, the current, anode distance, and gaseous atmosphere did not affect the probe-surface voltage at a given distance appreciably. It would seem, therefore, that 415 V represents the cathode drop for a discharge through water vapor to a liquid water cathode and this voltage appears to be confined to a very short distance near the liquid surface. Observations using a liquid ammonia solvent have given essentially similar results, although here the cathode drop is about 390 V. Figure 4 shows the distribution of potential as measured by a probe in a discharge at 0.025 A and 100 mm Hg pressure to a solution of 0.01 M NH_4NO_3 in liquid ammonia at $-78°C$ with the anode at distances of 0.5, 0.8, and 1.1 cm. The cathode and anode potential drops can readily be distinguished, and are separated by a linear voltage drop over the positive column. It is particularly note-

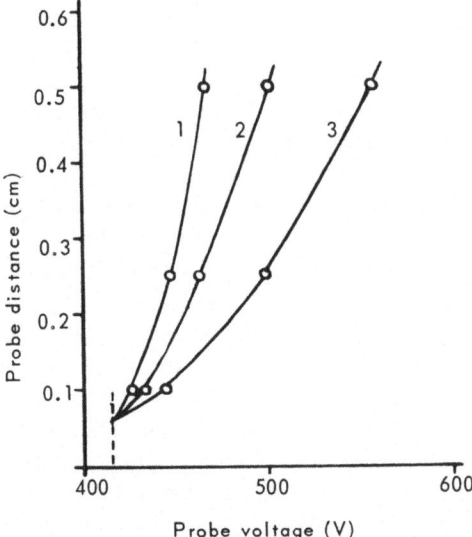

Figure 3. Measurement of potential drop near surface
of aqueous electrolyte at pressures of (1) 27, (2) 50,
and (3) 100 mm Hg.

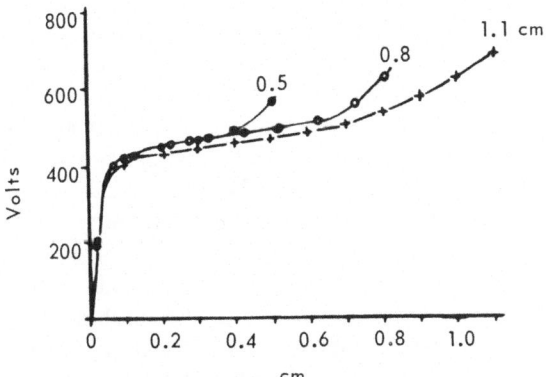

Figure 4 Variation of voltage distribution with anode-
surface distance in the glow-discharge electrolysis of
liquid ammonia (—●—) 0.5, (—○—) 0.8, and (—+—)
1.1 cm.

worthy that, while the total voltage across a discharge in GDE and hence the power dissipated can be markedly changed by alteration of experimental variables such as current, pressure, and anode-to-surface distance, the cathode drop itself remains much the same and this is of particular significance in considering the mechanism of GDE. Furthermore, from the point of view of possible practical application, the cathode drop seems to be a lower limit of applied voltage at which the characteristic chemical effects of GDE can be obtained even under optimum conditions of electrode spacing and pressure.

A parameter which would be expected to be of importance in GDE is the current density (CD) of the discharge at the liquid surface, and a measure of this can be obtained by dividing the current passing by the area of the glow-spot. Since the total area of the liquid surface is usually large, the condition is that of a normal glow-discharge in which the area of the glow may vary considerably with change in experimental factors. Increase of current causes the glow-spot to expand and the CD itself varies little with change of current; there is actually a slight decrease as the current increases. A rise of pressure decreases the area of the glow-spot and increases CD substantially, and the relation $jp^{-1} = $ const has sometimes been found, where j is CD and p the pressure; this contrasts with the normal glow-discharge between metal electrodes in a gas, where the corresponding relation at the cathode is $jp^{-2} = $ const. The area of the glow-spot in GDE is increased by moving the anode further from the liquid surface, and the CD is thereby decreased; the variation seems to be a linear one, but the effect is relatively small. Table 2 gives some illustrative values for the effect of experimental variables on CD; they refer to a discharge to an aqueous phosphate solution in a hydrogen atmosphere. The CD usually falls in the range 0.1–0.3 A cm^{-2}; with discharges to liquid ammonia solutions, the range is very similar.

When GDE is carried out with the electrode above the solution surface made a cathode, the nature of the discharge is changed completely. It appears as a narrow, cylindrical pencil of light to a very small glow-spot in the liquid surface, usually of about 2–3 mm diameter, so that the CD there is very high and may be about 1 A cm^{-2}. The cathode drop now appears close to the metal electrode, which becomes very hot, and the cathode glow spreads

Table 2
Influence of Experimental Variables on the Current Density at the Liquid Surface

Current, A	Pressure, mm Hg	Anode distance, cm	Glow-spot area, cm^2	Current density, $A\,cm^{-2}$
0.075	50	0.75	0.478	0.16
0.05	50	0.75	0.363	0.14
0.025	50	0.75	0.126	0.20
0.075	33	0.75	0.708	0.11
0.075	50	0.75	0.478	0.16
0.075	74	0.75	0.363	0.21
0.075	94	0.75	0.264	0.28
0.075	50	0.50	0.385	0.20
0.075	50	0.75	0.478	0.16
0.075	50	1.20	0.529	0.14
0.075	50	1.60	0.636	0.12

over its surface. Near the surface of the liquid, there occurs the anode drop in the discharge, usually of about 30 V. If GDE is carried out with 50-Hz alternating current, the discharge stops and restarts twice in each cycle; it thus consists of separate successive anodic and cathodic pulses each of which exhibits its normal dc features.

V. GLOW-DISCHARGE PHENOMENA IN CONVENTIONAL ELECTROLYSIS

In conventional electrolysis, conditions may arise near the electrodes such that glow-discharge through a gas or vapor phase spontaneously occurs if the applied voltage is sufficiently great. This was very convincingly demonstrated by Kellogg,[11] who sought an analogy in the electrolysis of aqueous solutions with the "anode effect" which sometimes occurs in molten salts. He electrolyzed dilute sulfuric acid and sodium hydroxide solutions with a sheet cathode and a platinum wire anode, gradually increasing the voltage applied to the cell. At first, the current increased according to Ohm's law, conventional electrolysis occurring with smooth evolution of gas bubbles from the anode. At about 30 V, however, the electrode entered a transition region in which the current fell with increasing voltage, the liquid near the anode becoming turbulent and emitting a hissing noise. When the voltage reached a critical value of ~ 45 V, evolution of gas bubbles largely ceased and the anode apparently became covered by a film of gas. This coincided with a large decrease of current and Kellogg termed the behavior the "aqueous anode effect." He showed that the phenomenon was connected with the volatility of the solvent, it could be obtained with either an anode or a cathode, and the transition region seemed to correspond to conditions in which the surface temperature of the electrode reached 100°C. He suggested, therefore, that the effect was due to local heating causing the formation of a sheath of water vapor around the electrode through which conduction probably occurred by local glow-discharges where the film thickness was least, but since his maximum voltage available was only 115 V, he did not in fact explore this further. Hickling and Newns[5a] observed a similar phenomenon when studying electrolysis in liquid ammonia, and they found that, with a thin platinum wire anode immersed in the solution, a full glow-discharge of greenish color could be developed around the electrode at voltages of 400–500 V and that, under these conditions, hydrazine was the main product formed in the liquid as is the case with GDE with the anode above the surface. This phenomenon of contact glow-discharge (CGDE) has been further studied in aqueous solutions by Hickling and Ingram.[8] They found that it is of quite general occurrence independent of the nature of the electrolyte, the electrode, or the electrode reaction; the only essential requirement is that the electrode be a wire of small radius. Figure 5 shows the typical current–voltage curve obtained with a 0.05 M Na_2HPO_4 electrolyte at a bulk temperature of 72°C with a platinum wire anode of diameter 0.0508 cm and immersed length 0.5 cm. In the linear section of the curve AB, 0–60 V, Ohm's law was obeyed and conventional electrolysis occurred with small bubbles of gas leaving the anode. At point B, with a power dissipation of about 26 W, smooth evolution of bubbles was interrupted and pulsing of steam and small flashes of light were seen at the anode. Between B and C, 60–270 V, was an unstable region in which current and voltage fluctuated wildly. At C, the current and voltage readings were suddenly stabilized. The flow of bubbles from the electrode surface had now

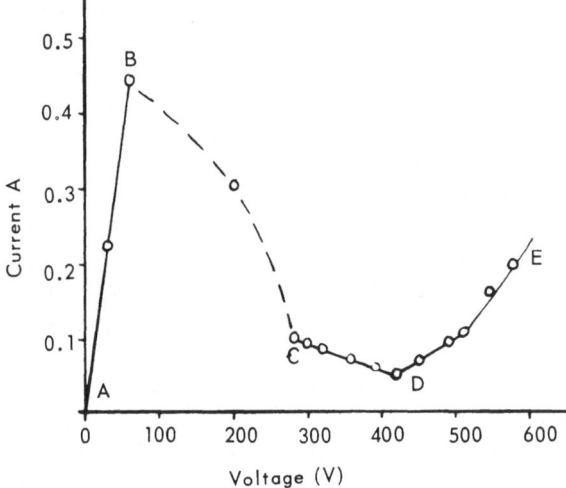

Figure 5. Current–voltage graph for contact glow-discharge electrolysis.

ceased and appeared to have been replaced by a smooth dark sheath of vapor which at times exhibited a kind of vortex motion. The electrode appearance and the negative slope of the current–voltage curve correspond to the electrode condition discovered by Kellogg, and this section of the characteristic is conveniently designated the Kellogg region. At point D, 420 V, the slope of the curve changes sign and this midpoint seems to indicate the start of a new section of the curve in which the glow, which had become apparent just before D, increased steadily with rising voltage; this appears to be a true glow-discharge region. Direct observation of the electrode suggested that, at the beginning of the Kellogg region, an extremely thin and mobile film covered the electrode surface. It had an average thickness of about 0.001 cm, but it was subject to frequent distortion, suggesting that liquid was repeatedly touching the metal at some point where the sudden rush of current generated a local pocket of vapor and the point of contact then moved to a new position on the metal surface. The average thickness of the sheath remained substantially constant throughout the Kellogg region and up to the midpoint voltage. It then began to increase rapidly in the discharge region proper; at 430 V, it was 0.015 cm and at 450 V,

0.025 cm. Examination of the chemical products of electrolysis confirmed this general interpretation; the products characteristic of GDE began to appear in the Kellogg region, but it was only above the midpoint voltage that their full yield was attained.

Kellogg[11] reported that the breakdown of conventional electrolysis could be obtained with the small immersed electrode cathodic as well as anodic. Hickling and Ingram[8] confirmed this and found that it occurred at a similar power dissipation, but the resulting condition seemed far less stable. Following the breakdown of conventional electrolysis, the electrode became covered with an intense glow, its color depending upon the nature of the metal ions in solution, and there was a loud sputtering and hissing similar to that caused when a hot rod is plunged into cold water. Quite high currents passed at voltages of 100–300 V and the wire electrode sometimes became white hot and melted. It seems likely that the intense heating of the metal in these circumstances is due to the cathode drop in the discharge occurring close to the metal surface, and this may well result in thermionic emission from the cathode and very high currents. As would be expected from what has been said above, the breakdown of conventional electrolysis has also been observed[8] using alternating current; there is, however, a marked tendency for the small electrode to burn out and for the metal to be dispersed in colloidal form in the solution.

The local heating at a wire electrode, which can lead to the breakdown of conventional electrolysis, seems to be inherent in the geometry of current flow to the electrode. At a thin wire electrode, the lines of current flow must converge rapidly and much of the Ohmic drop of potential in the electrolyte must occur within a short distance of the electrode surface. Thus, the equipotential surfaces will approximate to cylinders concentric with the electrode under study, and, for a concentric cylinder of radius r, the voltage gradient dV/dr will be given by the equation $I = -kA\,dV/dr$, where I is the current, k the conductivity of the electrolyte, and A the surfaces will approximate to cylinders concentric with the electrode Thus, $I = -2k\pi Lr\,dV/dr$, and since I must be constant for all values of r, $r\,dV/dr$ must be a constant, which is designated B. Thus, $B\,dr/r = dV$, and on integration between definite limits, $V_2 - V_1 = B\ln(r_2/r_1)$. Thus, on substituting for B in the current equation, $I = 2k\pi L(V_1 - V_2)/\ln(r_2/r_1)$. On inserting numerical values, the voltage drop and power dissipated close to the wire electrode can be calculated. Thus, Hickling and Ingram[8] calculated that, when breakdown of conventional electrolysis occurred under their standard conditions with a cell voltage of 60 V and total power dissipation of 26 W, a drop of 18 V occurred within a layer 1.75 mm thick around the wire anode and no less than 8 W was being dissipated in this very thin layer.

In CGDE with an immersed anode, electrical conditions are very similar to those in GDE with the anode above the surface, but for minimum electrode spacing. Thus, the midpoint voltage of 420 V is close to the cathode drop of 415 V found for aqueous electrolytes in GDE. Current densities are also not dissimilar; thus, for CGDE in aqueous electrolytes at atmospheric pressure, a value of about $0.6 \, A \, cm^{-2}$ is common. Hence, one would expect to find a similar yield of products on CGDE to those encountered in anodic GDE, and this seems to be the case.

From what has been said above, breakdown of conventional electrolysis might be expected whenever the lines of current flow converge in such a way that ohmic heating can cause local vaporization of the solvent. Thus, one might suppose that the phenomenon would have been frequently observed, and search of the literature shows that this is indeed the case, although usually the observation has been made in a particular context and its general significance has not been realized. A footnote in a paper by Fizeau and Foucault[12] as early as 1844 described an electrolysis of water with a battery of 80 voltaic couples between wire electrodes which became luminous; the luminosity was particularly apparent at the cathode and they regarded it as an arc discharge through hydrogen. Lagrange and Hoho[13] in 1893 applied this observation to the heat treatment of steel cathodes and laid the basis of the process of electrolytic heating of metals.[14] In this technological process, the metal to be treated is made the cathode in an aqueous electrolyte under conditions such that it becomes enveloped in vapor and it is heated to a high temperature in a reducing atmosphere, while interruption of the current leads to its rapid quenching; there seems no doubt that this is an example of CGDE. A further example of the same phenomenon is the Wehnelt electrolytic interrupter[15] for induction coils devised in 1899. This consisted of a cell containing a large plate cathode and a small platinum wire anode. Investigation by

Humphreys[16] showed that the interruption of current was associated with the development of a high resistance and visible glow at the small electrode, and he also showed that less oxygen was evolved at the anode than in normal electrolysis. Thus, CGDE seems to be a phenomenon of fairly general occurrence.

VI. CHEMICAL RESULTS OF GLOW-DISCHARGE ELECTROLYSIS

The general effect of GDE with an anode above the liquid surface is to bring about oxidation in the solution, and this seems to arise by breakup of solvent molecules to give active species which interact among themselves or with reactive solutes. Thus, a wide variety of chemical changes can be effected, and these will be classified according to the main type of reactant involved. The dominant variable governing the yields of products obtained is the quantity of electricity passed, and thus the yields can best be expressed in equiv F^{-1} or mole F^{-1}. These are, of course, integral yields obtained over a period of electrolysis, and it is often convenient to be able to specify a differential yield representing the instantaneous yield in equiv F^{-1} or in mole F^{-1} at a particular point in electrolysis when the concentrations of substances in the solution are accurately known. This differential yield, which can be obtained by drawing a tangent to the integral yield curve, will be denoted by G and it is particularly significant at the start of GDE, when it is designated G_0. It is analogous in some ways to the quantity used in radiation chemistry, where the yield is expressed in molecules per 100 eV, but it should be clearly realized that, as used here, it is a yield for a given quantity of electricity and this will only be proportional to the electrical energy put in if a constant voltage term is involved.

1. Water

The GDE of aqueous solutions in which the electrolyte is of an inert type results in the decomposition of water with the formation of hydrogen, oxygen, and hydrogen peroxide as products. The hydrogen peroxide is initially formed in amount proportional to the quantity of electricity passed, but the yield is higher than 1 equiv F^{-1}; thus, there is oxidation in excess of that possible by charge transfer and this excess is balanced by an equivalent amount

Table 3
Formation of Hydrogen Peroxide by GDE

Electrolyte (0.05 M)	Approx. pH	Initial $G_0(H_2O_2)$, equiv F^{-1}	Stationary concn. H_2O_2, N
NaOH	13–12	0	0
Na_2CO_3	11	0	0
$Na_2B_4O_7$	9	—	0.013
$NaHCO_3$	8	1.9	0.033
Na_2HPO_4	7	1.5	0.032
K_2SO_4	7–2	1.3	0.032
H_2SO_4	1	1.1	0.022

of hydrogen liberated into the gas phase. As the hydrogen peroxide accumulates in the solution, it itself is decomposed by the glow-discharge and eventually it reaches a stationary concentration when it is decomposing as fast as it is produced.

Table 3 gives some typical results for hydrogen peroxide formation in unstirred solutions of various electrolytes under standard discharge conditions.

In strongly alkaline solutions, no hydrogen peroxide can be detected, possibly due to its ready anodic decomposition by discharge of the HO_2^- ion, but in all other solutions, the formation of hydrogen peroxide and its decomposition seems to follow essentially the same path independent of the nature of the electrolyte. Over the pH range 8–2, the stationary concentration of hydrogen peroxide attained is substantially the same, although the initial yield of hydrogen peroxide seems to increase somewhat with rising pH. Moderate increase in concentration of electrolyte generally increases both the initial yield of hydrogen peroxide and the stationary concentration attained; thus, on increasing the concentration of a Na_2HPO_4 electrolyte from 0.05 to 0.5 M, $G_0(H_2O_2)$ increases from 1.5 to 1.9 equiv F^{-1} and the stationary concentration from 0.032 to 0.048 N. An increase in temperature also tends to raise the initial yield slightly, but diminishes somewhat the stationary concentration that can be attained. Stirring of the electrolyte is generally favorable to hydrogen peroxide production, probably by diminishing its concentration in the glow-spot. The presence of chloride ions in the electrolyte interferes considerably with the formation of hydrogen peroxide and the presence of 0.02 M NaCl completely inhibits its production, although the chloride ion itself is not appreciably consumed.

In contrast to the influence of these factors affecting the solution, factors affecting the discharge have very little effect on the yield of hydrogen peroxide. Thus, changing the current from 0.05 to 0.1 A has no significant effect, although the power dissipated in the discharge is increased from 28 to 60 W. Likewise, changes of the anode material, its shape, and its distance from the liquid surface are without appreciable effect. Using different gases in the space above the solution, e.g., air, nitrogen, oxygen, hydrogen, and nitrous oxide, has only a slight effect; hydrogen gives somewhat smaller and oxygen slightly higher yields than the other gases. Increase of gas pressure results in a small increase in the yield of hydrogen peroxide, possibly due to increase of CD in the glow-spot.

The trivial effect which experimental variables affecting the discharge have on the yield of hydrogen peroxide strongly suggests that this must arise by reactions occurring in the liquid rather than the gas phase, and quantity of electricity passed seems to be the most important factor. The general picture is summarized in Figure 6, which shows the integral and differential yields of hydrogen peroxide under various conditions. If it is assumed that the rate of formation of hydrogen peroxide is proportional to the current and its decomposition is dependent both on the current and upon its concentration, then

$$d(H_2O_2)/dt = G_0(I/F) - G_0(I/F)k[H_2O_2]$$

where (H_2O_2) is the integral yield, G_0 the initial differential yield, I the current, and k a coefficient for the decomposition reaction. On integration and application of the condition that there is no hydrogen peroxide present at the start of electrolysis, the equation becomes

$$(H_2O_2) = (V/k)[1 - \exp(-G_0kq/V)]$$

where q is the quantity of electricity and V the volume of the solution. This yield equation is found to represent closely the experimental results.[17]

CGDE gives very similar results to those obtained with GDE, and at atmospheric pressure with a 0.04 N H_2SO_4 electrolyte, a G_0

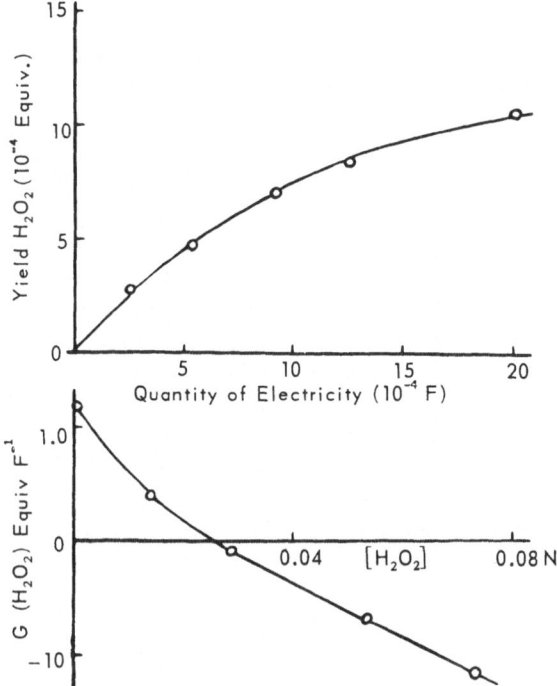

Figure 6. Integral and differential yields of hydrogen peroxide.

value for hydrogen peroxide as high as 1.8 equiv F^{-1} has been obtained. Spark electrolysis with solutions of inert electrolytes also furnishes very similar yields of hydrogen peroxide to those obtained in GDE and the effects of alkalinity, presence of chloride ion, and the nature of the gas atmosphere run closely parallel.[7]

2. Aqueous Solutions of Oxidizable Substrates

The oxidizing properties of anodic GDE are conveniently studied by having present in the conducting aqueous solution oxidizable substances such as ferrous, stannous, cerous, ferrocyanide, and azide ions. Oxidation is brought about in all such cases, and yields as high as 8 equiv F^{-1} can often be obtained in concentrated solutions, which is vastly in excess of any oxidation yield to be expected simply from charge transfer. In dilute solutions, there is usually hydrogen peroxide formation as well as oxidation of the substrate, but with increase in substrate concentration, peroxide formation diminishes and the oxidation yield approaches a limiting value.

The GDE of ferrous salts has been particularly investigated[5] since any hydrogen peroxide formed will itself react to bring about oxidation, and the total oxidation yield can readily be measured. Again, the dominant variable in determining the amount of oxidation is the quantity of electricity passed, with factors affecting the discharge having only a trivial effect, but the concentration of the ferrous salt has an important role. This is shown in Figure 7, where G_0(oxidation) is plotted against ferrous ion concentration for GDE of a solution in $0.8 N$ H_2SO_4 in the absence of oxygen. The initial oxidation yield starts at a value of about 2 equiv F^{-1} in very dilute ferrous solutions, but it increases very rapidly with rise in ferrous ion concentration, and it ultimately approaches a limiting value in relatively concentrated solutions. Over a period of years, this value has been measured by a number of different workers[5]; typical

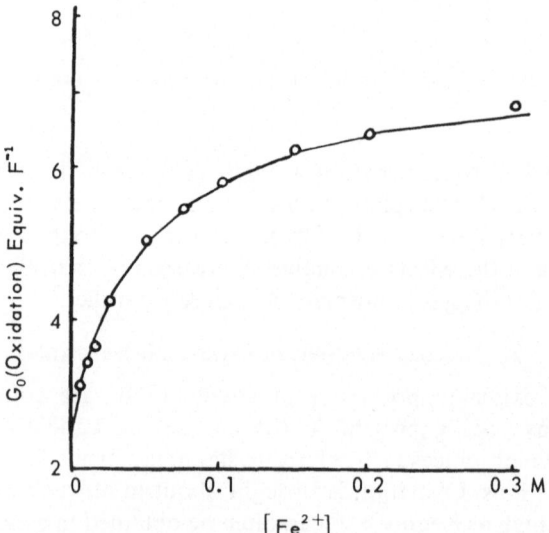

Figure 7. Influence of concentration on oxidation yields with ferrous ion.

figures obtained with slightly varying conditions are 7.2, 8.7, 7.9, 7.6, and 8.3 equiv F^{-1}. Thus, an average value of about 8 equiv F^{-1} seems to be indicated. In oxygen-saturated solutions of ferrous sulfate, the limiting G value is appreciably increased and a value as high as 12.5 equiv F^{-1} has been reported.[5] Original observations appeared to show that any inert atmosphere gave about the average limiting value of 8 equiv F^{-1}, but this has recently been questioned by Denaro,[18] who finds that, while hydrogen gives about this value, a nitrogen atmosphere gives an appreciably higher result. Just as in the case of hydrogen peroxide formation by GDE, increase of inert electrolyte concentration and temperature are found to favor an increased oxidation yield with ferrous sulfate.

The existence of a limiting oxidation yield at high ferrous ion concentrations seems unmistakably to indicate that the oxidation reaction in this case of GDE is competing kinetically with other processes. This conclusion is supported by observations with other oxidizable substrates. Thus, Figure 8 shows G_0(oxidation) and $G_0(H_2O_2)$ curves for GDE of azide, cerous, and ferrocyanide ions using a hydrogen atmosphere. The azide results[19] refer to solutions of sodium azide in neutral phosphate buffer, and it is seen that G_0(oxidation) starts at a very low value and increases with rising azide ion concentration, while formation of hydrogen peroxide, which reacts rather slowly with azide, is initially high but drops to zero when the azide concentration is moderately high. The limiting G_0(oxidation) value at high azide concentration is about 7 equiv F^{-1}, which is quite close to that found in the ferrous case. In the GDE of cerous ions, the situation is complicated by the fact that any hydrogen peroxide formed will reduce the ceric oxidation product. The yield curve in Figure 8 refers to solutions of cerous sulfate in 0.8 N H_2SO_4 and it is seen that a certain concentration of cerous sulfate is necessary before any oxidation will occur; at this concentration, the hydrogen peroxide yield sinks to zero, and at higher concentrations, G_0(oxidation) rises, ultimately approaching a value somewhat greater than 5 equiv F^{-1}. Since any hydrogen peroxide formed in the GDE will reduce the yield of ceric ion, the limiting value might be expected to be lower than with ferrous and azide ions, or at any rate to be attained more slowly. Figure 8 also shows the oxidation yield curve obtained for potassium ferrocyanide solutions[20] in 0.1 M NaOH on GDE in a hydrogen atmosphere. As has

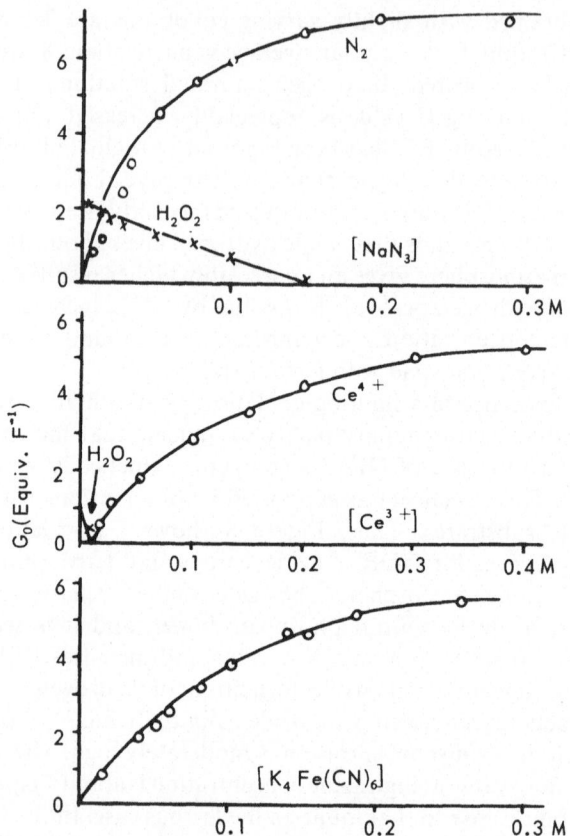

Figure 8. Influence of concentration on oxidation yields with
azide, cerous, and ferrocyanide ions; (circles) oxidation; (crosses)
H_2O_2 formation.

already been pointed out, hydrogen peroxide is not formed by GDE
in alkaline solutions, and G_0(oxidation) for formation of ferricyanide
shows a gradual increase with increasing ferrocyanide concentration
toward a limiting value of 5.6 equiv F^{-1}; in neutral solution, the
limiting value was somewhat higher, at 6.4 equiv F^{-1}.

The general conclusion to which experiments with oxidizable
substrates leads is that GDE with an anode above the aqueous
surface results in oxidation. The limiting oxidation yield depends

to some extent on the particular process studied, and, particularly, on how this is affected by hydrogen peroxide, which is always a possible product, but it appears to approach a maximum value of about 8 equiv F^{-1}.

3. Other Inorganic Reactions in Water

Peroxyacids and their salts can frequently be produced by GDE. Thus, even in dilute sulfuric acid solutions, some peroxysulfuric acid, H_2SO_5, is formed in addition to hydrogen peroxide,[18] and with concentrations of sulfuric acid greater than 1 M, substantial amounts of peroxydisulfuric acid, $H_2S_2O_8$, are produced as well as peroxysulfuric acid H_2SO_5.[21] Potassium peroxydiphosphate, $K_4P_2O_8$, has also been produced by GDE of phosphate solutions.[22] Yields of the peroxyacids are in general increased by the addition of potassium fluoride.

Very complex results are obtained in GDE of solutions of halides and oxyhalogen salts.[23] Some liberation of halogen may take place, and with iodides, this can predominate, but both formation and reduction of oxyhalogen salts may occur and the net process after a short time is often liberation of oxygen. Thus, with solutions of sodium chlorite, $NaClO_2$, a yield of oxygen of 34 equiv F^{-1} has been reported,[24] probably indicating a chain reaction.

4. Reactions of Organic Substances

The GDE of solutions of organic substances has not hitherto been extensively studied. Klemenc[25] examined the gaseous products from GDE of methanol, ethanol, and formic and acetic acids in water; they were mainly carbon dioxide, carbon monoxide, and hydrogen, accompanied sometimes by the formation of hydrocarbons, Gore and Hickling[26] have recently investigated the GDE of formic and acetic acids over a range of concentrations going from the dilute aqueous to the anhydrous acids. The main products from formic acid systems were found to be hydrogen, carbon dioxide, and carbon monoxide, but in aqueous solutions, oxalic acid was formed in substantial amounts, probably by dimerization of the ·COOH radical. Other products in small amounts included methane, methanol, and formaldehyde. With a large increase in acid concentration, the products became almost entirely gaseous and very large amounts of carbon monoxide were produced.

Acetic acid gave rather similar gaseous products, but containing more methane, while in solution, substances such as succinic acid $COOH \cdot CH_2 CH_2 \cdot COOH$, malic acid $COOH \cdot CH_2 CH(OH) \cdot COOH$, tricarballylic acid $COOH \cdot CH_2 CHCOOHCH_2 \cdot COOH$, and oxalic acid $COOH \cdot COOH$ were found; most of these products can be regarded as built up initially from the $\cdot CH_2 COOH$ radical. Again, solution products tended to diminish at very high acid concentrations. Thus, with formic and acetic acids, although GDE produces gaseous decomposition products which are very similar to those formed by passing a discharge between metallic electrodes in the acid vapor, synthetic reactions are possible in the liquid phase from radicals produced by abstraction of hydrogen atoms from the organic molecules. Another case of such synthesis which has been reported[27] is the production of oxamide $(CONH_2)_2$ by GDE of a solution of formamide $HCONH_2$ in dilute sulfuric acid.

The GDE of an aqueous solution of acrylonitrile will readily bring about polymerization, and it has been suggested as a general method of initiating polymerization in a liquid phase.[28] The polymerization of acrylamide in aqueous solution by GDE has recently been investigated.[18] Monomer disappears rapidly at a rate exceeding 10 mole F^{-1} in the concentration range 0.25–1 M, but the yield of polymer precipitated is very low, about 0.01 mole F^{-1}, with molecular weight varying between 15,000 and 200,000, depending upon experimental conditions. It seems likely that some products of much lower molecular weight are also being produced. The polymerization of organic vapors subjected to a glow-discharge has long been known and this has recently attracted attention as a possible means of coating metals with polymer films.[29] Strictly, however, it is outside the field of GDE, which is concerned primarily with reactions induced in the liquid phase.

5. Nonaqueous Reactions

GDE in liquid ammonia solutions has been extensively studied[30] and the results are analogous in many ways to those obtained in water. With inert electrolytes such as ammonium nitrate, hydrazine is the main product in solution, corresponding to hydrogen peroxide in water. The yield is quite high, the initial $G_0(N_2 H_4)$ value being about 2.5 mole F^{-1}, but it falls off slowly as hydrazine accumulates in the liquid and itself undergoes decom-

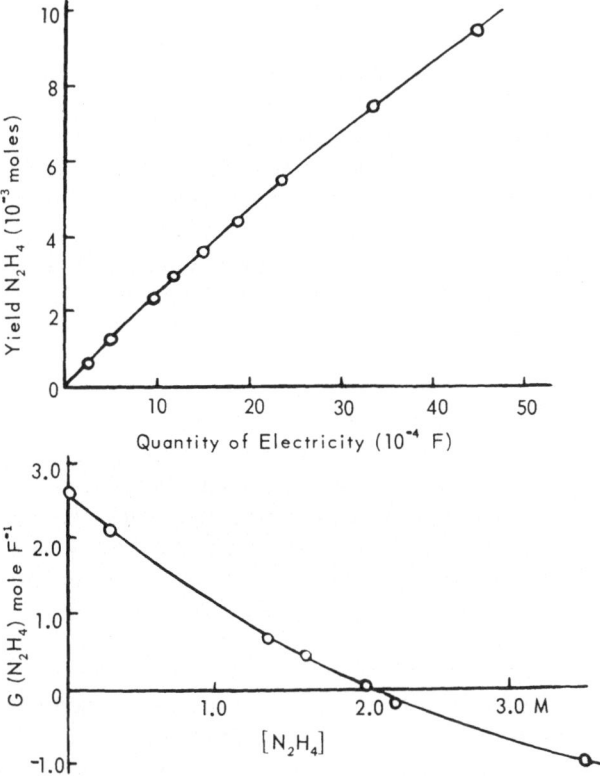

Figure 9. Integral and differential yields of hydrazine.

position by GDE. Eventually, a stationary concentration is reached which may be as high as 2 M under optimum conditions, and anhydrous hydrazine can be isolated from the solution by distillation. Figure 9 summarizes the integral and differential yields of hydrazine obtained with a 0.01 M NH_4NO_3 solution in liquid ammonia at $-78°C$ and under a pressure of 100 mm Hg when subjected to GDE at 0.025 A with the anode 0.5 cm above the liquid surface. Hydrazine formation appears to be largely independent of conditions in the discharge on the vapor side of the interface, and is not critically dependent upon the nature or concentration of the electrolyte provided the latter is chemically inert, but, in general, the highest yields are obtained in very dilute solution; a rise in temperature slightly increases the yield. Electrical conditions in the discharge are very similar to those with aqueous electrolytes. There is a cathode drop of about 390 V close to the liquid surface, and CD in the yellow glow-spot is usually in the range 0.1–0.3 A cm^{-2}.

Work on GDE in other nonaqueous solvents has hitherto been limited to the observations of Cady et al.[6b] who studied liquid sulfur dioxide containing small amounts of water to render it conducting. They found that sulfuric acid and sulfur were the main final products. With a glow-discharge anode, the yield of sulfuric acid was about 14 mole F^{-1}, and about 5 g-atom F^{-1} of sulfur was produced.

GDE with fused electrolytes might be expected to provide an interesting field of study. Sastri[6c] reported difficulty in striking a glow-discharge to fused electrolytes, but Hamilton and Ingram[30] have found that GDE may be conveniently carried out with both anhydrous chloride melts and molten bisulfates. The electrical characteristics of the discharge are similar to those with aqueous systems, and preliminary work on oxidation of suitable redox systems in these fused electrolytes has been carried out.

6. Chemical Effects of Other Discharges

The chemical effects of CGDE with an immersed anode are very similar to those of GDE with an anode above the surface and this is illustrated by the values in Table 4. When dealing with a volatile reactant in solution, CGDE may sometimes be adopted with advantage since decomposition of the vapor by the electrical discharge is then minimized. This has been done, for example, in working with formic and acetic acids.[26]

If GDE is operated with a cathode instead of an anode above the liquid surface, it might at first sight be expected that chemical reduction would occur in the liquid phase since the discharge process is equivalent to the introduction of electrons. Thon[6d] found that solutions of salts of silver, gold, and platinum could indeed be reduced to the metallic state, but, in general, with other metallic salts, only a precipitate of the metallic hydroxide was formed in the glow-spot. More recent investigation[31] has shown that in cathodic GDE the reactions produced are generally of the same type as in anodic GDE, but the yields are much less, and any reductions

Table 4
Comparison of Chemical Effects of GDE and CGDE

Chemical effect		GDE	CGDE
Oxidation of $FeSO_4$, limiting yield, equiv F^{-1}		8	7.9
Formation of H_2O_2 in dilute H_2SO_4:			
Initial yield, mole F^{-1}:	at 760 mm Hg, 18°C	2.1	1.8
	50 mm Hg, 18°C	1.1	2.5
	760 mm Hg, 50°C	1.6	1.8
Stationary concentration, N:	at 760 mm Hg, 18°C	0.115	0.155
	50 mm Hg, 18°C	0.03	0.085
	760 mm Hg, 50°C	0.09	0.11
Formation of N_2H_4 in liquid NH_3:			
Initial yield, mole F^{-1}		2.5	1.8
Stationary concentration, M		2.0	1.0

which occur seem to be due to hydrogen peroxide which is frequently formed. Thus, with dilute sulfuric acid, even in a hydrogen atmosphere, a small yield of hydrogen peroxide is obtained with a $G_0(H_2O_2)$ value of 0.04 equiv F^{-1}; with a ferrous sulfate electrolyte, oxidation occurs with a limiting value at high ferrous concentrations of approximately 0.28 equiv F^{-1}. Both these values are markedly increased in the presence of oxygen to 0.15 and 1.9 equiv F^{-1}, respectively. Using solutions of ferric salts, no certain evidence for reduction could be established; some reduction does, however, occur with ceric, dichromate, and ferricyanide solutions, but this may be partly due to hydrogen peroxide formation. It seems difficult to resist the view that in cathodic and anodic GDE the reactions are qualitatively similar, and that the smaller effect in cathodic GDE is connected with the much smaller potential drop near the liquid surface.

It has been pointed out earlier that when alternating current is used for GDE, the discharge usually stops and starts twice in each cycle and the current consists of separate successive anodic and cathodic pulses. From what has been said above, therefore, it would be expected that, chemically, the results would be the same qualitatively as in anodic GDE but that the yields would be reduced to about half because of the small contribution of the cathodic pulses.

This is found to be the case in practice. Thus, with a solution of sodium phosphate in water, the initial yield of hydrogen peroxide is about 0.7 equiv F^{-1} compared with 1.5 equiv F^{-1} for anodic GDE; similarly, in the formation of hydrazine in liquid ammonia, the initial yield of 2.5 mole F^{-1} for anodic GDE is reduced to 1.3 mole F^{-1} when alternating current is used.

Some other types of electric discharge to solutions have been used. Spark electrolysis at atmospheric pressure[7] gives chemical effects very closely analogous to those obtained in GDE. Silent electrical discharge in an ozonizer-type apparatus has also been tried[32] using acid solutions of ferrous sulfate and ceric sulfate; oxidation of ferrous ion and reduction of ceric ion occurs and the results have been interpreted in terms of radiolysis produced by slow electrons. The chemical effects produced by driving gaseous ions formed in a gas phase by radioactive substances into solution under the influence of an electric field has also been studied[33]; again, the effects are analogous to those in GDE.

VII. MECHANISM OF GLOW-DISCHARGE ELECTROLYSIS

From the experimental results on GDE which have accumulated, the following general conclusions seem to emerge.

1. In GDE, conditions existing in the gas or vapor phase seem to have very little influence either qualitatively or quantitatively on the chemical products formed in solution. This is particularly noteworthy when the electrical power dissipated in the discharge is considered. This can be varied substantially by changing the current or the electrode spacing, but in no such case is there much effect on the yields in solution. It might at first sight be thought that it is some chemically reactive species formed in the gas discharge which dissolves in the liquid phase and brings about the characteristic chemical effects. For example, the OH radical can usually be detected spectroscopically in the discharge when water vapor is present, and Klemenc[4] in 1938 seems to have regarded this as the effective agent in GDE. However, if a discharge is passed between metallic electrodes immediately above the surface of a suitable liquid, e.g., acidified ferrous sulfate solution, little or no oxidation

is produced. This seems unmistakably to indicate that the characteristic effects of GDE are associated with the passage of electricity into the solution, and this conclusion is very directly supported by the observations of Dewhurst *et al.*[6e] They found that, by interposing a negative grid between a glow-discharge anode and a ferrous sulfate solution, the normal oxidation observed could be almost entirely inhibited, suggesting that it must arise by entry of charged particles into the solution.

2. The initiation of chemical reaction in the liquid seems to be connected with the existence of a considerable electrical field near its surface which is constant under most conditions. Provided this characteristic cathode potential drop can be realized, then GDE occurs. The quantity of electricity passed is the experimental variable of dominant importance, and if a constant voltage drop is involved, this will also be proportional to the electrical energy expended in the thin layer adjacent to the liquid surface. Thus, it would appear that it is the amount of electrical energy put into the system due to the field near its surface which determines the kind and amount of chemical reaction.

3. The chemical reactions characteristic of GDE are very closely similar to those produced on the liquids concerned by ionizing radiation such as gamma rays and X-rays and alpha particles.[34] Such radiolysis processes are usually interpreted in terms of primary acts in which energy is transferred to solvent molecules, leading to their breakup and the formation of reactive entities in solution which interact or can be scavenged by suitable reactants. The initial process is much the same for different types of radiation but the final results depend upon their penetrating power and hence the distribution of spurs of reactivity in the solution. This penetration depends upon the initial energy of the particles and the rate at which they lose energy in the liquid; this latter factor is usually expressed as a linear energy transfer, or LET, value and the results of GDE are particularly like those of alpha particles of high LET value.

In the conventional view[35] of normal glow-discharge between metallic electrodes, it is considered that, adjacent to the cathode surface and separated from it by the cathode dark space, there exists a strong positive space charge due to the accumulation of gaseous positive ions; as a result of this, there is a considerable potential drop across a relatively small distance, which constitutes the normal cathode drop. Positive ions move across the cathode dark space and impinge upon the cathode, and in their passage, produce the electrons necessary for the maintenance of the discharge. At the cathode itself, the current is regarded as being carried almost entirely by positive ions. The cathode drop voltage is independent of gas pressure p and thickness d of the cathode drop region when these variables are considered separately, but is related to the product of these. This product pd is constant and corresponds to about 1 cm-(mm Hg) for common diatomic gases.

Application of this picture to the usual situation in GDE in which the anode is above the liquid surface would suggest that the current is carried to the solution by positive gaseous ions which are driven into the liquid from the gas phase and subsequently discharged. Since the local temperature under the glow-spot must be relatively high, the discharge must ultimately be one through the solvent vapor in all cases and this conforms to the observation that the nature of the gas present above the liquid does not usually have a marked effect in GDE. With aqueous solutions, therefore, the positive gaseous ions present are likely to be those found in ionized water vapor. We have no certain knowledge what these are at the pressures used in GDE, but observations made with the mass spectrometer at much lower pressures[36] have shown that the main positive ion is H_2O^+, and for simplicity, this will be assumed to be the species carrying the charge to the liquid surface. Close to the surface in the gas phase, there will be the cathode potential drop across a distance which, in the pressure range employed, should be between 0.01 and 0.1 cm. Thus, there will exist a considerable electric field near the surface in which the gaseous ions will be accelerated and they will enter the liquid with substantial energies.

The cathode drop observed in GDE with aqueous solutions is 415 V, and hence the maximum energy that the gaseous ions entering the solution could have would be 415 eV if they passed through the cathode drop without any loss of energy. In practice, much of this energy must be dissipated by collisions in the vapor phase. At first sight, it would be expected that the average energy would depend upon the mean free path of the ions in the discharge. Calculations based on this premise lead to rather low energies, less than 10 eV. However, direct experimental investigation of the energies of positive ions passing through perforated cathodes in gas discharges have shown that the problem is not a simple one. Thus, Chaudrhi and Oliphant,[37] working at pressures of less than 1 mm Hg, found that positive ions were present with all energies from zero up to the full energy corresponding to the total cathode drop. At any given pressure, the number of ions possessing any given energy was approximately proportional to the discharge current, and the maximum on the energy distribution curve, which, in a typical experiment, occurred at about 300 eV, was independent of discharge voltage and current but varied with pressure. It would be expected that, with increase in pressure, the energy should fall, and their results indicated that this was so. However, as pressure rises, the thickness of the cathode drop region decreases and this may have a compensating effect. More recent work[38] suggests that the factor determining the energy of a positive ion striking the cathode is the ratio between the mean free path of the ion for symmetrical charge transfer with ambient gas molecules and the cathode dark-space distance, and pressure may have little effect since pd is relatively constant. One can get an idea of the energies which the positive gaseous ions in GDE would have to have in order to produce the chemical effects observed by comparing the yields with those obtained using ionizing radiations. Thus, with the deaerated acidified ferrous sulfate system commonly used as a dosimeter,[34] $G(Fe^{3+})$ varies from about 8 molecules per 100 eV for gamma rays or X-rays of low LET value to about 4 molecules per 100 eV for alpha particles of a high LET value of above 10 eV A^{-1}. The corresponding value found in GDE is about 8 equiv F^{-1}, and so, if the chemical effects of GDE are produced in a manner similar to that in radiolysis, this would imply that the gaseous ions entering the solution would on the average each require an energy of 100–200 eV. This seems to be feasible and it suggests that the ions may well enter the liquid with energies adequate to bring about dissociation of solvent molecules. If the value of 100 eV is adopted as being of the right order of magnitude, the G values in GDE expressed in mole F^{-1}, become directly comparable with G values in radiation chemistry expressed in molecules per 100 eV.

The penetration of the gaseous ions into the liquid phase would be expected to be very small. In water, slow alpha particles have

LET values of about $10 \, \text{eV} \, \text{A}^{-1}$, and for these gaseous H_2O^+ ions, the value will presumably be higher than this. Thus, their range in the liquid phase is not likely to exceed about $10 \, \text{Å}$. This implies that the primary reaction zone in the liquid will be a very thin disk, frequently of area about $0.5 \, \text{cm}^2$ and thickness of perhaps $10 \, \text{Å}$. Within this primary reaction zone, the bombarding ions will each ionize or activate several solvent molecules by collision, and one molecule per ion will be broken up by charge transfer; in aqueous solutions, the primary processes can be depicted as follows:

Collision

$$
\text{ionization:} \quad H_2O \rightsquigarrow \begin{array}{l} H_2O^+ \xrightarrow{\ H_2O\ } OH + H_3O^+ \\ + \\ e^- \xrightarrow{\ H_2O\ } e^-_{aq} \xrightarrow{\quad\quad} H + OH^- \end{array} \tag{1}
$$

$$
\text{activation:} \quad H_2O^* \rightsquigarrow OH + H \tag{2}
$$

Charge transfer

$$
H_2O^+ + H_2O \rightarrow OH + H_3O^+ \tag{3}
$$

These primary processes lead in all cases to the production of OH radicals and hydrated electrons and, sometimes, H atoms. However, because the reactive species are packed very closely together in the primary reaction zone, there is a good chance that molecular yields of H_2O_2 and H_2 may arise either by dimerization of the free radicals or conceivably by interaction of their precursors, e.g., according to the following possibilities:

$$
2H_2O^+ \rightarrow H_2O_2 + 2H^+
$$
$$
2H_2O^* \rightarrow H_2O_2 + H_2 \tag{4}
$$

and

$$
2e^-_{aq} \rightarrow H_2 + 2OH^- \tag{5}
$$

Interaction of the primary products to re-form water will also, of course, always take place. Therefore, diffusing out from the primary reaction zone will be a mixture of OH, e^-_{aq} or H, H_2O_2, and H_2, and, as these species diffuse into the body of the solution, they may

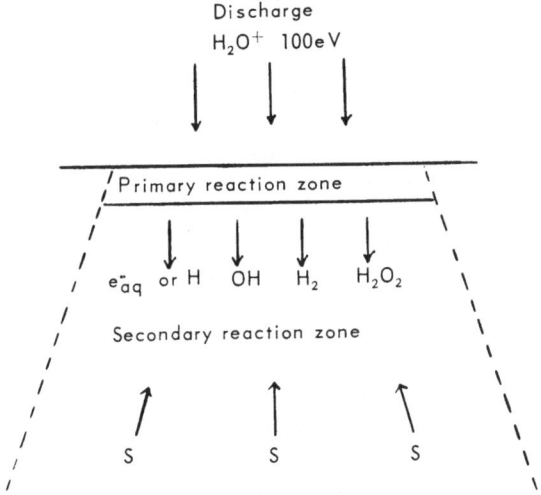

Figure 10. Reaction zones.

interact with one another or be scavenged by reactants in the solution. The situation thus envisaged is diagrammatically represented in Figure 10 and it has many features in common with the picture usually adopted to represent radiolysis[39] in which the primary collision and ionization acts are considered to occur in a spur of activity (corresponding to the path of the particle) from which emerges a yield of radical and molecular products. In GDE, the primary reaction zone corresponds to the spur; within it, there will be many separate hot spots of activity, each produced by an incoming ion, but, as the radicals and molecular products diffuse away from their initial positions, they will merge into a single downward and sideways drift into the solution. Thus, the primary reaction zone is a kind of spur, but one which is located in a fixed position and constantly renewed by the introduction of new gaseous ions.

The radicals and products which diffuse into the body of the electrolyte will interact among themselves and with any reactive substrate in the solution. A bewildering multiplicity of reactions has been considered in radiolysis,[39] but if we select from these only those reactions which are known to proceed at high velocities,[40] the following main reactions seem to be relevant to GDE:

Interaction of radicals and products:

$$H + OH \rightarrow H_2O \tag{6}$$

$$OH + OH \rightarrow H_2O_2 \tag{7}$$

$$H + H \rightarrow H_2 \tag{8}$$

$$OH + H_2O_2 \rightarrow H_2O + HO_2$$

followed by $OH + HO_2 \rightarrow H_2O + O_2$ $\qquad(9)$

Interaction with substrate S:

$$OH + S \rightarrow S^+ + OH^-$$

or $OH + SH \rightarrow S + H_2O$ $\qquad(10)$

$H_2O_2 \rightarrow$ possible oxidation and reduction   (11)
reactions with substrate and its
oxidation products

$$H + SH \rightarrow S + H_2 \tag{12}$$

This scheme permits many of the features of GDE to be understood. Thus, in aqueous solutions of inert electrolytes, the chief
product in solution should be hydrogen peroxide arising by reaction (7) and possibly in part by (4). This itself can be decomposed
by reaction (9), and so, as its concentration in solution increases, it
will be decomposed by the glow-discharge at an increasing rate until
ultimately a stationary concentration is reached. If the GDE is
started with a high concentration of hydrogen peroxide present,
then reaction (9) will predominate and again a stationary state will
be approached when the rate of decomposition has decreased until
it is equal to the rate of formation. This is in agreement with what is
observed experimentally. Increase of CD might be expected to
favor production of hydrogen peroxide by reactions (7) and (4) and
hence to increase both the initial yield and the stationary concentration. Raising the gas pressure increases the CD and the yield of
hydrogen peroxide does increase somewhat. A more marked effect
has been found[26] using a condensed glow-discharge. With this
technique, a condenser is charged to a high voltage and periodically
discharged between the anode and the surface of the solution; each

discharge is very short and produces a very high CD. By this method, with a dilute sulfuric acid electrolyte, $G_0(H_2O_2)$ was increased from 1.1 to 2.0 equiv F^{-1} and the stationary concentration was raised from 0.03 to 0.11 N under otherwise identical conditions. The yield of hydrogen peroxide by GDE is somewhat increased when an oxygen atmosphere is used, and this may be due to reduction of oxygen in solution by hydrogen atoms, e.g.,

$$H + O_2 \rightarrow HO_2$$
$$2HO_2 \rightarrow H_2O_2 + O_2 \tag{13}$$

The prejudicial effect which chloride ion seems to exert on hydrogen peroxide formation may be due to the removal of OH radicals by the reaction

$$Cl^- + OH \rightarrow Cl + OH^- \tag{14}$$

When reaction with substrates is considered, it can readily be seen that the main type of reaction likely to occur is oxidation brought about by OH radicals and H_2O_2 as indicated in (10) and (11), and this fits what has been observed in GDE using oxidizable substrates such as ferrous, azide, cerous, and ferrocyanide ions. As the concentration of such reactants is increased, they will be able to penetrate further into the reaction zone until, at high concentrations, all the oxidizing species are being scavenged and a limiting oxidation yield will then be approached. The role of the H atoms in GDE is somewhat obscure. At first sight, it might be expected that they would bring about reduction, but usually no such reaction can be detected; for example, in the GDE of Fe^{2+} ions, the presence of Fe^{3+} has no influence on the yields. In radiation chemistry, an oxidizing function has often been ascribed to them. This was formerly interpreted as a reaction with hydrogen ions leading to the oxidizing species H_2^+ which then accepts an electron, e.g.,

$$H + H^+ \rightarrow H_2^+$$
$$H_2^+ + Fe^{2+} \rightarrow Fe^{3+} + H_2 \tag{15}$$

but it can more plausibly be considered[39] as abstraction of a hydrogen atom from a water molecule in the hydration sphere of the ferrous ion, thus

$$Fe \cdot H_2O^{2+} + H \rightarrow FeOH^{2+} + H_2 \tag{16}$$

It could well be that H atoms act similarly in GDE, but it is note-worthy that the limiting oxidation yield with different ions in acid, neutral, and alkaline solutions shows a certain constancy when the action of hydrogen peroxide is allowed for, which seems unlikely if it is largely due to H atoms. Reactions which can plausibly be attributed to H atoms are (a) the increase in ferrous oxidation in the presence of oxygen, which may be due to the formation of the HO_2 radical by reaction (13), where HO_2 may then oxidize Fe^{2+} ions; and (b) hydrogen-atom-abstraction reactions of type (12), which seem to occur with some organic substances; for example, in the GDE of formic and acetic acids in aqueous solution,[26] the yield of hydrogen gas is higher than is normally obtained from the breakup of water molecules. What seems clear experimentally in GDE is that there is a limiting oxidation yield for inorganic ions of some 7–8 equiv F^{-1} which probably arises predominantly by the scavenging of $OH\cdot$ radicals and H_2O_2 molecules; most of this oxidation probably occurs in the secondary reaction zone, although in exceptional cases it may be possible for the scavengers to penetrate into the primary zone and react with precursors of these species. It is apparent that there is need for further careful work with selective scavengers in GDE to determine radical and molecular yields, and to identify the species involved in different reactions.

When the substrate concentration is not sufficient for complete scavenging, the yields in GDE will depend upon the competition among reactions (6)–(12) under the conditions of the experiment. If it is supposed, for simplicity, that the OH radical is the main oxidizing species, that a constant yield of n equiv F^{-1} of OH emerges from the primary zone, and that the velocities of the competing reactions are v_6, v_7, v_8, etc., then (neglecting the molecular yield of H_2O_2, which may be small), G values can be written down as follows, assuming that hydrogen peroxide reacts completely with ferrous and ceric ions:

1. Inert electrolytes, $G(H_2O_2) = n(v_7 - v_9)/(v_6 + v_7 + v_9)$
2. Ferrous ion, $G(\text{oxidation}) = n(v_{10} + v_7)/(v_6 + v_7 + v_{10})$
3. Cerous ion, $G(\text{oxidation}) = n(v_{10} - v_7)/(v_6 + v_7 + v_{10})$

The velocity terms will of course involve the concentrations of the radicals concerned in the reactions and these will change as they advance in the secondary reaction zone. Thus, the kinetic situation is extremely complex and requires for its solution a complete knowledge of the concentration profiles of all the reacting species in the zone. Such knowledge is not yet available and a general solution to the problem cannot be given. However, the results of some very crude approximations in particular cases are interesting. For example, the velocity of reaction (10) must involve the concentration of substrate, and if it is supposed in a particular system that the radical concentrations can be regarded as constant, then an expression for the dependence of G on substrate concentration can be written down. Thus, in the ferrous case, the result

$$G(\text{oxidation}) = n([Fe^{2+}] + A)/([Fe^{2+}] + A + B)$$

The quantities A and B will depend upon the concentrations of the reactive radicals and cannot, therefore, be independent of $[Fe^{2+}]$. Nevertheless, an expression of this type fits the experimental yields quite closely, as shown by Hickling and Linacre,[5b] if n is taken to be 7 and A and B are given empirical values. The agreement is, of course, in part artificial since A and B are given appropriate values to secure a fit at certain points, but the expression is not without utility. Comparison of the equations for $G_0(H_2O_2)$ and $G_0(\text{oxidation})$ for ferrous ion show that, at very low ferrous concentrations, $v_{10} \to 0$ and the oxidation yield under these conditions should approximate to the hydrogen peroxide yield provided v_9 is small. Thus, $G_0(\text{oxidation})$ when $[Fe^{2+}] \to 0$ should give an upper limit for $G_0(H_2O_2)$ in inert electrolytes; this limit should be about $2\,\text{equiv}\,F^{-1}$ and $G_0(H_2O_2)$ yields up to this value have been obtained in practice.

Although in the foregoing discussion attention has been directed mainly to the oxidation of inorganic ions, the results of GDE in most systems can be qualitatively understood in terms of the reactions of the OH radical, a subsidiary role being allotted to the other species present. Thus, with organic substrates, the initial process seems usually to be the abstraction of a H atom by the OH radical to give an organic radical. This may itself dimerize to a stable product, as, for example, in the production of oxamide from formamide in aqueous solution[27]:

$$HCONH_2 + OH \to {^{\cdot}CONH_2} + H_2O$$
$$2\,{^{\cdot}CONH_2} \to (CONH_2)_2$$

(17)

Alternatively, the organic radical may enter into decomposition reactions and in some cases the products could undergo hydroxylation. Thus, with formic acid,[26] the initial step seems to be the formation of the carboxyl radical followed by dimerization to oxalic acid or decomposition according to the following scheme:

$$OH + HCOOH \rightarrow \dot{C}OOH + H_2O$$

or

$$2\,\dot{C}OOH \rightarrow (COOH)_2$$

and

$$2\,\dot{C}OOH \rightarrow HCOOH + CO_2 \tag{18}$$

$$OH + \dot{C}OOH \rightarrow CO_2 + H_2O$$

$$H + \dot{C}OOH \rightarrow H_2 + CO_2$$

With acetic acid, the initial product of GDE seems to be the $\dot{C}H_2COOH$ radical, which may dimerize to succinic acid. This, in turn, may lose a H atom, providing further possibilities of radicals linking up, or direct combination with the OH radical may lead to hydroxyacids being formed. Thus,

$$OH + CH_3COOH \rightarrow \dot{C}H_2COOH + H_2O$$

(and perhaps

$$H + CH_3COOH \rightarrow \dot{C}H_2COOH + H_2)$$

$$2\,\dot{C}H_2COOH \rightarrow \begin{array}{l} CH_2COOH \\ | \\ CH_2COOH \end{array}$$

$$OH + \begin{array}{l} CH_2COOH \\ | \\ CH_2COOH \end{array} \rightarrow \begin{array}{l} \dot{C}HCOOH \\ | \\ CH_2COOH \end{array} + H_2O \tag{19}$$

$$\dot{C}H_2COOH + \begin{array}{l} \dot{C}HCOOH \\ | \\ CH_2COOH \end{array} \rightarrow \begin{array}{l} CH_2COOH \\ | \\ CHCOOH \\ | \\ CH_2COOH \end{array} \quad \text{(tricarballylic acid)}$$

$$OH + \begin{array}{l} \dot{C}HCOOH \\ | \\ CH_2COOH \end{array} \rightarrow \begin{array}{l} HOCHCOOH \\ | \\ CH_2COOH \end{array} \quad \text{(malic acid)}$$

Many other possibilities are apparent with a scheme of this kind.

Most of the results obtained in the GDE of solutions in liquid ammonia can be understood if it is supposed that the main reactive species is the NH_2 radical, corresponding to the OH radical in aqueous solutions. Here, the current may be carried to the liquid by NH_3^+ ions accelerated in the cathode drop of about 390 V and these will enter the liquid with energies similar to those involved in aqueous systems. Thus, the initial reactions in the primary zone due to the entry of the energetic gaseous ions may well be given by the following:

Collision

$$NH_3^+ \xrightarrow{NH_3} NH_2 + NH_4^+$$

ionization: $NH_3 \rightsquigarrow +$ (20)

$$e^- \xrightarrow{NH_3} e_s^- \rightarrow H + NH_2^-$$

activation: $NH_3^* \rightsquigarrow NH_2 + H$ (21)

Charge transfer

$$NH_3^+ + NH_3 \rightarrow NH_2 + NH_4^+$$ (22)

Molecular yields of hydrazine may result from the interaction of the close-packed radicals or their precursors, and thus a mixture of NH_2, e_s^- or H, N_2H_4, and H_2 will diffuse out into the secondary reaction zone. Possible reactions here will be re-formation of ammonia and dimerization to hydrazine and hydrogen:

$$NH_2 + H \rightarrow NH_3$$ (23)

$$NH_2 + NH_2 \rightarrow N_2H_4$$ (24)

$$H + H \rightarrow H_2$$ (25)

and decomposition of hydrazine itself by NH_2 radicals:

$$NH_2 + N_2H_4 \rightarrow N_2H_3 + NH_3$$

followed by   (26)

$$NH_2 + N_2H_3 \rightarrow N_2 + H_2 + NH_3$$

From this scheme, it is apparent that hydrazine should accumulate in the solution until its rate of decomposition approaches its rate of formation, so that a stationary concentration is established. On this basis, and by making drastic simplifying assumptions similar to those considered for aqueous reactions, the competitive kinetics lead to an equation of the form

$$G = 0.5n(A - [N_2H_4])/(A + B + [N_2H_4])$$

for the yield of hydrazine in moles F^{-1} in solutions of various hydrazine concentrations, where n is the initial yield of NH_2 in moles F^{-1}. Values of A and B can be found empirically from the experimental observations, and, with n given the value of 12, the equation fits the results fairly closely.[41] The equation can be integrated to give the concentration of hydrazine which will exist after the passage of Q faradays of electricity starting with no hydrazine initially present, and this equation is

$$0.5nQ/V = (2A + B) \log_e\{A/(A - [N_2H_4])\} - [N_2H_4]$$

where V is the volume of the solution; this equation also satisfactorily represents the experimental data.

VIII. GLOW-DISCHARGE ELECTROLYSIS AND RADIATION CHEMISTRY

GDE is undoubtedly a form of electrolysis in which some of the electrical energy put into a system can bring about chemical reaction by a process additional to ordinary charge transfer. An attempt has been made above to formulate a mechanism for this in terms of gaseous ions being accelerated in a constant cathode drop communicating their energy to solvent molecules and bringing about their decomposition by a process related to radiolysis. Hence, GDE then takes its place as a minor section of radiation chemistry and the facts, concepts, and ideas developed in the latter field can be applied to it. This seems, in general, quite a fruitful procedure and by it, many of the results of GDE and of other electrical discharges to solutions which fall into a similar category can be explained. It is, however, worth emphasizing some of the differences between the energetic particles in GDE and the more usual kinds of ionizing radiation. First, while in GDE the energy *per particle* is extremely low (probably of the order of 10^2 eV as compared to 10^4–10^7 eV for the quantum of most ionizing radiation), the *dose rate* can be extremely high. For example, with a current of 0.075 A, the number of singly charged gaseous ions reaching the solution surface per minute is 2.8×10^{19}, and assuming them to have an average energy of 100 eV, this corresponds to a dose rate of $2.8 \times 10^{21} \, \text{eV min}^{-1}$. This is considerably higher than the dose rate normally used in steady radiolysis (usually of the order 10^{16}–$10^{20} \, \text{eV cc}^{-1} \, \text{min}^{-1}$), although much less than that achieved in pulse radiolysis. In a typical GDE experiment, the energy put into the solution is about $10^{21} \, \text{eV cc}^{-1}$, while in radiolysis experiments, doses of about $10^{18} \, \text{eV cc}^{-1}$ are common. Thus, the amount of chemical change which can be effected in GDE is much greater than that in radiolysis, and high concentrations of substrate can be used; also, under these conditions, impurities seem to have much less effect. Possibly connected with the high dose rate and LET value is the fact that concentration of substrate seems to have a greater influence in GDE than in radiolysis. Thus, with gamma rays and X-rays, concentration of the substrate often has little influence on the yields unless the concentration is extremely low, and in this "concentration-independent" region, it may be supposed that the substrate is scavenging all the radicals rather thinly dispersed between the strungout spurs of activity in the solution. With denser radiations, such as alpha particles, the yields depend more on concentration, and scavenging probably extends into the spurs themselves. GDE seems to be an extreme extrapolation of this latter case. The dose here is confined to a very small primary reaction zone, and in the steady-state condition which is set up, the profile of the free-radical concentrations will show a continuous variation from a high value at the surface of the solution to zero at a short distance inside. Thus, all reactions will occur in a heterogeneous region and will result from the penetration of the substrate by the concentration gradient of radicals. Thus, a "concentration-independent" region is not likely to occur until very high concentrations of substrate are reached, and the distinction which is often made in radiation chemistry between radical and molecular yields loses much of its significance in GDE.

Closely related to the point made above is the fact that, in GDE, yields of dimers of radicals tend always to be high, probably because of the very high local concentration of radicals in the primary reaction zone. Thus, in the formation of hydrazine in liquid ammonia, GDE gives an initial yield of 2.5 mole F^{-1}; in contrast, gamma radiolysis gives $G(N_2H_4) = 0.13$ molecule per 100 eV.[42] Attempts to calculate radical concentrations in GDE are probably not very meaningful since the concentration will vary greatly with distance, but, by adopting a very simple reaction scheme for the formation of hydrogen peroxide in dilute sulfuric acid and using known rate constants, Denaro[18] has calculated that, for a homogeneous reaction zone, [OH] would be about 10^{-5} M.

Attempts to relate absolute yields of initial species obtained in radiolysis with those apparently obtained in GDE are not entirely successful. For example, for the radiolysis of pure deaerated water by high-energy electrons, the following yields per 100 eV have been quoted[43]:

$$H_2O \rightsquigarrow e_{aq}^-(2.5) + OH(2.3) + H(0.6)$$

$$+ H_2(0.45) + H_2O_2(0.85) + H_3O^+(2.5)$$

If similar yields occurred in GDE and only the oxidizing species OH and H_2O_2 were involved in bringing about oxidation, as has been suggested earlier, this would imply a maximum oxidation yield of 4.0 equiv F^{-1} if the gaseous ions have average energies of 100 eV. The actual yield observed is nearer 8 equiv F^{-1} and, although 1 equiv F^{-1} will arise by charge transfer, this still leaves a discrepancy of 3 equiv F^{-1} to be explained. Again, in the radiolysis of liquid ammonia,[42] the following primary yields have been quoted:

$$NH_3 \rightsquigarrow H(1.75) + NH_2(1.75) + H_2(0.12) + N_2H_4(0.12)$$

According to this, if all the NH_2 radicals combined to give N_2H_4, the yield would only be 1.0 molecule per 100 eV, whereas in GDE, it is 2.5 mole F^{-1}. The situation could therefore arise that, in GDE, the average energy of the positive ions entering the solution is appreciably greater than the 100 eV which has been adopted as a convenient rough estimate.

IX. APPLICATIONS OF GLOW-DISCHARGE ELECTROLYSIS

The foregoing treatment of GDE leads to the conclusion that the method provides a simple way of generating radicals derived from solvent molecules at a very high local concentration. The conditions are highly favorable for dimerization of these radicals, or they may be used for interaction with a suitable substrate. It seems likely, therefore, that GDE may have possibilities for industrial application. It is important, however, to bear certain limitations in mind. First, the consumption of electrical energy is considerable. To get the characteristic effects of GDE, the cathode drop voltage is essential and this is usually about 400 V. Thus, even if the yield of desired product is high in the sense that several moles of it are formed for each faraday of electricity passed, the electrical energy expended is still an order of magnitude greater than that in conventional electrolytic processes. Second, although GDE provides a means of introducing a considerable amount of energy into a liquid system, it is an unselective type of process and may lead to a variety of disintegration reactions with great multiplicity of products; this is particularly so when organic systems are used. Third, it may often happen that the desired product is itself decomposed by GDE unless it can be removed rapidly from the reaction zone. Thus, the situation for successful industrial application would seem to be one in which the desired product is costly and difficult to make by other methods, and where GDE gives a good yield of it without many alternative products arising.

An important case in which it might be hoped that these conditions would be partly met is the production of hydrazine. This is usually manufactured chemically by the reaction of hypochlorite and excess ammonia in aqueous solution according to the Raschig process; the reaction requires careful control and leads to a dilute crude liquor which needs elaborate purification and concentration before a concentrated solution of hydrazine hydrate can be obtained. Direct decomposition of ammonia by electrical discharge to give hydrazine is an attractive alternative, but in spite of much work, it has been found difficult to obtain substantial yields in the gas phase.[44] However, by GDE of liquid ammonia, hydrazine is very readily produced, as has been described, and anhydrous hydrazine can be obtained from the resulting liquid directly by distillation. This process would seem to offer industrial possibilities, and it has been patented.[45] It has not, however, so far been commercially exploited, which would seem to indicate that its economics are unfavorable.

REFERENCES

[1] J. Gubkin, *Ann. Physik* **32**[III] (1887) 114.
[2] K. Klüpfel, *Ann. Physik* **16**[IV] (1905) 574.
[3] A. Makowetsky, *Z. Elektrochem.* **17** (1911) 217.

[4]A. Klemenc et al., Z. Elektrochem 20 (1914) 485; 37 (1931) 742. Z. Physik. Chem. 130 (1927) 378; 154 (1931) 385; 166 (1933) 343; 27B (1935) 369; 179 (1937) 1; 182 (1938) 91; 40B (1938) 252; 183 (1938) 217, 297; Z. Anorg. Allgem. Chem. 240 (1939) 167. Monatsh. 75 (1944) 42; 76 (1946) 38; 78 (1948) 243; 81 (1950) 122; 82 (1951) 708; 869, 1041; 84 (1953) 365, 498, 1053; 85 (1954) 47. Chimia (Aarau) 6 (1952) 177. Z. Elektrochem. 56 (1952) 198, 634, 917; 57 (1953) 615.

[5]S. Glasstone and A. Hickling, J. Chem. Soc. (1934) 1772. R. A. Davies and A. Hickling, J. Chem. Soc. (1952) 3595. A. R. Denaro and A. Hickling, J. Electrochem. Soc. 105 (1958) 265. A. Hickling and M. D. Ingram, J. Electroanal. Chem. 8 (1964) 65.

[5a]A. Hickling and G. R. Newns, Proc. Chem. Soc. (1959) 368; J. Chem. Soc. (1961) 5177, 5186.
A. Hickling and M. D. Ingram, J. Electroanal. Chem. 8 (1964) 65.

[5b]A. Hickling and J. K. Linacre, J. Chem. Soc. (1954) 711.

[6]W. R. Cousins, Z. Physik. Chem. 4B (1929) 440. W. Braunbek, Z. Physik 91 (1934) 184. F. Fichter and K. Kestenholz, Helv. Chim. Acta 23 (1940) 209. A. Muta et al., J. Electrochem. Soc. Japan 17 (1949) 74, 113, 202, 235, 265, 298; 18 (1950) 17, 82. M. Haïssinsky and A. Coche, J. Chim. Phys. 51 (1954) 581. E. H. Brown, W. D. Wilhide, and K. L. Elmore, J. Org. Chem. 27 (1962) 3698. A. Banege-Nia, F. Basquin, and G. Morand, Compt. Rend. 258 (1964) 4521. A. Banege-Nia, D. Kaspar, and G. Morand, Compt. Rend. 258 (1964) 5213.

[6a]Z. Sternberg, in Proc. Int. Conf. Ionization Phenomena Gases, 3rd, Venice (1957), p. 1061.

[6b]G. H. Cady, H. J. Emeleus, and B. Tittle, J. Chem. Soc. (1960) 4138.

[6c]B. S. R. Sastri, J. Sci. Ind. Res. (India) 19B (1960) 317.

[6d]N. Thon, Compt. Rend. 197 (1933) 1114.

[6e]A. Dewhurst, J. F. Flagg, and P. K. Watson, J. Electrochem. Soc. 106 (1959) 366.

[7]P. De Beco, Compt. Rend. 207 (1938) 623; 208 (1939) 797. Bull. Soc. Chim. France 12 (1945) 779, 789, 795. N. A. Goryunova and V. I. Pavlov, Zh. Obshchei Khim. 23 (1953) 1253.

[8]A. Hickling and M. D. Ingram, Trans. Faraday Soc. 60 (1964) 783.

[9]K. O. Hough and A. R. Denaro, J. Sci. Instr. 43 (1966) 488.

[10]D. E. Couch and A. Brenner, J. Electrochem. Soc. 106 (1959) 628.

[11]H. H. Kellogg, J. Electrochem. Soc. 97 (1950) 133.

[12]H. Fizeau and L. Foucault, Compt. Rend. 18 (1844) 860.

[13]M. M. Lagrange and M. Hoho, Compt. Rend. 116 (1893) 575.

[14]M. Hoho, Elec. Rev. 104 (1929) 185. I. Z. Yasnogorodskii, Avto. Trakt. Prom. 6 (1954) 21; Chem. Abs. 48 (1954) 12586. S. Owaku and K. Kuroyanagi, J. Japan. Inst. Metals 20 (1956) 63. T. Sato and H. Mii, Rep. Gov. Ind. Res. Inst. Nagoya 5 (1956) 313, 415, 586; 6 (1957) 179, 338, 610.

[15]A. Wehnelt, Elektrotech. Z. 20 (1899) 76.

[16]W. J. Humphreys, Phys. Rev. 9 (1899) 33.

[17]A. Klemenc, Monatsh. 76 (1946) 38. R. A. Davies and A. Hickling, J. Chem. Soc. (1952) 3595.

[18]A. R. Denaro, private communication.

[19]A. R. Denaro and A. Hickling, J. Electrochem. Soc. 105 (1958) 265.

[20]A. R. Denaro and P. A. Owens, Electrochim. Acta 13 (1968) 157.

[21]A. Klemenc, Z. Anorg. Allgem. Chem. 240 (1939) 167.

[22]F. Fichter and K. Kestenholz, Helv. Chim. Acta 23 (1940) 209.

[23]A. Klemenc and H. F. Hohn, Z. Physik. Chem. 154A (1931) 385; 166A (1933) 343.

[24]A. Klemenc, Monatsh. 81 (1950) 122.

[25]A. Klemenc, Z. Elektrochem. 56 (1953) 694.

[26] G. H. Gore and A. Hickling, unpublished work.

[27] E. H. Brown, W. D. Wilhide, and K. L. Elmore, *J. Org. Chem.* **27** (1962) 3698.

[28] J. F. Woodman, United States Patent (1953) 2,632,729.

[29] A. Bradley and J. P. Hammes, *J. Electrochem. Soc.* **110** (1963) 15. A. R. Denaro, P. A. Owens, and A. Crawshaw, *European Polymer J.* **4** (1968) 93.

[30] A. Hamilton and M. D. Ingram, private communication.

[31] A. Hickling and J. V. Shennan, unpublished work.

[32] A. Yokohata and S. Tsuda, *Bull. Chem. Soc. Japan* **39** (1966) 46, 53.

[33] V. I. Pavlov, *Compt. Rend. Acad. Sci. U.S.S.R.* **43** (1944) 236, 383, 385.

[34] A. O. Allen, *The Radiation Chemistry of Water and Aqueous Solutions*, Van Nostrand Company, New York, 1961. A. J. Swallow, *Radiation Chemistry of Organic Compounds*, Pergamon Press, London, 1960.

[35] A. Von Engel, *Ionized Gases*, 2nd ed., Clarendon Press, Oxford, 1965.

[36] H. A. Barton and J. H. Bartlett, *Phys. Rev.* **31** (1928) 823. H. D. Smyth and D. W. Mueller, *Phys. Rev.* **43** (1933) 116. M. M. Mann, A. Hustrulid, and J. T. Tate, *Phys. Rev.* **58** (1940) 340.

[37] R. M. Chaudrhi and M. L. Oliphant, *Proc. Roy. Soc.* **137A** (1932) 662.

[38] W. D. Davis and T. A. Vanderslice, *Phys. Rev.* **131** (1963) 219.

[39] E. J. Hart, *Ann. Rev. Nuclear Sci.* **15** (1965) 125.

[40] M. Anbar and P. Neta, *Int. J. Appl. Radiation Isotopes* **16** (1965) 227.

[41] A. Hickling and G. R. Newns, *J. Chem. Soc.* (1961) 5177.

[42] D. Cleaver, E. Collinson, and F. S. Dainton, *Trans. Faraday Soc.* **56** (1960) 1640.

[43] D. C. Walker, *Quart. Rev. Chem. Soc.* **21** (1967) 79.

[44] F. K. McTaggart, *Plasma Chemistry in Electrical Discharges*, p. 199, Elsevier Publishing Co., London, 1967.

[45] A. Hickling and G. R. Newns, British Patent, (1962) 896,113.

Index